DISCRETE DYNAMICAL SYSTEMS and DIFFERENCE EQUATIONS with *Mathematica*

DISCRETE DYNAMICAL SYSTEMS and DIFFERENCE EQUATIONS with *Mathematica*

Mustafa R.S. Kulenović
Orlando Merino
Department of Mathematics
University of Rhode Island
Kingston, Rhode Island

CRC Press is an imprint of the
Taylor & Francis Group, an **informa** business
A CHAPMAN & HALL BOOK

CRC Press
Taylor & Francis Group
6000 Broken Sound Parkway NW, Suite 300
Boca Raton, FL 33487-2742

First issued in paperback 2019

ISBN-13: 978-1-58488-287-9 (hbk)
ISBN-13: 978-0-367-39630-5 (pbk)

Library of Congress Cataloging-in-Publication Data

Catalog record is available from the Library of Congress

Visit the Taylor & Francis Web site at
http://www.taylorandfrancis.com

and the CRC Press Web site at
http://www.crcpress.com

M. R. S. Kulenović and O. Merino
Department of Mathematics
University of Rhode Island
Kingston, RI 02881-0816, USA

Discrete Dynamical Systems and Difference Equations with *Mathematica*

CRC PRESS
Boca Raton Ann Arbor London Tokyo

To

Gabi and, Rešad and Ema

Contents

Preface

This book is both an introductory survey on theory and techniques of discrete dynamical systems and difference equations, and a manual for the use of the software package *Dynamica.*

This book is intended for students of mathematics, life sciences, physics, economics, engineering and other areas. The prerequisites for Chapter 1 are two courses in calculus. The rest of the book requires knowledge of a few basic facts in linear algebra and multi dimensional calculus.

The theory of discrete dynamical systems and difference equations developed greatly during the last twenty five years of the twentieth century, following the publication of the seminal paper "Period Three Implies Chaos", by J. Yorke and Y. Li in 1975. In 1987 R. Devaney published "*An introduction to Chaotic Dynamical Systems*", the first book on the subject. Applications of difference equations also experienced enormous growth in many areas, for example in Biology following R. Mays 1976 article "Simple Mathematical Models with Very Complicated Dynamics".

In 1985, the software program *Phaser* by H. Kocak appeared and made a great impact. *Phaser* became the standard software tool for studying difference equations numerically and visually. The main strength of *Phaser* was its graphical interface, which allowed users without programming experience to quickly plot trajectories. This was already a big advance, as it was quick and interactive. Another important software is *Dynamics*, by J. Yorke s group, which appeared in 1994. *Dynamics* is a program written in C that, in addition to plotting trajectories, has other capabilities such as calculating of Lyapunov exponents, plotting bifurcation diagrams, and finding basins of attraction.

Recent advances in the technology of Computer Algebra Systems (CAS) now allow the use of symbolic calculation to study difference equations. For example, linearized stability analysis of systems with parameters, calculation of invariants, finding Lyapunov functions (based on invariants), finding symbolic periodic solutions, can all be treated with a CAS. We developed the *Mathematica* based package *Dynamica* as a collection of tools for use in the study of discrete dynamical systems and difference equations. No programming experience is needed to use *Dynamica.* Also, creating technical reports is no more difficult than using a word processor.

Dynamica implements a series of tools and techniques of algebraic, numerical, and graphical nature. These include: finding equilibrium and periodic points, classifying the stability character of equilibrium and periodic points, semicycle analysis of solutions, calculation and visualization of invariants (first integrals), calculation and visualization of Lyapunov functions, plotting bifurcation diagrams, visualization of stable and unstable manifolds, calculation and visualization of Lyapunov numbers, and calculation of Box Dimension. *Dynamica* is easy to learn and use. In very little time the new user may be ready to study complex difference equations of which not much (maybe almost nothing!) is known.

This book can be used as the main text, for the students who wish not to emphasize proofs, or as a supplement for an undergraduate junior or senior course in discrete dynamical systems and difference equations. It can also be useful to graduate students and researchers studying higher order dynamics. The book is to be used with the software *Dynamica* witten in *Mathematica.*

One set of *Dynamica* sessions presented in the book serve as a tutorial of the programs and the different techniques. The necessary concepts and results are provided in the text, as well as references to research publications, and suggested problems. However, the presentation style emphasizes use of the software and not theoretical discussions.

The second set of *Dynamica* sessions presented in the book consists of case studies of well known difference equations, which come under name such as Henon, Lyness, Todd, and others. An interesting feature of the sessions is that they show how to prove, with help of *Dynamica*, many results published recently.

Many applications of discrete dynamical systems and difference equations have appeared recently in the areas of biology, economics, physics, resource management, and others. We present as exercises and research projects many of these applications that appeared in the last ten years and are presented for the first time in a book. *Dynamica* offers a set of tools that can be used to gain insight and explore exciting aspects of these systems. In addition, the code of *Dynamica* can be customized by the user with some programming experience. Finally, *Dynamica* can be used to produce superb technical documents on the analysis of discrete dynamical systems as well as superior graphics that can be easily implemented in some other applications. For example all the graphics in this book has been produced by using *Dynamica* and *Mathematica.* The *Dynamica* package and the notebooks that correspond to particular chapters can be downloaded from

www.math.uri.edu/Dynamica

Chapter 1 presents a short exposition of the dynamics of one-dimensional maps. It provides many effective tests for the stability of both hyperbolic and nonhyperbolic fixed and periodic points, as well as the symbolic dynamics for one-dimensional maps. In addition, it provides some tools for checking chaos theoretically, numerically, and visually.

Chapter 2 gives a detailed exposition of the dynamics of two-dimensional maps. It provides some effective tests for the stability of hyperbolic fixed and periodic points. In particular, it contains very recent reserch results about the stability of hyperbolic fixed points for a special class of maps with invariant sets and the global attractivity results for dissipative maps. This chapter presents the stability of nonhyperbolic fixed points for area preserving maps. In addition we present some very recent theoretical results about the semicycle analysis and provide the corresponding computer exercises that implements this method. This chapter contains an extensive list of two-dimensional maps which appears in recent biological and economical modeling.

Chapter 3 presents a short review of the dynamics of n- dimensional maps. It lists some basic results such as The Stable Manifold Theorem, Hartman-Grobman Theorem, and Lyapunov Theorem on Stability, and provide illustrative examples for some basic concepts in the theory of discrete dynamical systems. It also provides some recent results on dissipative maps.

Chapter 4 provides a new and fast developing area of discrete dynamical systems and difference equations that gives essential tools for finding the solutions of difference equations in exact form, and for investigating the short term and long term behavior of the solutions. The theory is based on the existence of an expression, called *invariant* or *first integral*, that remain constant or invariant along solutions of a difference equation and which reveals the behavior of solutions of the equation. The major result presented is the discrete version of Dirichlet theorem. In many cases, the *Dynamica* functions that we developed may be used to find rational invariants symbolically and to construct a corresponding Lyapunov function, which proves stability.

Chapter 5 presents a short exposition of the dynamics of three-dimensional maps. The theory for three-dimensional maps is mainly a specialization of the general theory presented in Chapter 3. This reflects the lack of specific results in the theory of dynamics of three-dimensional maps at this point in time. The emphasis in this chapter is on computer simulation rather than on the theory.

Chapter 6 gives a concise and self contained review on a special class of fractals generated by affine transformations, known as iterated function system. This chapter includes the presentation of an effective formula for the computation of the fractal dimension. The corresponding *Dynamica* functions can be used to create new iterated function systems, and to either compute or estimate their fractal dimensions.

Every chapter in this book contains several exercises and suggested projects related to some equations that have appeared recently in the literature.

Acknowledgements

This book is the outgrowth of lecture notes and seminars as well as numerous computer experiments that were performed at the University of Rhode Island during the last six years. We are thankful to Professors D. Clark, E. A. Grove, G. Ladas, and J. Montgomery and to our graduate students K. Cunningham, H. El-Metwally, C. Kent, L. McGrath, N. Prokup, M. Radin, C. T. Teixeira, S. Valicenti, and K. P. Wilkinson for their enthusiastic participation in using and testing the software and useful suggestions that helped to improve the exposition as well as the code.

Chapter 1

Dynamics of One-Dimensional Dynamical Systems

1.1 Introduction

In this chapter we investigate and visualize dynamics of the first-order difference equation

$$x_{n+1} = f(x_n), \quad n = 0, 1, \ldots \tag{1.1}$$

where $f : R \to R$ is a given function. To do this, we shall survey the theory, discuss examples, and present computer sessions with the software package *Dynamica*.*

When we study the *dynamics* of a difference equation, we attempt to do the following actions: determine equilibrium points and periodic points, analyze their stability and asymptotic stability, and determine aperiodic points and chaotic behavior. In this chapter we give rigorous definitions of all these notions for (1.1).

We now introduce some terminology. We shall also refer to (1.1) as a one-dimensional **dynamical system**. The function f is called the **map** associated with (1.1). A **solution** of equation (1.1) is a sequence $\{\phi_n\}_{n=0}^{\infty}$ that satisfies equation (1.1) for all $n = 0, 1, \ldots$. If an **initial condition** $x_0 = d$ is given, the problem of solving the difference equation (1.1) so that the solution satisfies the initial condition is called the **initial value problem** (abbreviated as IVP).

The **general solution** of equation (1.1) is a sequence $\{\phi_n\}_{n=0}^{\infty}$ that satisfies equation (1.1) for all $n = 0, 1, \ldots$ and involves a constant C that can be determined once an initial value is prescribed. A **particular solution** of equation (1.1) is a sequence $\{\phi_n\}_{n=0}^{\infty}$ that satisfies equation (1.1) for all $n = 0, 1, \ldots$.

We present the linear theory in Sections 1.2 and 1.3 because it serves to introduce and motivate some of the basic notions for (1.1). Also, solutions in the linear case may be obtained in an explicit, closed form, thus facilitating the presentation of the material. Stability of equilibrium points is discussed in

**Dynamica is available for download from www.math.uri.edu/Dynamica and www.mathsource.com.*

Sections 1.4 and 1.5. Section 1.6 is an introduction to bifurcations and period doubling. Section 1.7 is a computer session with the *Dynamica* package. Symbolic dynamics for one-dimensional maps are discussed in Section 1.8, which also includes a computer session. We finish the chapter with a discussion on dissipative maps in Section 1.9 and parametrization and Poincaré functional equation in Section 1.10.

We warn the reader about the use throughout this book of the notation $\{x_n\}$ (rather than $\{\phi_n\}$) to refer to solutions of equation (1.1). This is done here for practical reasons (mainly to simplify the exposition), but also because it is standard practice in books and research articles in the field of Difference Equations.

1.2 Linear Difference Equations with Constant Coefficients

An equation of the form

$$x_{n+1} = ax_n + b, \quad n = 0, 1, \dots \tag{1.2}$$

where $a \in \mathbf{R}\backslash\{0\}$ and $b \in \mathbf{R}$ is called **first-order linear difference equation.** When $b = 0$, equation (1.2) is called **homogeneous**, while equation (1.2) is called **nonhomogeneous** when $b \neq 0$. The main result of this section is the following theorem.

THEOREM 1.1
Let $a, b \in \mathbf{R}$ with $a \neq 0$. There exists a unique solution of equation (1.2) with initial condition $x_0 = d \in \mathbf{R}$. The solution is given by

$$x_n = \begin{cases} d + bn & \text{if } a = 1 \\ \left(d - \dfrac{b}{1-a}\right)a^n + \dfrac{b}{1-a} & \text{if } a \neq 1 \end{cases}, \quad n = 0, 1, \dots. \tag{1.3}$$

REMARK 1.1 Suppose $b \neq 0$. Note that when $a = 1$, every solution of equation (1.2) is unbounded. Note also that when $a \neq 1$, equation (1.2) has the constant solution

$$x_n = \frac{b}{1-a}, \quad n = 0, 1, \dots. \tag{1.4}$$

Such a solution is called an *equilibrium* solution of equation (1.2). Every other solution of equation (1.2) with $|a| > 1$ is unbounded and converges to the equilibrium solution (1.4) if $|a| < 1$. ∎

Example 1.1

(a) The equation

$$x_{n+1} - 2x_n = 0, \quad n = 0, 1, \ldots$$

has the general solution

$$x_n = C2^n,$$

where C is an arbitrary constant. The corresponding IVP has the solution

$$x_n = d2^n.$$

(b) The equation

$$x_{n+1} - \frac{1}{2}x_n = 2, \quad n = 0, 1, \ldots$$

has the general solution

$$x_n = C\left(\frac{1}{2}\right)^n + 4,$$

where C is an arbitrary constant. The corresponding IVP has the solution

$$x_n = (d-4)\left(\frac{1}{2}\right)^n + 4.$$

(c) The equation

$$x_{n+1} - x_n = 5, \quad n = 0, 1, \ldots$$

has the general solution

$$x_n = C + 5n,$$

where C is an arbitrary constant. The corresponding IVP has the solution

$$x_n = d + 5n.$$

(d) The equation

$$x_{n+1} + x_n = -4, \quad n = 0, 1, \ldots$$

has the solution

$$x_n = C(-1)^n - 2,$$

where C is an arbitrary constant. The corresponding IVP has the solution

$$x_n = (d+2)(-1)^n - 2.$$

▮

1.3 Linear Difference Equations with Variable Coefficients

The next result in this section gives an explicit formula for the solutions of the first-order, linear, nonhomogeneous equation with variable coefficients

$$x_{n+1} = a_n x_n + b_n\,, \quad n = 0, 1, \ldots. \tag{1.5}$$

THEOREM 1.2

Let $\{a_n\}_{n=0}^{\infty}$ and $\{b_n\}_{n=0}^{\infty}$ be sequences of real numbers. There exists a unique solution of equation (1.5) with initial condition $x_0 = d$. Such a solution is given by

$$x_n = \left(\prod_{i=0}^{n-1} a_i\right) d + \sum_{k=0}^{n-1}\left(\prod_{i=k+1}^{n-1} a_i\right) b_k, \quad n = 1, 2, \ldots, \tag{1.6}$$

where by definition,

$$\prod_{i=0}^{-1} a_i = 1, \prod_{i=n}^{n-1} q_i = 1.$$

Example 1.2 The solution of the *IVP*

$$x_{n+1} + \frac{2n+1}{2n+3} \cdot x_n = r_n, \quad n = 0, 1, \ldots$$

$$x_0 = d.$$

is

$$x_n = \frac{d + (n^2 + 2n) r_n}{2n+1}.$$

■

Mathematica has a powerful built-in package named `RSolve` for computing solutions to difference equations using the generating functions method. The form of the equations given in `RSolve` is similar to the form used in the built-in package `DSolve` for solving differential equations. One can give a single equation, equations with initial conditions, or several equations as input. The solution is returned in the form of a list of replacement rules.

When an initial condition is specified, the range of integers n for which the difference equation is valid is inferred from the initial condition. When the initial condition is not specified, the equation is assumed to be valid for $n \geq 0$.

This loads the package.

```
In[1]:= << DiscreteMath`RSolve`
```

The general solution of this difference equation is a power. The constant `C[1]` may be specified with the option `RSolveConstants`.

```
In[2]:= RSolve[a[n + 1] == 5 a[n], a[n], n]
Out[2]= {{a[n] → 5^n C[1]}}
```

This input specifies an initial condition.

```
In[3]:= RSolve[{a[n + 1] == 5a[n],
              a[0] == 2}, a[n], n]
Out[3]= {{a[n] → 2 5^n}}
```

These are the first 8 terms in the solution.

```
In[4]:= Table[(a[n]/.%)[[1]],
              n, 0, 7]//Expand
Out[4]= {2, 10, 50, 250, 1250, 6250,
        31250, 156250}
```

Here are the terms from a_{10} to a_{14} in the solution.

```
In[5]:= Table[(a[n]/.%3)[[1]],
              {n, 10, 14}]//Expand
Out[5]= {19531250, 97656250, 488281250,
        2441406250, 12207031250}
```

`RSolve` can solve any linear constant coefficient first-order equation. It can also solve certain linear variable coefficient equations.

`RSolve` can solve some first-order homogeneous equations that have coefficients that are rational functions of n.

```
In[6]:= RSolve[{T[0] == 4,
          T[n] == 1/(n + 1) - n/(n + 1) T[n - 1]},
          T[n], n]
Out[6]= {{T[n] → (1 + 7 (-1)^n)/(2 (1 + n))}}
```

Other equations that can be solved with `RSolve` are those for which the following two conditions hold: (a) the solution grows slower than $k^n n!$ for any constant k, and, (b) `DSolve` can solve the associated differential equation.

1.4 Stability

In this section we are especially interested in the topic of stability of equilibrium points. The symbol $f^r(d)$ denotes the r-th iterate of a function f starting at the point d. In other words, $f^2(d) = f(f(d))$ and $f^3(d) = f(f(f(d)))$, etc. We need the following definitions.

DEFINITION 1.1

(i) The **positive orbit of** $x_0 = d$ *for a dynamical system (1.1) is the sequence*

$$\gamma^+(d) := \{x_0, x_1, x_2, \ldots\} = \{d, f(d), f(f(d)), \ldots\}.$$

(ii) A point $\bar{x} \in R$ *is an* **equilibrium point** *for the dynamical system (1.1) or a* **fixed point for map** f *if* $f(\bar{x}) = \bar{x}$.

(iii) A point $x^* \in R$ *is said to be an* **eventually equilibrium point** *for the difference equation (1.1) or an* **eventually fixed point** *for* f *if there exists a positive integer* r *and a fixed point* $\bar{x}$ *of* f *such that*

$$f^r(x^*) = \bar{x}, \quad \text{and} \quad f^{r-1}(x^*) \neq \bar{x}.$$

(iv) A point $p \in R$ *is called a* **periodic point** *of period* k *(1.1) if* $f^k(p) = p$. *The point* p *is called a* **periodic point of minimal period** k *or* **prime period** k **point** *if* $f^k(p) = p$, *and* k *is the smallest positive number for which this holds. If* p *is a periodic point, then* $\gamma^+(p)$ *is called the* **periodic orbit**. *In this case, it is customary to represent* $\gamma^+(p)$ *as the finite set* $\{x_0, x_1, x_2, \ldots x_k\}$. *Orbits that are not periodic are said to be* **aperiodic**

(v) A point $p^* \in R$ *is said to be an* **eventually periodic point** *of minimal period* k *for the difference equation (1.1) or an eventually periodic point for the map* f *if there exists a positive integer* r *and a periodic point* p *of minimal period* k *such that*

$$f^r(p^*) = p, \quad \text{and} \quad f^{r-1}(p^*) \neq p.$$

Let us illustrate this definition with some examples:

Example 1.3 The following is an example of logistic difference equation that will be studied in more detail in Section 1.16:

$$x_{n+1} = f(x_n) = 4x_n(1 - x_n) \tag{1.7}$$

Equation (1.7) has two fixed points, 0 and 3/4, and the following eventually fixed points:

eventually fixed point	reason
$\frac{1}{4}$	$f(\frac{1}{4}) = \frac{3}{4}$,
$\frac{1}{2} + \frac{1}{4}\sqrt{3}$	$f(f(\frac{1}{2} + \frac{1}{4}\sqrt{3})) = \frac{3}{4}$,
$\frac{1}{4}$	$f(f(\frac{1}{2})) = 0$,
$\frac{1}{2} + \frac{1}{4}\sqrt{2}$	$f(f(f(\frac{1}{2} + \frac{1}{4}\sqrt{3}))) = 0$.

Also, equation (1.7), has the period-two solution

$$\left\{ \frac{5}{8} - \frac{1}{8}\sqrt{5},\ \frac{5}{8} + \frac{1}{8}\sqrt{5} \right\},$$

and two eventually period-two solutions with initial points

$$\frac{1}{2} + \frac{1}{8}\sqrt{10 - 2\sqrt{5}} \quad \text{and} \quad \frac{1}{2} - \frac{1}{8}\sqrt{10 - 2\sqrt{5}},$$

because

$$f\left(f\left(\frac{1}{2} \pm \frac{1}{8}\sqrt{10 - 2\sqrt{5}}\right)\right) = \frac{5}{8} + \frac{1}{8}\sqrt{5}.$$

Finding the eventually fixed points is an interesting computational problem that leads to the equation

$$f^r(x) = \bar{x},$$

where r is a positive integer greater than 1, and $\bar{x}$ is a fixed point of f. For equation (1.7), with $\bar{x} = \frac{3}{4}$ and $r = 2$ we obtain the algebraic equation

$$f(f(x)) = 16x(1 - x)(1 - 4x(1 - x)) = \frac{3}{4}$$

Similarly, finding the eventually periodic points leads to the equation

$$f^r(x) = p,$$

where r is a positive integer greater than 1, and p is a fixed point of f. For equation (1.7) with $p = \frac{5}{8} \pm \frac{1}{8}\sqrt{5}$ and $r = 2$ we obtain the algebraic equation

$$f(f(x)) = 16x(1 - x)(1 - 4x(1 - x)) = \frac{5}{8} \pm \frac{1}{8}\sqrt{5}.$$

Mathematica can be used to solve such equations. ▮

Example 1.4 A piecewise linear version of the logistic equation is the *tent equation* given by

$$x_{n+1} = T_2(x_n), \quad n = 0, 1, \ldots,$$

where

$$T_2(x) = \begin{cases} 2x, & \text{if } x \le \frac{1}{2} \\ 2(1 - x), & \text{if } x > \frac{1}{2}. \end{cases}$$

This map may be written in the form $T_2(x) = 1 - 2\left|x - \frac{1}{2}\right|$. There are two equilibrium points $\bar{x}_1 = 0$ and $\bar{x}_2 = \frac{2}{3}$. Moreover, the point $\frac{1}{2}$ is an eventually fixed point since

$$T_2\left(T_2\left(\frac{1}{2}\right)\right) = 0.$$

In fact, it is easy to show that every point of the form $k2^{-n}$, where k, and n are positive integers such that $k2^{-n} \in (0, 1]$, is an eventually fixed point.

For more on the tent map, see Section 1.8. ▮

Example 1.5 The difference equation

$$x_{n+1} = \begin{cases} \frac{1}{2} x_n & \text{if } x_n \text{ is an even positive number} \\ 3x_n + 1 & \text{if } x_n \text{ is an odd positive number,} \end{cases} \tag{1.8}$$

where $n = 0, 1, \ldots$ and $x_0 = d$ is a positive integer, has no equilibrium points. This equation has a period-three solution $\{1, 4, 2\}$ and many eventually period-three solutions. For example, the solution that starts at 8 has the corresponding orbit

$$\{8, 4, 2, 1, 4, 2, 1, \ldots\},$$

and the solution that starts at 7 has the corresponding orbit

$$\{7, 22, 11, 34, 17, 52, 26, 13, 40, 20, 10, 5, 16, 8, 4, 2, 1, 4, 2, 1, \ldots\},$$

etc. In fact, it is well known that there are infinitely many eventually periodic solutions, see [BL] and [Lg]. One of the most difficult open problems in the theory of difference equations is proving that every solution of equation (1.8) is eventually periodic to the period-three solution $\{1, 2, 4\}$. This problem is known as the $3x + 1$ **conjecture** or **Collatz conjecture**, see [BL] and [Lg]. ▮

An interesting fact is that the phenomenon of eventually equilibrium points and eventually periodic points does not have a counterpart in differential equations, whenever the solutions are unique.

1.4.1 Visualization of Orbits and Time Series

A type of plot that is frequently used to visualize the solutions to the one-dimensional difference equation (1.1) is called a **stair-step diagram** or **staircase diagram**. The staircase diagram is a plot in a rectangular coordinate system of

a. the graph of the function $y = f(x)$,

b. the identity line $y = x$, and,

c. a polygonal line that results from joining the points

$$(x_0, x_1),\ (x_1, x_1),\ (x_1, x_2),\ (x_2, x_2),\ (x_2, x_3),\ (x_3, x_3),\ \ldots$$

The line segments of the polygonal line create the impression of stairs (see Figure 1.1.) Also, it is important to observe that the equilibrium points of (1.1) are the points of intersection of the graph of f with the diagonal $y = x$.

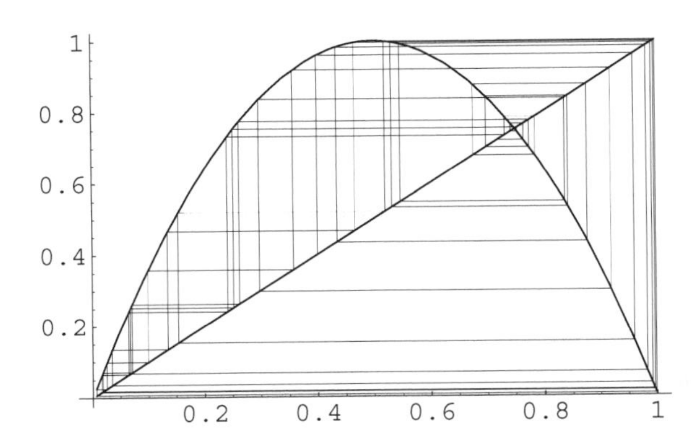

FIGURE 1.1: Staircase diagram of $\gamma^+(0.4)$ for $x_{n+1} = 4x_n(1 - x_n)$.

Another plot used for visualization of the solutions of one-dimensional difference equation (1.1), is called **time series**. It consists of a representation of the variable x_n as a function of n. Typically the horizontal axis represents n and the vertical axis represents x_n, see Figures 1.2 and 1.3.

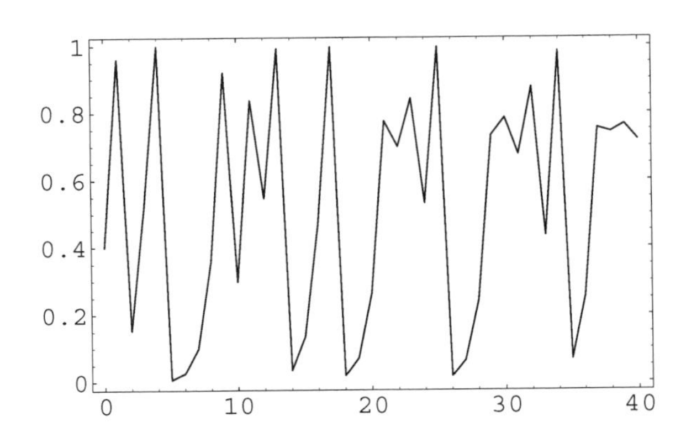

FIGURE 1.2: Time series plot of the orbit $\gamma^+(0.4)$ for $x_{n+1} = 4x_n(1 - x_n)$.

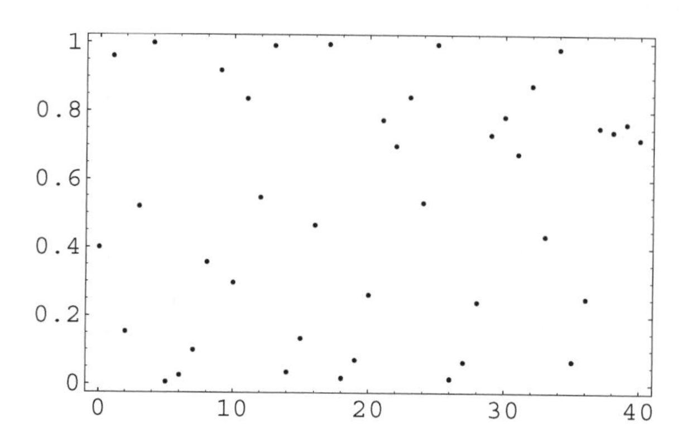

FIGURE 1.3: Plot of points of the orbit $\gamma^+(0.4)$ for $x_{n+1} = 4x_n(1 - x_n)$.

One of the major tasks addressed throughout this book is the study of the behavior of orbits near fixed points and periodic orbits. In the language of difference equations, this is the study of the behavior of solutions near equilibrium points and periodic solutions. The corresponding theory is known as **stability theory**. In addition to elucidating local behavior of difference equations, stability theory very often leads to global results for an equation. In other words, stability theory describes the behavior of all solutions of the equation, see [E2], [KoL], and [KL]. We start by introducing basic notions of stability.

DEFINITION 1.2

(a) *An equilibrium solution or fixed point $\bar{x}$ of (1.1) is said to be* **stable** *if for any $\epsilon > 0$ there exists $\delta > 0$ such that whenever $|x_0 - \bar{x}| < \delta$, the points x_n in the orbit $\gamma^+(x_0)$ satisfy $|x_n - \bar{x}| < \epsilon$.*

(b) *An equilibrium solution $\bar{x}$ of (1.1) is said to be* **unstable** *if it is not stable.*

(c) *An equilibrium solution $\bar{x}$ of (1.1) is said to be* **asymptotically stable** *or a* **sink**, *or an* **attracting fixed point** *of the function f if it is stable and, in addition, there exists $r > 0$ such that for all x_0 satisfying $|x_0 - \bar{x}| < r$, the iterates x_n satisfy $\lim_{n\to\infty} x_n = \bar{x}$.*

(d) *An equilibrium solution $\bar{x}$ of (1.1) is a* **global attractor** *on an interval I if*

$$x_0 \in I \quad \text{implies} \quad \lim_{n\to\infty} x_n = \bar{x}.$$

(e) An equilibrium solution $\bar{x}$ of (1.1) is **globally asymptotically stable** *if it is stable and is a global attractor.*

(f) An equilibrium solution $\bar{x}$ of (1.1) is a **source**, *or a* **repeller** *of the function f in (1.1) on an interval I if there exists $r > 0$ such that for all $x_0 \in I$ with $0 < |x_0 - \bar{x}| < r$, there exists $N \geq 1$ such that $|x_N - \bar{x}| \geq r$.*

It is known from the theory, see [D1], [E1], [E2], [KoL], that, under certain conditions, the stability type of an equilibrium $\bar{x}$ of (1.1) is the same as the stability type of the equilibrium point of the corresponding linearized equation

$$y_{n+1} = f'(\bar{x})y_n. \tag{1.9}$$

A simple but not rigorous argument is that for x close to the value of an equilibrium point $\bar{x}$ we have

$$f(x) = f(\bar{x}) + f'(\bar{x})(x - \bar{x}) + HOT,$$

where HOT denotes the higher-order terms in $x - \bar{x}$. Taking into account that $\bar{x}$ is an equilibrium, neglecting HOT, and replacing $x = u_n$, we obtain the approximate equation

$$u_{n+1} = \bar{x} + f'(\bar{x})(u_n - \bar{x}).$$

Finally, setting $y_n = u_n - \bar{x}$, we obtain equation (1.9).

The stability of (1.9) is evident from Theorem 1.1 and/or the corresponding staircase diagrams from which the following **linearized stability result** follows.

THEOREM 1.3

([Linearized Stability]) Let $f : R \to R$ be continuously differentiable in a neighborhood of the equilibrium point $\bar{x}$. An equilibrium point $\bar{x}$ of f is asymptotically stable if $|f'(\bar{x})| < 1$, and it is unstable if $|f'(\bar{x})| > 1$.

We now introduce the notion of hyperbolic equilibrium points of (1.1).

DEFINITION 1.3 *An equilibrium point $\bar{x}$ of (1.1) is* **hyperbolic** *if f is differentiable at $\bar{x}$ and $|f'(\bar{x})| \neq 1$.*

From the linearized stability Theorem 1.3, if the equilibrium point $\bar{x}$ of (1.1) is hyperbolic, then it must be either asymptotically stable or unstable, and the stability type is determined from the size of $f'(\bar{x})$. However, we should emphasize that the behavior of orbits near an equilibrium point is different depending on whether $f'(\bar{x}) > 0$ or $f'(\bar{x}) < 0$. In fact, the following linearized oscillation theorem that parallels the linearized stability theorem 1.3 holds, see [KL1]. Let us first define oscillation of a sequence.

DEFINITION 1.4 (Oscillation)

(a) A sequence $\{x_n\}_{n=0}^{\infty}$ is said to have **eventually** *some property P, if there exists an integer $N \geq k$ such that every term of $\{x_n\}_{n=N}^{\infty}$ has this property.*

(b) A sequence $\{x_n\}$ is said to **oscillate about zero** *or simply to* **oscillate** *if the terms x_n are neither eventually all positive nor eventually all negative. Otherwise the sequence is called* **nonoscillatory**. *A sequence $\{x_n\}$ is called* **strictly oscillatory** *if for every $n_0 \geq 0$, there exist $n_1, n_2 \geq n_0$ such that $x_{n_1} x_{n_2} < 0$.*

(c) A sequence $\{x_n\}$ is said to **oscillate about** $\bar{x}$ *if the sequence $x_n - \bar{x}$ oscillates. The sequence $\{x_n\}$ is called* **strictly oscillatory about** $\bar{x}$ *if the sequence $x_n - \bar{x}$ is strictly oscillatory.*

THEOREM 1.4 (Linearized Oscillation Theorem)

Let $f : R \to R$ be a continuous function such that $uf(u) > 0$ for $u \neq 0$, and $f'(0) = 1$. If there exists $r > 0$ such that either $f(u) \leq u$ for $u \in [0, r]$ or $f(u) \geq u$ for $u \in [-r, 0]$, then every solution of

$$x_{n+1} - x_n + pf(x_n) = 0 \tag{1.10}$$

oscillates if and only if every solution of the corresponding linearized equation

$$y_{n+1} - y_n + py_n = 0$$

oscillates, that is, if and only if $p \geq 1$.

The stability type of a nonhyperbolic equilibrium point cannot be determined from the linearization of the map. We illustrate this later in Examples 1.7 to 1.10.

Since a periodic point p of minimal period k is an equilibrium point of the map f^k, the notion of the stability of p follows from the definition of an equilibrium point, and the linearized stability result can be applied to f^k to determine the stability type of p.

DEFINITION 1.5 (Stability of Periodic Points)

A periodic point p of equation (1.1) of minimal period k is said to be **stable, asymptotically stable, unstable,** *or a* **global attractor** *if p is, respectively, stable, asymptotically stable, an unstable equilibrium point, or a global attractor of f^k.*

In the special case of a **period-two solution** $(p, f(p))$ of (1.1) we obtain that this solution is stable if

$$|f'(p)f'(f(p))| < 1,$$

and unstable if

$$|f'(p)f'(f(p))| > 1.$$

This result follows from Theorem 1.3 and the chain rule applied to the function $f^2(x) = f(f(x))$.

The number $\lambda = |f'(p)f'(f(p))|$ is called a **multiplier** of the orbit. Likewise, the multiplier of a periodic orbit of any period n can be defined. Again, applying the chain rule to the function $f^k(x)$, we conclude that the multiplier is given by

$$\lambda = f'(p)f'(f(p))\cdots f'(\ldots(f(p)\ldots).$$

In addition to determining the attracting fixed point and the attracting periodic point A we want to locate the maximal set that is attracted to A. Such a set is called the **basin of attraction** of A and is denoted by $\mathcal{B}(A)$.

DEFINITION 1.6 *Let $\bar{x}$ be an asymptotically stable fixed point of a map f. The* **basin of attraction** *$\mathcal{B}(\bar{x})$ of $\bar{x}$ is defined as the maximal set J that contains $\bar{x}$ and is such that*

$$f^n(x) \to \bar{x}, \quad \text{as} \quad n \to \infty \quad \text{for every } x \in J.$$

The basin of attraction of an attracting periodic point of period p is defined in analogous fashion, with the map f replaced by the p-th iterate f^p.

Example 1.6 The map $f(x) = 2x(1-x)$ has an attracting fixed point $\bar{x} = 1/2$ with a basin of attraction $\mathcal{B}(1/2) = (0, 1)$. ∎

We shall see later that the basins of attraction may have very complicated structures even for very simple looking maps. As a rule in the case of chaotic maps the basins of attraction are as complicated as Cantor sets. (See Section 1.8.) Finding the basin of attraction of a fixed point or a periodic point is in general a difficult task. Here we present one of the basic topological properties of basins of attraction. To do so we define the important concept of the invariant set.

DEFINITION 1.7 *A set M is said to be* **invariant** *under a map f if $f(M) \subset M$, that is, if for every $x \in M$ the elements of the orbit $\gamma^+(x)$ belong to M.*

This concept will be generalized in Chapter 3 to the case of R^k. The next result is proved in [E2].

THEOREM 1.5

Let $\bar{x}$ be an attracting fixed point of a map f. Then the basin of attraction $\mathcal{B}(\bar{x})$ is an invariant open interval.

In the general case of R^k one can show that the basin of attraction of an attracting fixed point is an invariant open set. See Example 1.6.

1.5 Stability in the Nonhyperbolic Case

When $\bar{x}$ is a nonhyperbolic equilibrium point of (1.1) two cases are possible:

$$(a) \quad f'(\bar{x}) = 1$$
$$(b) \quad f'(\bar{x}) = -1. \tag{1.11}$$

In view of Theorem 1.4, in case (a) the map f is increasing in some neighborhood of $\bar{x}$, which implies that every orbit that starts in this neighborhood is a monotonic sequence. If we assume that f has a continuous second derivative, then a more precise statement can be given about the stability of $\bar{x}$.

1.5.1 The Case $f'(\bar{x}) = 1$

Here we consider the case when $f'(\bar{x}) = 1$. Let us start with examples that illustrate possible behaviors.

Example 1.7 Consider the equation

$$x_{n+1} = x_n + x_n^3, \quad n = 0, 1, \ldots . \tag{1.12}$$

The only equilibrium point of this equation is zero. Assume that $x_0 > 0$. Then $x_1 = x_0 + x_0^3 > x_0$, and by using induction one can prove that $x_n > x_{n-1}$ for $n = 1, 2, \ldots$. The sequence either converges to a fixed point or diverges to infinity. Since the only fixed point is zero, we have that $\{x_n\}$ diverges to ∞.

On the other hand, assume $x_0 < 0$. Then, $x_1 = x_0 + x_0^3 < x_0$, and by using induction one can prove that $x_n < x_{n-1}$ for $n = 1, 2, \ldots$. This implies that $\{x_n\}$ diverges to $-\infty$. Thus, in this case, the equilibrium is a source. ∎

Example 1.8 Consider the equation

$$x_{n+1} = x_n - x_n^3, \quad n = 0, 1, \ldots . \tag{1.13}$$

The only equilibrium point of this equation is zero. Assume that $x_0 \in (0, 1/\sqrt{3})$. Then $x_1 = x_0 - x_0^3 < x_0$ and $x_1 \in (0, 1/\sqrt{3})$. By using induction one can prove that $\{x_n\}$ is a decreasing sequence of numbers from $(0, 1)$. Consequently, by the monotone convergence principle, $\{x_n\}$ is convergent to zero.

On the other hand, assume $x_0 \in (-1/\sqrt{3}, 0)$. Then, $x_1 = x_0 - x_0^3 > x_0$ and by using induction one can prove that $\{x_n\}$ is an increasing sequence of numbers from $(-1/\sqrt{3}, 0)$. Consequently, by the monotone convergence principle, $\{x_n\}$ is convergent zero. Thus, in this case, the equilibrium is a sink. ∎

Example 1.9 Consider the equation

$$x_{n+1} = x_n + x_n^2, \quad n = 0, 1, \dots . \tag{1.14}$$

The only equilibrium point of this equation is zero. Assume that $x_0 > 0$. Then $x_1 = x_0 + x_0^2 > x_0$ and by using induction one can prove that $x_n > x_{n-1}$ for $n = 1, 2, \dots$. Consequently, $\{x_n\}$ diverges to ∞.

Now, assume $x_0 \in (-1/2, 0)$. Then, $x_1 = x_0 + x_0^2 > x_0$ and $x_1 \in (-1/2, 0)$. By using induction one can prove that $\{x_n\}$ is an increasing sequence of numbers from $(-1/2, 0)$. Consequently, by the monotone convergence principle, $\{x_n\}$ is convergent to zero. ∎

Example 1.10 Consider the equation

$$x_{n+1} = x_n - x_n^2, \quad n = 0, 1, \dots . \tag{1.15}$$

The only equilibrium point of this equation is zero. Assume that $x_0 < 0$. Then $x_1 = x_0 - x_0^2 < x_0$ and by using induction one can prove that $\{x_n\}$ is a decreasing sequence. Consequently, $\{x_n\}$ diverges to $-\infty$.

Now, assume $x_0 \in (0, 1/2)$. Then, $x_1 = x_0 - x_0^2 < x_0$ and $x_1 \in (0, 1/2)$. By using induction one can prove that $\{x_n\}$ is an decreasing sequence of numbers from $(0, 1)$. Consequently, by the monotone convergence principle $\{x_n\}$ is convergent to the unique equilibrium that is zero. ∎

The examples we just discussed motivate the following definition.

DEFINITION 1.8 (Semi-Stable Equilibrium Points)

1. *The equilibrium point $\bar{x}$ of (1.1) is called* **semi-stable from below** *if there exists $r > 0$ such that the following statements are true.*

   (i) If $\{x_n\}_{n=0}^{\infty}$ is a solution of (1.1) with $\bar{x} - r < x_0 < \bar{x}$, then

   $$x_0 < x_1 < \cdots < x_n < \cdots \quad \text{and} \quad \lim_{n\to\infty} x_n = \bar{x}.$$

   (ii) If $\{x_n\}_{n=0}^{\infty}$ is a solution of (1.1) with $\bar{x} < x_0 < \bar{x} + r$, then there exists $N \geq 1$ such that

   $$\bar{x} < x_0 < \cdots < x_{N-1} < \bar{x} + r \leq x_N.$$

2. *The equilibrium point* $\bar{x}$ *is called* **semi-stable from above** *if there exists* $r > 0$ *such that the following statements are true.*

   (i) *If* $\{x_n\}_{n=0}^{\infty}$ *is a solution of (1.1) with* $\bar{x} < x_0 < \bar{x} + r$, *then*

$$x_0 > x_1 > \cdots > x_n > \cdots \quad \text{and} \quad \lim_{n\to\infty} x_n = \bar{x}.$$

   (ii) *If* $\{x_n\}_{n=0}^{\infty}$ *is a solution of (1.1) with* $\bar{x} - r < x_0 < \bar{x}$, *then there exists* $N \geq 1$ *such that*

$$x_N \leq \bar{x} - r < x_{N-1} < \cdots < x_0 < \bar{x}.$$

3. *The equilibrium point* $\bar{x}$ *is called* **semi-stable** *if it is either semi-stable from above or from below.*

We are now ready for the first result of this section, which deals with the case when $f''(\bar{x}) \neq 0$. The proofs of the results on semi-stability can be found in [Sf], pp. 160–172.

THEOREM 1.6 (Semi-Stable Case – Second Derivative Test) *Assume that* f *has a continuous second derivative at an equilibrium point* $\bar{x}$. *Suppose that* $f'(\bar{x}) = 1$ *and that* $f''(\bar{x}) \neq 0$. *Then* $\bar{x}$ *is a semi-stable equilibrium solution of (1.1). In particular, the following statements are true.*

1. *If* $f''(\bar{x}) < 0$, *then* $\bar{x}$ *is a semi-stable-from- above equilibrium solution of (1.1).*

2. *If* $f''(\bar{x}) > 0$, *then* $\bar{x}$ *is semi-stable-from-below equilibrium solution of (1.1).*

If $f'(\bar{x}) = 1$ and $f''(\bar{x}) = 0$, then the existence of the third derivative of f at $\bar{x}$ may resolve the stability question.

THEOREM 1.7 (Semi-Stable Case – Third Derivative Test) *Assume that* f *has a continuous third derivative at an equilibrium point* $\bar{x}$. *If* $f'(\bar{x}) = 1$, $f''(\bar{x}) = 0$, *and* $f'''(\bar{x}) \neq 0$, *then* $\bar{x}$ *is stable if* $f'''(\bar{x}) < 0$, *and* $\bar{x}$ *is unstable if* $f'''(\bar{x}) > 0$.

Example 1.11 Let us apply Theorems 1.6 and 1.7 to give alternative solutions to the problems of stability of equations (1.12)–(1.15).

In the case of equation (1.12), the corresponding function f is given by $f(x) = x + x^3$, and the equilibrium point is $\bar{x} = 0$. All conditions of Theorem 1.7 are satisfied and the equilibrium point is unstable.

In the case of equation (1.13), the corresponding function f is given by $f(x) = x - x^3$ and the equilibrium point is $\bar{x} = 0$. All conditions of Theorem 1.7 are satisfied and the stability of the equilibrium point follows.

In the case of equation (1.14), the corresponding function f is given by $f(x) = x + x^2$ and the equilibrium point is $\bar{x} = 0$. All conditions of Theorem 1.6 are satisfied and the equilibrium point is semi-stable from below.

In the case of equation (1.15), the corresponding function f is given by $f(x) = x - x^2$ and the equilibrium point is $\bar{x} = 0$. All conditions of Theorem 1.6 are satisfied and the equilibrium point is semi-stable from above. ∎

1.5.2 The Case $f'(\bar{x}) = -1$

In this case the map f is not monotone but rather oscillatory, and it flips from a point close to $\bar{x}$ to the other side of $\bar{x}$. If the equilibrium point $\bar{x}$ becomes unstable, an orbit cannot approach $\bar{x}$. However, if the iterates remain bounded, it is possible that the odd iterates (staying on the same side of $\bar{x}$) converge to a limit point, say, p and the even iterates (staying on the same side of $\bar{x}$) converge to a limit point $f(p)$. If this happen, then $f(f(p)) = p$ with $f(p)$ different from p (here $f(p)$ and p are on different sides of q). Then, p is a periodic point of period two. This change in global behavior of solutions of (1.1) is called period-doubling or flip bifurcation. A **bifurcation diagram** showing the limiting set versus a parameter is frequently used to graphically illustrate this phenomenon.

Example 1.12 Consider the equation

$$x_{n+1} = -x_n, \quad n = 0, 1, \dots . \tag{1.16}$$

The unique equilibrium point is 0. Every solution of this equation, except the equilibrium point, is periodic with period two. ∎

Example 1.13 Consider the equation

$$x_{n+1} = -x_n + x_n^2, \quad n = 0, 1, \dots . \tag{1.17}$$

The equilibrium points of this equation are $\bar{x}_1 = 0$ and $\bar{x}_2 = 2$. Let us investigate the behavior of even-indexed terms x_{2k}. Equation (1.17) gives

$$x_{2k+1} = -x_{2k} + x_{2k}^2$$

and

$$x_{2k+2} = -x_{2k+1} + x_{2k+1}^2 = -(-x_{2k} + x_{2k}^2) + (-x_{2k} + x_{2k}^2)^2.$$

Thus we obtain

$$x_{2k+2} = x_{2k} - 2x_{2k}^3 + x_{2k}^4.$$

Set $y_k = x_{2k}$ $y_k = x_{2k}$ in this equation to obtain

$$y_{k+1} = y_k - 2y_k^3 + y_k^4 = f(y_k). \tag{1.18}$$

Since $f'(0) = 1, f''(0) = 0, f'''(0) = -8$, Theorem 1.7 implies that $\bar{x}_1 = 0$ is a sink. Likewise, we conclude that $\bar{x}_2 = 2$ is a source.

Similarly we conclude that $\{x_{2k+1}\}$ satisfies equation (1.18), hence it is convergent to the zero equilibrium. Consequently, both even–indexed terms and odd-indexed terms are convergent to zero, hence $\bar{x}_1 = 0$ is a sink, while $\bar{x}_2 = 2$ is a source. ▮

Example 1.14 Consider the equation

$$x_{n+1} = -x_n - x_n^3, \quad n = 0, 1, \ldots . \tag{1.19}$$

The unique equilibrium point of this equation is $\bar{x}_1 = 0$. Let us investigate the behavior of even-indexed terms x_{2k}. Equation (1.19) gives

$$x_{2k+1} = -x_{2k} - x_{2k}^3$$

and

$$x_{2k+2} = -x_{2k+1} - x_{2k+1}^3 = -(-x_{2k} - x_{2k}^3) + (-x_{2k} - x_{2k}^3)^3.$$

Thus we obtain

$$x_{2k+2} = x_{2k} + x_{2k}^3 + (x_{2k} + x_{2k}^3)^3.$$

Set $y_k = x_{2k}$ in this equation to obtain

$$y_{m+1} = y_m + y_m^3 + (y_m + y_m^3)^3 = f(y_m). \tag{1.20}$$

Since $f'(0) = 1, f''(0) = 0, f'''(0) = 12$, then Theorem 1.7 implies that zero equilibrium is a source.

Similarly, we conclude that $\{x_{2k+1}\}_{k=0}^{\infty}$ satisfies (1.20), hence the zero equilibrium is a source. ▮

We have the following results, see [D1], [E2], and [Sf].

THEOREM 1.8 (Stability – Case $f'(\bar{x}) = -1$)
Let $f(x)$ be a three times continuously differentiable function in a neighborhood of an equilibrium point $\bar{x}$ such that $f'(\bar{x}) = -1$. Set

$$S_f(\bar{x}) = 2f'''(\bar{x}) + 3(f''(\bar{x}))^2 \tag{1.21}$$

If $S_f(\bar{x}) > 0$, then $\bar{x}$ is asymptotically stable, and if $S_f(\bar{x}) < 0$, then $\bar{x}$ is unstable.

The last result can be restated in terms of the so-called **Schwarzian derivative** of f.

DEFINITION 1.9 *Let $f(x)$ be a three times continuously differentiable function at a point x such that $f'(x) \neq 0$. The* **Schwarzian derivative** *of f at the point x is defined as*

$$\mathcal{S}f(x) = \frac{f'''(x)}{f'(x)} - \frac{3}{2}\left(\frac{f''(x)}{f'(x)}\right)^2.$$

COROLLARY 1.1

Let $f(x)$ be a three times continuously differentiable function in a neighborhood of an equilibrium point $\bar{x}$. Suppose also that either $f'(\bar{x}) = 1$ and $f''(\bar{x}) = 0$, or $f'(\bar{x}) = -1$. Then the following statements are true.

1. *If $\mathcal{S}f(\bar{x}) < 0$, then $\bar{x}$ is a sink of (1.1).*
2. *If $\mathcal{S}f(\bar{x}) > 0$, then $\bar{x}$ is a source of (1.1).*

Example 1.15 Let us apply Theorem 1.8 to give alternative solutions to the problems of stability of equations (1.17) and (1.19).

In the case of equation (1.17), the corresponding function f is given by $f(x) = -x + x^2$ and the equilibrium point is $\bar{x} = 0$. A simple computation gives $S_f(0) = 12$. Thus, all conditions of Theorem 1.6 are satisfied, hence the equilibrium point is a sink.

Similarly, in the case of equation (1.19), the corresponding function f is given by $f(x) = -x-x^3$ and the equilibrium point is $\bar{x} = 0$. Simple computation gives $S_f(0) = -12$. All conditions of Theorem 1.6 are satisfied, hence the equilibrium point is a source. ▮

1.6 Bifurcations

In this section we consider an interesting case of nonhyperbolic equilibrium point $\bar{x}$ when $f'(\bar{x}) = -1$. As we have seen in this case it is possible to get all solutions to be periodic, see (1.16). Let us motivate our main results by the following example.

Example 1.16 (Discrete Logistic Equation)

In the early 1970s, May [M1] and others began to study the equations used by biologists to model fluctuations in certain species such as insects and fish. Perhaps the most famous nonlinear discrete equation used to model a single species is the *logistic equation*, given by

$$x_{n+1} = f_p(x_n) = px_n(1 - x_n), \quad n = 0, 1, \ldots \tag{1.22}$$

where $p > 0$ is a parameter and $0 \leq x_n \leq 1$ represents the scaled population size at time n. The parameter p measures the reproduction rate. In [M1] the author considers the population of blowflies. Notice that equation (1.22) may be written as

$$x_{n+1} - x_n = px_n(1 - x_n) - x_n = x_n(p - 1 - x_n),$$

which implies that the corresponding differential equation has the form

$$y' = py(p - 1 - y). \tag{1.23}$$

The last equation is a well-known logistic differential equation that has a simple general solution of the form

$$y(t) = \frac{p-1}{\left(1 + (p-1)C_1 e^{-pt(-1+p)}\right)}.$$

One can consider equation (1.22) as the Euler discretization of the corresponding differential equation (1.23). In the case of equation (1.22) there are known explicit solutions in the cases where $p = 2$ and $p = 4$. There is no known formula for the solutions of equation (1.22) for any other value of parameter p (there is no proof that such a formula does not exist either). To find the equilibrium points, it is necessary to solve the equation

$$f_p(x) = px(1 - x) = x,$$

which gives two solutions: $\bar{x_1} = 0$ and $\bar{x_2} = 1 - \frac{1}{p}$. By using Theorem 1.3 one may establish the stability character of both of these equilibrium points. The multiplier at the equilibrium point $\bar{x_1} = 0$ is $f_p'(0) = p$. Thus $\bar{x_1} = 0$ is locally asymptotically stable for $0 < p < 1$ and unstable for $p > 1$. Next, the multiplier at the equilibrium point $\bar{x_2} = 1 - \frac{1}{p}$ is $f_p'(1 - \frac{1}{p}) = 2 - p$. Thus $\bar{x_2}$ is locally asymptotically stable for $1 < p < 3$ and unstable for $p > 3$.

To find period-two solutions, it is necessary to solve the following equation:

$$f_p(f_p(x)) = p^2x(1 - x)(1 - px(1 - x)) = x. \tag{1.24}$$

Two solutions of equation (1.24) are known, namely, the equilibrium points $\bar{x_1} = 0$ and $\bar{x_2} = 1 - \frac{1}{p}$, because every equilibrium point is also a periodic point of any period. Therefore, equation (1.24) can be simply factored and reduced to a quadratic equation with roots

$$x_{2,1} = \frac{\frac{1}{2}p + \frac{1}{2} + \frac{1}{2}\sqrt{(p-3)(p+1)}}{p} \quad \text{and} \quad x_{2,2} = \frac{\frac{1}{2}p + \frac{1}{2} - \frac{1}{2}\sqrt{(p-3)(p+1)}}{p}.$$

Thus there are two points of prime period-two when $p > 3$. So, as the parameter p changes its values, the qualitative behavior of the solutions also changes. Let us denote the value of the parameter p at the k-th point where the change occurs as b_k. We have the following table.

parameter interval	type of behavior	critical value of parameter
$0 < p < 1$	zero equilibrium is asymptotically stable	$b_0 = 1$
$1 < p < 3$	positive equilibrium is asymptotically stable	$b_1 = 3$
$3 < p < 1 + \sqrt{6}$	period-two solution is asymptotically stable	$b_2 = 1 + \sqrt{6}$.

The values b_k are called the **bifurcation values** of the parameter. As we continue this process, we can find that the next bifurcation value b_3 does not correspond to the appearance of a prime period-three solution, but it corresponds to a prime period-four solution. The next bifurcation value b_4 corresponds to a prime period-eight solution. In addition, one can prove that for $p \in (b_2, b_3)$, the prime period-four solution is stable while the period-two solution becomes unstable. Similarly, one can prove that for $p \in (b_3, b_4)$, the prime period-eight solution is stable while the period-four solution becomes unstable. As this process continues, one can see that there is a sequence of bifurcation values of parameter $\{b_k\}_{k=0}^{\infty}$ with the following property: for $p \in (b_k, b_{k+1})$ the prime period-2^k solution is stable, while the periodic solutions of all periods $2, \ldots, 2^{k-1}$ become unstable.

This phenomenon is known as the **period-doubling bifurcation route to the chaos**. This means that as the parameter p increases beyond b_1, the equilibrium point branches into a period-two solution; at the value b_2, the period-two solution branches into a period-four solution, etc. The sequence of period-doubling bifurcations ends at the value which is approximately $b = 3.56994\ldots$, where equation (1.22) has the periodic solutions of all periods as well as some nonperiodic solutions. The last situation is often described as **chaotic behavior** or **chaos**. The last period that can arise in this bifurcation process is period three; this is the origin of the title of [LY]. There are three important features of the period-doubling bifurcation route to chaos. First, periods appear in the order $2, 4, 8, \ldots, 2^k, \ldots$ finishing with 3; the order is known as Sharkovsky's order, which is described in this section. Second, there is only one periodic solution which is stable in each of the intervals (b_k, b_{k+1})-this is an important result known as Singer's theorem which is given in this section. Third, the sequence $\{b_k\}_{k=0}^{\infty}$ has the remarkable property

$$\lim_{k\to\infty} \frac{b_k - b_{k-1}}{b_{k+1} - b_k} = \delta \approx 4.66920.$$

The constant δ is called **Myrberg's number** or **Feigenbaum's number**. It was first discovered by Myrberg in the series of papers [My1]–[My3] and was rediscovered by Feigenbaum in [F1] and [F2]. ∎

We present next a simple heuristic derivation of the Myrberg-Feigenbaum constant. Let us assume that the sequence of bifurcation values $\{b_n\}$ converges to a limiting value b_∞ and that convergence is geometric with the ratio δ, that is,

$$b_n - b_\infty \approx C\delta^{-n}, \quad n \to \infty, \tag{1.25}$$

where C is a constant. This assumption is known as a **scaling hypothesis**, see [Mt] and [D]. Equation (1.25) implies that

$$b_n - b_{n+1} = (b_n - b_\infty) - (b_{n+1} - b_\infty) \approx C\delta^{-n} - C\delta^{-n-1} = C\delta^{-n}(1 - \delta^{-1}). \tag{1.26}$$

Then,

$$\delta \approx \frac{b_n - b_{n+1}}{b_{n+1} - b_{n+2}}, \quad n \to \infty. \tag{1.27}$$

We can also estimate r_∞ by observing that

$$0 \approx b_n - b_\infty - \delta(b_{n+1} - b_\infty) \approx b_n - b_\infty - \frac{b_n - b_{n+1}}{b_{n+1} - b_{n+2}}(b_{n+1} - b_\infty).$$

Solving this approximate equation for b_∞ yields

$$b_\infty = \frac{b_n b_{n+2} - b_{n+1}^2}{b_n - 2b_{n+1} + b_{n+2}}.$$

Now we give rigorous results behind some of the phenomena described in Example 1.16.

1.6.1 Sharkovsky's Theorem and Period Doubling

In 1975, Li and Yorke [LY] published the seminal paper "Period Three Implies Chaos" in the *American Mathematical Monthly* in which they proved that if a continuous map f has a period-three solution, then it must have solutions of any period k. A precise formulation of a special case of this theorem is the following.

THEOREM 1.9 (Period Three Implies Chaos)
Let $f : I \longrightarrow I$ be a continuous map on an interval I. If this map has a periodic point p_3 of **minimal period** *3 (i.e., $f^3(p_3) = p_3$ and $f^m(p_3) \neq p_3$ for $m = 1, 2$), then for every $k = 1, 2, \ldots$ there is a periodic point p_k of minimal period k (i.e., $f^k(p_k) = p_k$ and $f^m(p_k) \neq p_k$ for $m = 1, 2, \ldots$).*

The proof of this theorem is technical and can be found in [BlC], [E2], [LY], and [Ro].

Theorem 1.9 is a special case of a theorem published in 1964 by the Ukranian mathematician A. N. Sharkovsky [Sh]. Sharkovsky introduced a new ordering of the positive integers $\triangleright$ in which the number 3 appears first. He proved that if $k \triangleright m$ and f has a k-periodic point, then it must have an m-periodic point. To state this result more precisely, let us introduce **Sharkovsky's ordering** as follows:

$$3 \triangleright 5 \triangleright 7 \triangleright \ldots \triangleright 2 \times 3 \triangleright 2 \times 5 \triangleright 2 \times 7 \ldots \triangleright 2^n \times 3 \triangleright 2^n \times 5 \triangleright 2^n \times 7 \ldots \triangleright 2^n \triangleright 2^2 \triangleright 2 \triangleright 1.$$

We first list all the odd integers except 1, then 2 times the odd integers, 2^2 times the odd integers, etc. This is followed by powers of 2 in descending order ending in 1. The notation $k \triangleright m$ means that k appears before m in Sharkovsky's ordering.

THEOREM 1.10 (Sharkovsky's Theorem)

Let I be an interval in R (finite or infinite). Let $f : I \longrightarrow I$ be a continuous map. If f has a periodic point of period k, then it must possess a periodic point of period m for all m with $k \triangleright m$.

The proof of this theorem is highly technical and can be found in [BlC], [E2], and [Ro].

REMARK 1.2 Theorem 1.10 gives a very simple and intuitive geometric method for checking complicated behavior in similar cases. In fact, if we plot the bisector $y = x$ together with the graph of the map f and the graph of the third iterate $f^3(x) = f(f(f(x)))$, and if the graphs of the bisector and f^3 intersect in some points that are not the equilibrium points, then there is at least one periodic point of period three and consequently by Theorem 1.10 or Theorem 1.9 there are periodic points of all other periods. Similarly, one can check visually the existence of other periodic points of reasonably small periods. See Figures 1.4 and 1.5 for illustrations of this method. ∎

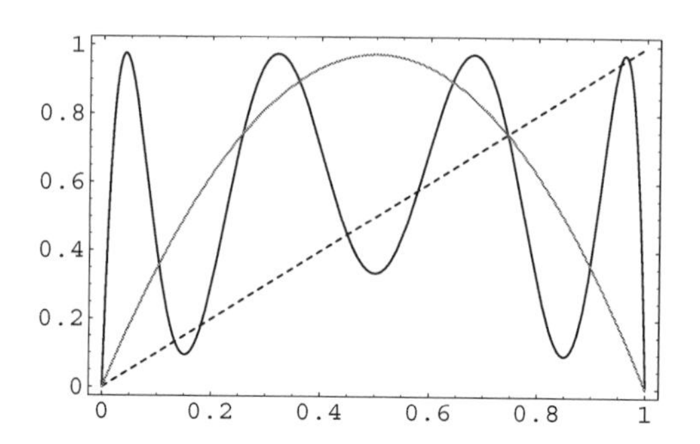

FIGURE 1.4: Plot of the third iterate f^3 of $f(x) = 3.9x(1-x)$. The function f^3 is shown here as a dark line, f is shown as a gray line, and the identity function is shown as dashed line. Two intersections correspond to fixed points of f. The remaining six intersections indicate that there are two prime period-three solutions $\{p_1, q_1, r_1\}$ and $\{p_2, q_2, r_2\}$. However, the plot does not provide sufficient information to distinguish between elements of the orbits.

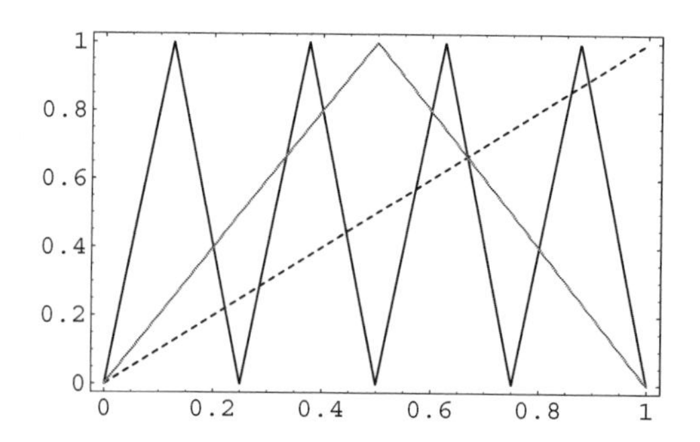

FIGURE 1.5: Plot of the third iterate T_3^2 of the tent map T_3 in the first quadrant. The function T_2^3 is shown here as a dark line, T_2 is shown as a gray line, and the identity function is shown as dashed line. Just as in the case of Figure 1.4 it is clear that there are two period-three orbits.

Sharkovsky's theorem does not extend to two or higher dimensions nor are there any known results analogous to this theorem. It is worth mentioning that one can construct a continuous map with a periodic point of period 5 but not with period 3, or with a periodic point of period 11 but not with period

5, etc. The following general result has been proved in [E2] and [E3]:

THEOREM 1.11 (A Converse of Sharkovsky's Theorem)
For any positive integer r, there exists a continuous map $f_r : I \longrightarrow I$ on the closed interval I such that f_r has a point of minimal period r but no points of minimal periods s, for all positive integers s that precede r in the Sharkovsky's ordering, i.e., $s \rhd r$.

Our next goal is to answer the following question: How many attracting periodic points can a differentiable map possess? A partial answer to the question is a well-known theorem by D. Singer [Si]. Recall that the symbol $Sf(x)$ denotes the Schwarzian derivative of function f.

THEOREM 1.12 (Singer's Theorem)
Let $f : I \longrightarrow I$ be a map defined on the closed interval I such that $Sf(x) < 0$ for all $x \in I$. If f has n critical points in I, then for every positive integer k, the map f has at most $n + 2$ attracting period-k solutions.

REMARK 1.3 Using Theorem 1.12 in a sophisticated way, see [E2], pp. 67, one can show that the logistic map $f_p(x) = px(1 - x)$, for $0 < p \le 4$ and $x \in [0, 1]$, has at most one attracting periodic solution (the equilibrium point is considered to be a periodic solution of period 1). ∎

Now we present a theoretical result that explains the appearance of bifurcation values b_k where period-doubling takes place of the logistic equation (1.22). In the statement of this result it is customary to assume that the equilibrium point is at the origin. This can always be achieved for a given equation

$$x_{n+1} = f(x_n), \quad n = 0, 1, \ldots,$$

with equilibrium point $\bar{x}$ by means of the simple transformation

$$y_n = x_n - \bar{x}.$$

The transformed equation

$$y_{n+1} = f(y_n + \bar{x}) - \bar{x}, \quad n = 0, 1, \ldots,$$

has zero as the equilibrium point that corresponds to the equilibrium point $\bar{x}$.

The logistic equation (1.22) and the corresponding map f_p also depend on the parameter p. Thus we have to consider a **one-parameter family of maps** to see how parameter changes affect the dynamic behavior of the solutions of the corresponding equation. The rigorous result is the following.

THEOREM 1.13 (Period-Doubling Bifurcations)
Let $f(x)$ be a three times continuously differentiable function with a fixed point at the origin satisfying the following conditions:

$$f(0) = 0, \quad f'(0) = -1, \quad f''(0) \neq 0.$$

If there exists a one-parameter family of functions $F(p, x)$ continuous in (p, x) and continuously differentiable with respect to x such that

$$F(0, x) = f(x), \quad F(a, 0) = 0, \quad \frac{\partial F}{\partial x}(a, 0) = -(1 + a),$$

then there exists a neighborhood of $(p, x) = (0, 0)$ in which,

a. *for values of p such that $(f^2)'''(0) < 0$, there exists a unique periodic orbit of minimal period 2 of the function $F(p, x)$;*

b. *for values of p such that $(f^2)'''(0) > 0$, there is no periodic orbit of minimal period 2.*

Furthermore, for this value of p, the period-two orbit is asymptotically stable (respectively, unstable) if the origin is an unstable (respectively, asymptotically stable) fixed point.

Here we discussed only period-doubling bifurcations. There exist several other types of bifurcation known as saddle node or tangent bifurcation, transcritical bifurcation, and pitchfork bifurcation, which are elaborated in textbooks such as [D1], [E2], [HK], and [Ml].

1.7 *Dynamica* Session

In this section the *Dynamica* package is introduced. Basic commands are used in Subsection 1.7.1 to produce orbits and their plots. Then, stability and multipliers are discussed in Subsection 1.7.2, bifurcation diagrams in Section 1.7.3, Lyapunov numbers in Subsection 1.7.4, and box dimension in Subsection 1.7.5.

1.7.1 Orbits and Periodic Points

After showing how to produce orbits and plots, we briefly treat an example of a difference equation defined in terms of a parameter, and look at changes in the behavior of solutions when the parameter is varied. Also, we shall see that in simple cases symbolic initial values may be used to generate orbits. In addition, we present here an example in which numerical problems are encountered, and where periods of solutions are found with *Dynamica.*

1.7.1.1 Generating Orbits and Their Plots with *Dynamica*

First we get acquainted with *Dynamica*'s basic commands for generating orbits and their corresponding plots.

Our first example is a dynamical system defined in terms of the constant A.

```
In[1]:= << Dynamica'
Dynamica Version 1.0

In[2]:= eqn = x[n + 1] == x[n]^2 + A;
```

The corresponding quadratic map may be obtained with the command DEToMap

```
In[3]:= q = DEToMap[eqn];

In[4]:= q[x]
Out[4]= x^2 + A
```

A particular value of the constant may be set in order to perform numerical simulations.

```
In[5]:= A = -0.5;
```

This is how one may obtain the 14-th iterate of the orbit $\gamma^{+}(0.1)$.

```
In[6]:= Iterate[q, 0.1, 14]
Out[6]= -0.363795
```

The first 14 terms of the orbit $\gamma^{+}(0.1)$ can be produced with the *Dynamica* command Orbit.

```
In[7]:= orbl = Orbit[q, 0.1, 14]
Out[7]= {0.1, -0.49, -0.2599, -0.432452,
         -0.312985, -0.40204, -0.338364,
         -0.38551, -0.351382, -0.376531,
         -0.358225, -0.371675, -0.361858,
         -0.369059, -0.363795}
```

We may use *Mathematica*'s TableForm function to display the orbit in tabular form. Here, the output suggests the existence of an asymptotically stable fixed point.

```
In[8]:= orbl // TableForm
         0.1
         -0.49
         -0.2599
         -0.432452
         -0.312985
         -0.40204
         -0.338364
Out[8]=  -0.38551
         -0.351382
         -0.376531
         -0.358225
         -0.371675
         -0.361858
         -0.369059
         -0.363795
```

The `Time Series plot` of the orbit shows that it converges to an equilibrium point.

```
In[9]:= TimeSeriesPlot[q, 0.1, 20];
```

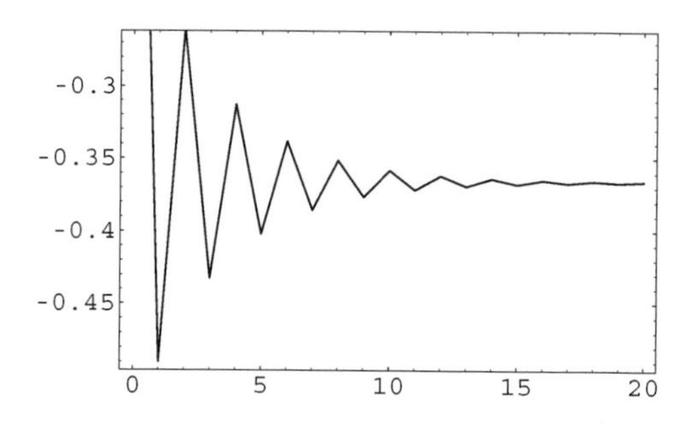

Here is another plot of the orbit, produced with the command `OrbitPlot`. Note that joining the points yields the time series plot.

```
In[10]:= OrbitPlot[q, 0.1, 20];
```

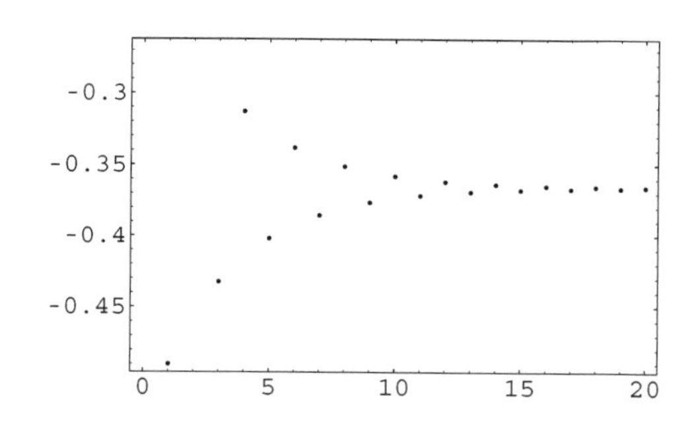

1.7.1.2 Sensitive Behavior of Solutions of Difference Equation

In this subsection we see that, at least for the map $f(x) = x^2 + A$, "relatively small" changes of the parameter may yield different behaviors of solutions.

By looking at the orbit $\gamma^+(0.1)$ for $A = -1.22$ we may see that there is an asymptotically periodic solution with period 2.

```
In[11]:= A = -1.22;

In[12]:= orb2 = Orbit[q, 0.1, 60];

In[13]:= Take[orb2, {50, 60}];
Out[13]= {0.188187, -1.18459, 0.183243,
         -1.18642, 0.187597, -1.18481,
         0.183768, -1.18623, 0.18714,
         -1.18498, 0.184174}
```

The time series plot shows the asymptotic period-two character of the solution. Note that this time the input to `TimeSeriesPlot` is a list of points.

```
In[14]:= TimeSeriesPlot[orb2];
```

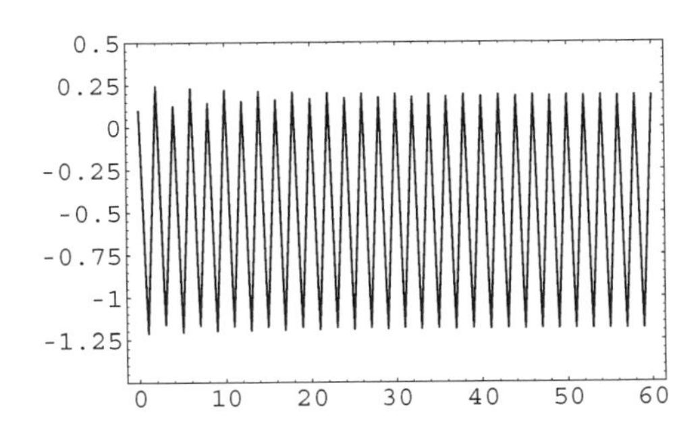

Dynamica's `OrbitPlot` command gives the same information. It also accepts lists of points as input.

```
In[15]:= OrbitPlot[orb2];
```

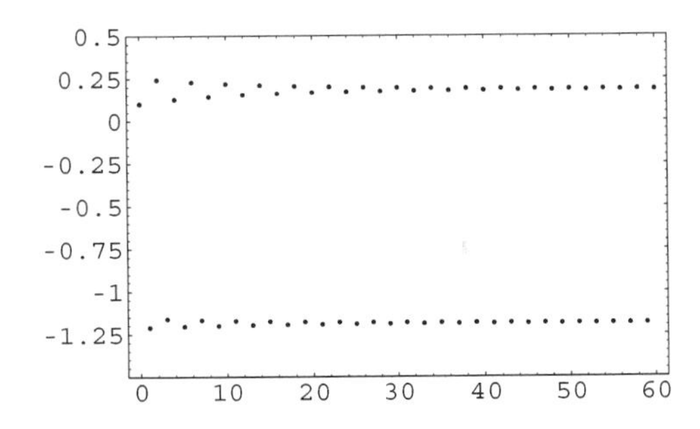

Another test, now with $A = -1.3$, suggests the existence of an asymptotically periodic point with period four. Shown here are the iterates $x_{50}, \ldots, x_{60}$.

```
In[16]:= A = -1.3;

In[17]:= orb3 = Orbit[q, 0.1, 60];

In[18]:= Take[orb3, {50, 60}]
Out[18]= {-1.29962, 0.389019, -1.14866,
          0.0194303, -1.29962, 0.389019,
          -1.14866, 0.0194303, -1.29962,
          0.389019, -1.14866}
```

Asymptotic behavior may be difficult to visualize with this time series plot.

```
In[19]:= TimeSeriesPlot[orb3];
```

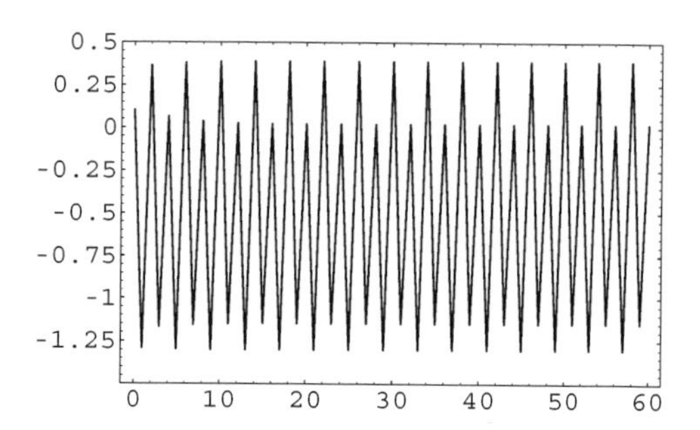

Plotting the orbit with `OrbitPlot` helps to see the asymptotic behavior of the sequence.

```
In[20]:= OrbitPlot[orb3];
```

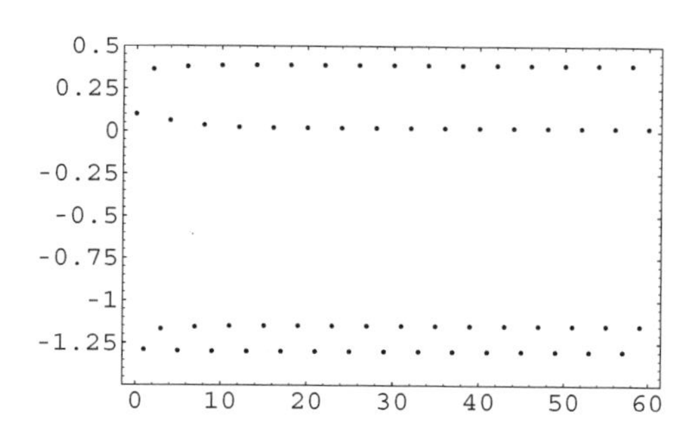

`FindMinimalPeriod` may be used to locate minimal periods of eventually periodic orbits.

```
In[21]:= FindMinimalPeriod[Orbit[q, 0., 200]]
Minimal period = 4
Periodic orbit =
  {-1.29962, 0.389019, -1.14866,
   0.0194303}
```

Finally, we perform one more test, this time with $A = -1.38$. A pattern seems to emerge when iterates $x_{50}, \ldots, x_{60}$ are displayed.

```
In[22]:= A = -1.38;

In[23]:= orb4 = Orbit[q, 0.1, 60];

In[24]:= Take[orb4, {50, 60}]
Out[24]= {-1.3667, 0.487872, -1.14198,
          -0.0758786, -1.37424, 0.508542,
          -1.12138, -0.122496, -1.36499,
          0.48321, -1.14651}
```

Plotting 60 points suggests that there is a possible period-four solution which is asymptotically stable.

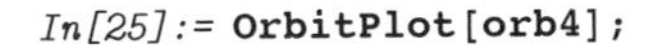

```
In[25]:= OrbitPlot[orb4];
```

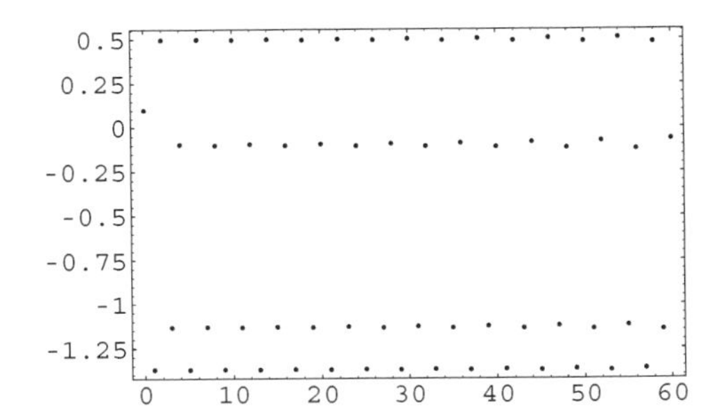

Plotting the first 600 points gives a surprise: the solution converges to a period-eight solution!

```
In[26]:= OrbitPlot[q, 0.1, 600];
```

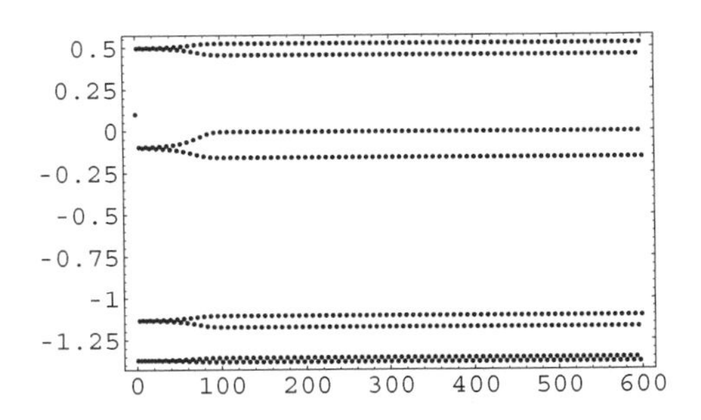

A histogram may be used to show the eventually periodic character of a solution. A histogram of the first 3000 terms of $\gamma^{+}(0.0)$ shows several spikes, indicating that the solution converges to the periodic solution of period-eight.

```
In[27]:= Histogram[Orbit[q, 0.1, 3000]];
```

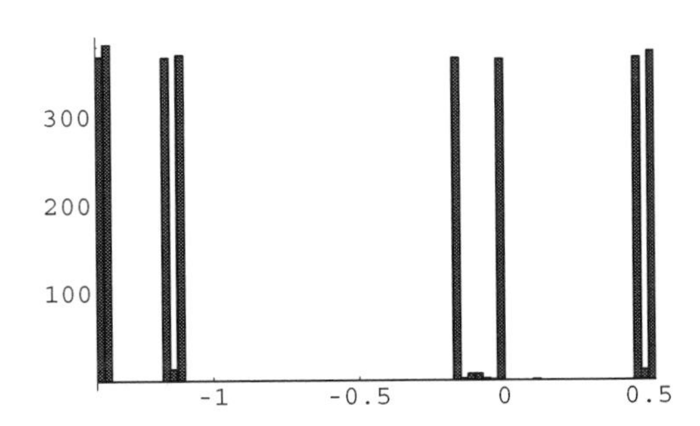

1.7.1.3 Initial Symbolic Values

In some simple cases, *Dynamica* may be used to establish that all solutions are periodic.

The `Orbit` function may also be used with symbolic initial values. Here is an example in which $\gamma^+(a)$ is periodic with period 2 for $a \neq 0$.

```
In[28]:= f = MapToDE[x[n + 1] == 1/x[n]];

In[29]:= Orbit[f, a, 4]
Out[29]= {a, 1/a, a, 1/a, a}
```

For this example, every solution is periodic with period 4, for all initial points $a \notin \{-1, 0, 1\}$.

```
In[30]:= g = MapToDE[x[n + 1] == (1 + x[n])/(1 - x[n])];

In[31]:= Simplify[Orbit[g, a, 7]]
Out[31]= {a, (1 + a)/(1 - a), -1/a, (-1 + a)/(1 + a), a,
          (1 + a)/(1 - a), -1/a, (-1 + a)/(1 + a)}
```

1.7.1.4 Numerical Error Buildup May Lead to Wrong Conclusions

The next example shows how `FindMinimalPeriod` works in the case when the orbit is generated by the *dbl*, which when applied to x gives the decimal part of $2x$. In symbols, $dbl(x) = 2x \pmod 1$.

We begin by defining the map and by generating an orbit with initial point $\frac{15}{13}$.

```
In[32]:= dbl[x_] := Mod[2 x , 1]

In[33]:= orb5 = Orbit[dbl, 15/13, 20]
Out[33]= {15/13, 4/13, 8/13, 3/13, 6/13, 12/13, 11/13,
          9/13, 5/13, 10/13, 7/13, 1/13, 2/13, 4/13,
          8/13, 3/13, 6/13, 12/13, 11/13, 9/13, 5/13}
```

One can also use *Dynamica*'s function `FindMinimalPeriod` to find the minimal period of a given orbit.

```
In[34]:= FindMinimalPeriod[orb5]
Minimal period = 12
Periodic orbit =
 {4/13, 8/13, 3/13, 6/13, 12/13, 11/13, 9/13, 5/13, 10/13,
  7/13, 1/13, 2/13}
```

Let us perform another round of computations, only this time with floating point arithmetic (instead of exact arithmetic).

```
In[35]:= orb6 = Orbit[dbl, 15./13., 20]
Out[35]= {1.15385, 0.307692, 0.615385,
          0.230769, 0.461538, 0.923077,
          0.846154, 0.692308, 0.384615,
          0.769231, 0.538462, 0.0769231,
          0.153846, 0.307692, 0.615385,
          0.230769, 0.461538, 0.923077,
          0.846154, 0.692308, 0.384615}
```

The difference due to round off is initially negligible, but it seems to grow with the iteration number.

```
In[36]:= orb5 - orb6
Out[36]= {-2.22045 × 10^-16, -2.22045 × 10^-16,
   -4.44089 × 10^-16, -9.4369 × 10^-16,
   -1.88738 × 10^-15, -3.77476 × 10^-15,
   -7.66054 × 10^-15, -1.53211 × 10^-14,
   -3.05866 × 10^-14, -6.11733 × 10^-14,
   -1.22458 × 10^-13, -2.4486 × 10^-13,
   -4.89719 × 10^-13, -9.79439 × 10^-13,
   -1.95888 × 10^-12, -3.91781 × 10^-12,
   -7.83562 × 10^-12, -1.56712 × 10^-11,
   -3.13426 × 10^-11, -6.26852 × 10^-11,
   -1.2537 × 10^-10}
```

The growth of the error is best shown on a `Log plot`.

```
In[37]:= LogListPlot[Abs[orb5 - orb6]]
```

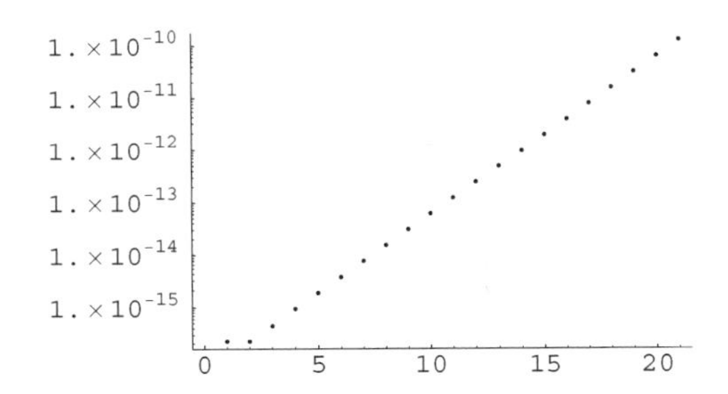

Error buildup causes `FindMinimalPeriod` to fail to detect an orbit. Thus, in some cases, there is the possibility of arriving at wrong conclusions because of error buildup.

```
In[38]:= FindMinimalPeriod[orb6]
No periodic orbit was detected
```

Next we consider an example from p. 26 in [D2]. We shall see once more that computer calculations may give the wrong answer.

The orbit $\gamma^{+}(0.4)$ seems to be eventually asymptotically stable.

```
In[39]:= orb7 = Orbit[dbl, 0.4, 60]
Out[39]= {0.4, 0.8, 0.6, 0.2, 0.4, 0.8, 0.6,
         0.2, 0.4, 0.8, 0.6, 0.2, 0.4, 0.8,
         0.6, 0.2, 0.4, 0.8, 0.6, 0.2, 0.4,
         0.8, 0.6, 0.2, 0.4, 0.8, 0.6, 0.2,
         0.4, 0.8, 0.6, 0.2, 0.4, 0.8, 0.6,
         0.200001, 0.400002, 0.800003,
         0.600006, 0.200012, 0.400024,
         0.800049, 0.600098, 0.200195,
         0.400391, 0.800781, 0.601563,
         0.203125, 0.40625, 0.8125,
         0.625, 0.25, 0.5, 0., 0.
         , 0., 0., 0., 0., 0., 0.}
```

A plot of the orbit $\gamma^{+}(0.4)$ also suggests asymptotic stability.

```
In[40]:= OrbitPlot[orb7];
```

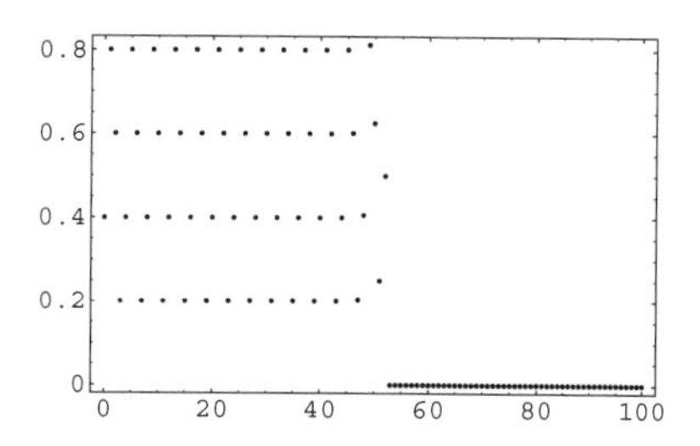

Here is the the same orbit $\gamma^{+}(\frac{2}{5})$, only now calculated with exact arithmetic.

```
In[41]:= orb8 = Orbit[dbl, 4/10, 20]
```

Out[41]= $\left\{\frac{2}{5}, \frac{4}{5}, \frac{3}{5}, \frac{1}{5}, \frac{2}{5}, \frac{4}{5}, \frac{3}{5}, \frac{1}{5}, \frac{2}{5}, \frac{4}{5}, \frac{3}{5}, \frac{1}{5}, \frac{2}{5}, \frac{4}{5}, \frac{3}{5}, \frac{1}{5}, \frac{2}{5}, \frac{4}{5}, \frac{3}{5}, \frac{1}{5}, \frac{2}{5}\right\}$

The number 0 doesn't seem to be an attractor anymore! In fact we get a period four solution. Why do we get different answers? (See Devaney [D2], p. 26 for a hint)

```
In[42]:= FindMinimalPeriod[orb8]
Minimal period = 4
Periodic orbit =
```

$\{\frac{2}{5}, \frac{4}{5}, \frac{3}{5}, \frac{1}{5}\}$

1.7.2 Stability and Multipliers

Stability of the orbit can be checked by using the `Multiplier` function which returns the product of the derivatives along the orbit. Of course this requires that the input function is one that *Mathematica* can differentiate.

Recall that a multiplier of absolute value less than 1 indicates an attractor.

1.7.2.1 Staircase Diagrams

A periodic orbit of period 4 that is asymptotically stable.

```
In[43]:= A = -1.3;
         orb9 = Orbit[q, 0, 3, 1000];
         Multiplier[q, orb8]
Out[43]= 0.180543
```

StaircaseDiagram plots the graph of a function, bisector and staircase (or stairstep) diagrams of up to three orbits. The orbits should be computed using Orbit. Note the rapid convergence to a period-four solution, with points that show in the figure as the intersection of the thick lines with the identity function.

```
In[44]:= StaircaseDiagram[q, orb9];
```

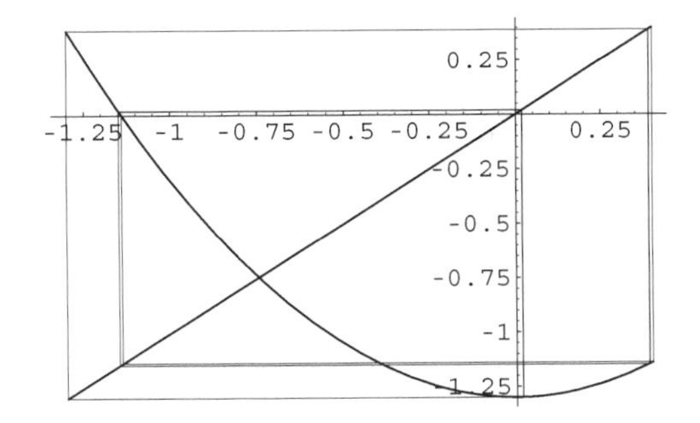

Adding another orbit produces a figure similar to the previous one.

```
In[45]:= StaircaseDiagram[q,
           {orb9, Orbit[q, 1.7, 1000]}]
```

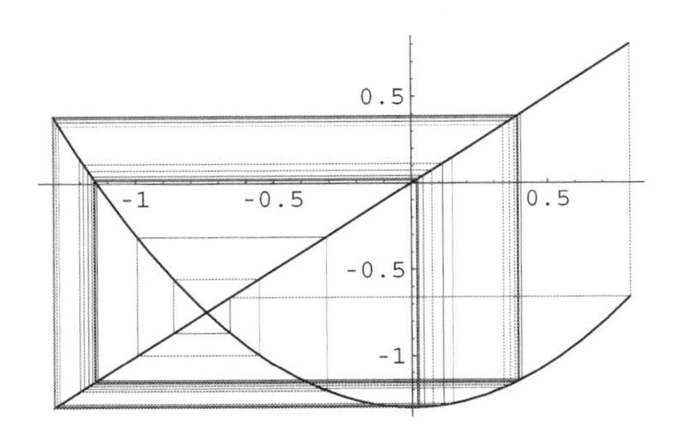

Let us first find the equilibrium points of the equation.

```
In[46]:= eqp = Solve[q[x] == x, x]
Out[46]= {{x → -0.74499}, {x → 1.74499}}
```

Here, the equilibrium points are named fp[1] and fp[2]

```
In[47]:= {fp[1], fp[2]} = x /. %
Out[47]= {-0.74499, 1.74499}
```

The solution is periodic with period 4, with semicycles of length 1. The positive semicycle is the number of consecutive terms of a solution greater than or equal to the equilibrium, and the negative semicycle is the number of consecutive terms smaller than the equilibrium of the equation. For a detailed discussion of semicycles see Section 3.9.

```
In[48]:= TimeSeriesPlot[q, 0., 40,
           AxesOrigin → {0, fp[1]}];
```

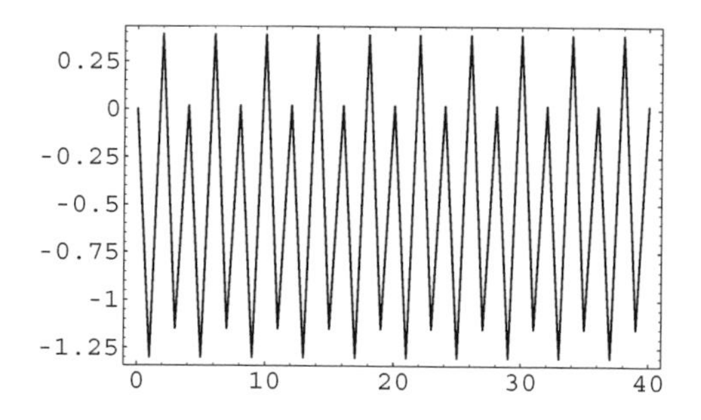

An orbit plot clearly shows the period-four solution.

```
In[49]:= OrbitPlot[q, 0.1, 40,
           AxesOrigin → {0, fp[1][[1]]}];
```

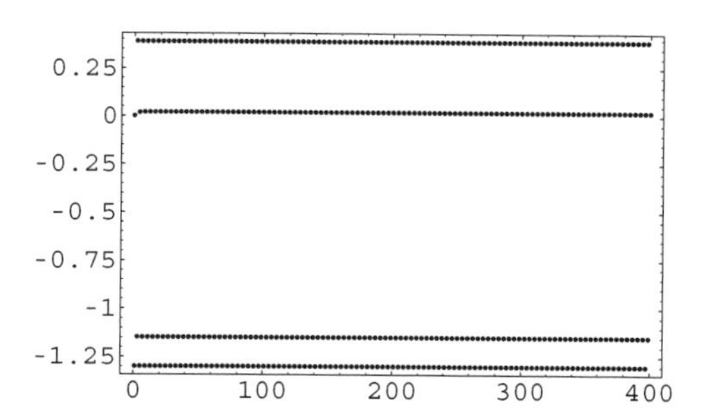

1.7.2.2 Rate of Attraction of Orbits

Finally, we illustrate how the multiplier of an attracting periodic solution determines the **rate of attraction** of nearby orbits. The distance to the equilibrium point should go to zero exponentially fast.

Here is an orbit for $A = -0.4$.

```
In[50]:= A = -0.4;

In[51]:= orb11 = Orbit[q, 0., 40];
```

A time series plot of the sequence shows rapid convergence of the sequence of points.

```
In[52]:= TimeSeriesPlot[orb11,
              PlotRange → All];
```

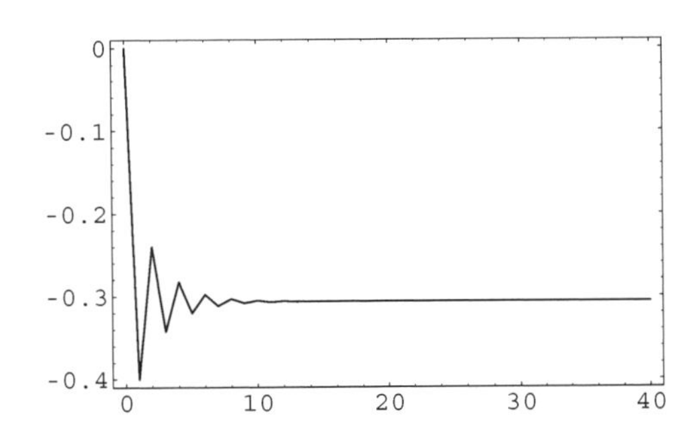

Here is a plot of the orbit.

```
In[53]:= OrbitPlot[orb11,
              PlotRange → All];
```

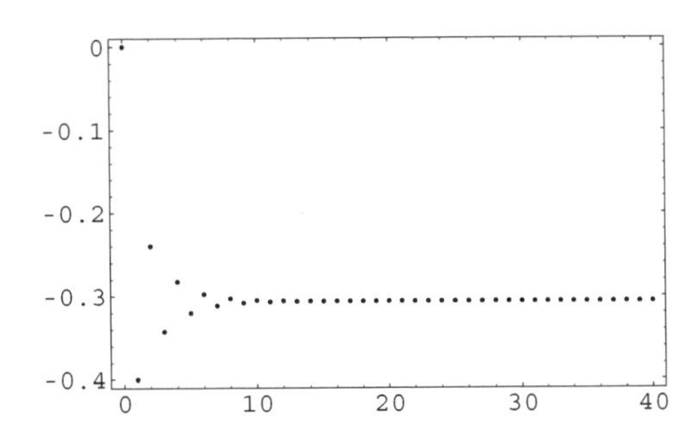

Now we shall see how fast the iterates approach the equilibrium point by using a computational method.

One can get a reasonable approximation of the equilibrium point by iterating the map many times.

```
In[54]:= fixedpoint = Iterate[q, 0., 10000]
Out[54]= -0.306226
```

Then one subtracts this value from each entry in the orbit to obtain a list of differences, which may be used to get an idea of the **speed of convergence**.

```
In[55]:= diff = orb - fixedpoint;
```

A log-plot of the differences may be used to show that they approach zero as an exponential function of n.

```
In[56]:= LogListPlot[Abs[diff]];
```

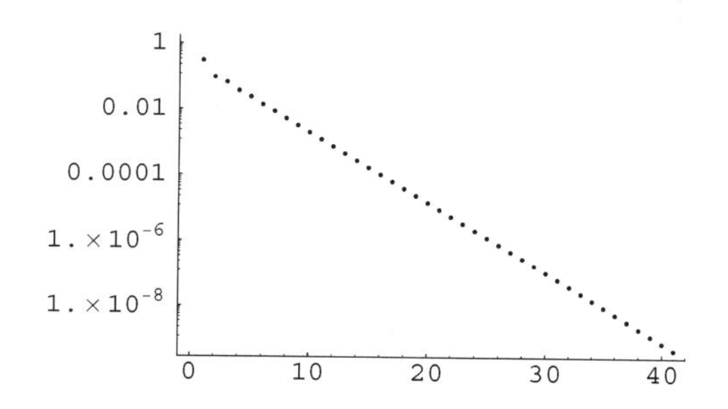

This list of ratios shows that each difference is approximately 0.612 times the previous one.

```
In[57]:= ratios = Table[
           diff[[i + 1]]/diff[[i]],
           {i, 1, 30}]
Out[57]= {-0.306226, -0.706226, -0.546226,
          -0.648626, -0.588988, -0.626271,
          -0.603797, -0.617677, -0.609224,
          -0.614418, -0.611243, -0.61319,
          -0.611999, -0.612729, -0.612282,
          -0.612556, -0.612388, -0.612491,
          -0.612428, -0.612466, -0.612443,
          -0.612457, -0.612448, -0.612454,
          -0.61245, -0.612452, -0.612451,
          -0.612452, -0.612451, -0.612452}
```

A plot displays the rapid convergence of the ratios to their limit.

```
In[58]:= ListPlot[ratios, Frame → True];
```

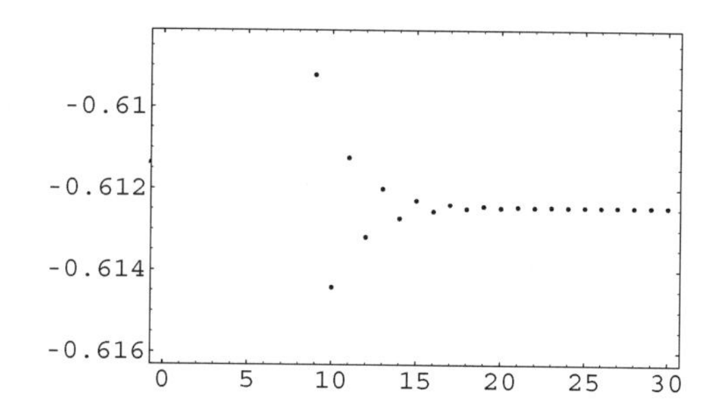

The multipliers function may be used to compute the constant ratio.

```
In[59]:= Multiplier[q, {fixedpoint}]
Out[59]= -0.612452
```

Several of the functions that we have used so far can be combined in a single "cell" of *Mathematica* to be run together. This is one example.

```
A = -1.43;

Histogram[Orbit[q, 0., 300]];

Multiplier[q, Orbit[q, 0, 7, 1000]]

StaircaseDiagram[q, Orbit[q, 0, 8, 1000],
Orbit[q, 1.7, 20]]

TimeSeriesPlot[q, {0.}, 40]
```

1.7.3 Bifurcation Diagrams

In this subsection we continue to consider difference equations that depend on a parameter. We have already seen that, at least for the logistic difference equation, different values of the parameter may yield qualitatively different solutions. We are especially interested in the changed nature of solutions as the parameter is varied, and to investigate such changes we shall use the **bifurcation diagram** plot. The word "bifurcation" refers to a sudden qualitative change in the nature of solutions that occurs as the parameter is varied. In the simplest case of only one parameter, the parameter value at which a bifurcation occurs is called a **bifurcation parameter value**. The bifurcation diagram shows many sudden qualitative changes in the attractor as well as in the periodic orbits.[†]

One of the first discovered bifurcations is a period-doubling bifurcation occurring in the case of the one-dimensional *logistic equation*:

$$x_{n+1} = p x_n (1 - x_n), \tag{1.28}$$

where p is a positive parameter. Among many authors who discovered this phenomenon, Myrberg [My1]–[My3] in 1958, 1959, 1963 probably was the first one. He found that as a parameter was varied an equilibrium attractor could bifurcate into an attracting period-two orbit, which could again double to an attracting period-four orbit, followed by an infinite sequence, period 8, 16, etc. (see [ASY], [D1], [E2], [HK], [Ho], [K], [Ml], [R], [Sf], etc).

[†]Bifurcation plots are generated by repeating the following procedure. Here n_1, n_2, N_{steps}, p_{min}, and p_{max} are given numbers.

a. Choose a value of the parameter p, starting with the initial value p_{min}.
b. Choose an initial point seed x_0.
c. Calculate the orbit of x_0.
d. Ignore the first $n_1 - 1$ iterates and plot the orbit from x_{n_1} to x_{n_2}.
e. Increment p by $(p_{max} - p_{min})/N_{steps}$ and begin the procedure again.

First define the logistic map in terms of a parameter.

```
In[60]:= Clear[p]

In[61]:= logisp = x[n + 1] == px[n] (1 - x[n]);

In[62]:= logispmap = DEToMap[logisp];
```

In this plot the horizontal axis represents the parameter p and the vertical axis the solution x_n. The number of intersections with a vertical line at any point gives the period of an attracting solution for the corresponding value of the parameter. One can see the appearance of period-two, period-four, and period-eight solutions. This is the beginning of *period-doubling route to chaos*. It also shows a "white window" for parameter values slightly greater than 3.8, which suggests the existence of period-three solutions near 3.8.

```
In[63]:= BifurcationPlotND[logispmap,
           {p, 2.5, 4.1}, {0.2},
           FirstIt → 400,
           Iterates → 150,
           Steps → 300];
```

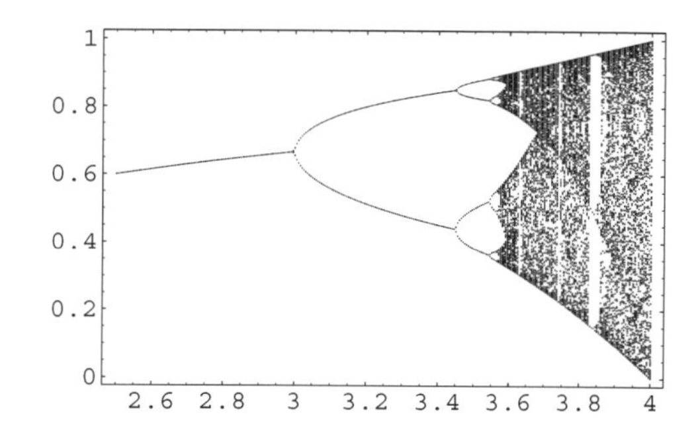

The bifurcation values for the periodic solutions of periods 2, 4, 8, and 16 can be located precisely on this plot with $3.4 \leq p \leq 3.6$. Some white vertical windows located between $p = 3.58$ and $p = 3.59$ are clearly visible. These windows are closely connected to Sharkovsky's ordering, see [E2], pp. 42–44. To estimate the periods of the attracting periodic solution more zooming may be applied. Sharkovsky's theorem may be used to predict the periods of all solutions for values of the parameter smaller than the bifurcation value that occurs in the interval $(3.58, 3.59)$.

```
In[64]:= BifurcationPlotND[logispmap,
           {p, 3.4, 3.6}, {0.2},
           FirstIt → 400,
           Iterates → 150,
           Steps → 300];
```

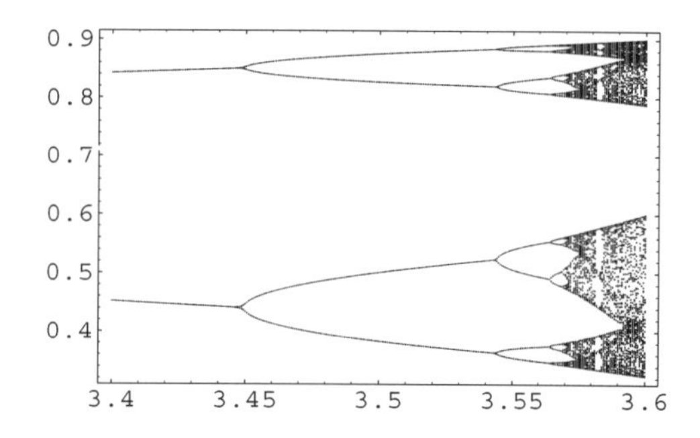

1.7.4 Lyapunov Numbers

Another way to measure the complexity of solutions behavior of difference equations is with calculation of the Lyapunov number. For example, it is well known [ASY] that chaotic dynamics is characterized by an exponential divergence of initially close points.

In the case of the one-dimensional discrete map of an interval (a, b) into itself,

$$x_{n+1} = f(x_n),$$

the "Lyapunov exponent" is a measure of the divergence of two orbits starting with slightly different initial conditions x_0 and $x_0 \pm \delta_0$. As we have shown earlier, if x_0 is a point of a period-k orbit and if we start the orbit from a nearby point $x_0 \pm \delta_0$, then after one iteration the distance between the two is approximated by

$$\delta_1 \approx |f'(x_0)| \, \delta_0 = M_0 \delta_0,$$

where M_0 is the magnification factor for the first step. At the second step,

$$\delta_2 \approx |f'(x_1)| \, \delta_1 = M_1 \delta_1 = M_1 M_0 \delta_0,$$

where M_1 is its magnification factor.

Continuing in this manner, we conclude that the total magnification factor over one cycle of the period-k orbit is the product

$$M_0 M_1 \ldots M_{k-1}.$$

Since this product is an accumulation of magnification factors, it makes sense to consider some average of it. The most convenient is the geometric average

$$\left(M_0 M_1 \ldots M_{k-1}\right)^{1/k},$$

which by taking logarithms leads to the arithmetic average:

$$\begin{aligned} \lambda &= \ln \left(M_0 M_1 \ldots M_{k-1}\right)^{1/k} \\ &= \tfrac{1}{k} \left(\ln M_0 + \ldots + \ln M_{k-1}\right) \\ &= \tfrac{1}{k} \left(\ln |f'(x_0)| + \ldots + \ln |f'(x_{k-1})|\right). \end{aligned}$$

Intuitively, the condition for stability of a periodic orbit is that the average magnification factor is less than 1, which is equivalent to

$$\lambda < 0 \text{ stable}$$
$$\lambda > 0 \text{ unstable.}$$

We shall use this approach to extend this process to arbitrary orbits.

DEFINITION 1.10 (Lyapunov Exponent)
Let f be a smooth map on R and let x_0 be a given initial point. The **Lyapunov exponent** $\lambda(x_0)$ *of a map f is given by*

$$\lambda(x_0) = \lim_{k\to\infty} \frac{1}{k}\left(\ln\left|f'(x_0)\right| + \ldots + \ln\left|f'(x_{k-1})\right|\right),$$

provided the limit exists. In the case when any of the derivatives are zero, set $\lambda(x_0) = -\infty$. The **Lyapunov number** $L(x_0)$ *is defined as the exponent of the Lyapunov exponent, whenever the latter exists:*

$$L(x_0) = e^{\lambda(x_0}.$$

For an arbitrary orbit, the existence of the above limit is an open question.

DEFINITION 1.11 (Asymptotically Periodic Orbit)
An orbit $\{x_1, x_2, x_3, \ldots\}$ is called asymptotically periodic if there exists a periodic orbit $\{y_1, y_2, y_3, \ldots\}$ such that

$$\lim_{n\to\infty} \left|x_n - y_n\right| = 0.$$

Now we are ready to define the chaotic orbit in the sense of Yorke [ASY].

DEFINITION 1.12 (Chaotic Orbit in the Sense of Yorke)
Let f be a map of R and let $\{x_0, x_1, x_2, \ldots\}$ be a bounded orbit of f. The orbit is chaotic if

1. $\gamma^+(x_0) = \{x_0, x_1, \ldots\}$ *is not asymptotically periodic,*
2. *no Lyapunov exponent is* 0, *and*
3. $\lambda(x_0) > 0$.

As we can see from the definition, an important interpretation of the Lyapunov exponent is the measure of information loss during the process of iteration.

Example 1.17 (Lyapunov Exponents for the Tent Map)
It is usually the case that Lyapunov exponents are very difficult to calculate exactly. One of the exceptions is the general tent map

$$T_p(x) = \begin{cases} px, & \text{if } x \le \frac{1}{2} \\ p(1-x), & \text{if } x > \frac{1}{2}. \end{cases}$$

Clearly,

$$T_p'(x) = \begin{cases} p, & \text{if } x < \frac{1}{2} \\ -p, & \text{if } x > \frac{1}{2}, \end{cases}$$

and $T_p'(1/2)$ is undefined. Thus, for any orbit of the tent map that does not contain the point $1/2$, we have $\lambda(x_0) = \ln p$. This formula holds for all x_0 that are not eventually equal to $1/2$. ∎

Example 1.18 (Lyapunov Exponents for the Logistic Map)
Lyapunov exponents for some maps can be calculated sometimes using a certain equivalency between the map and another map from which the Lyapunov exponent can be calculated easily. Let us denote by T the tent map for which $p = 2$, and denote by L the logistic map with $p = 4$, i.e., $L(x) = 4x(1-x)$. Let us consider the transformation

$$y = C(x) = \sin^2\left(\frac{\pi x}{2}\right), \quad x \in [0, 1]. \tag{1.29}$$

This is a one-to-one transformation of the interval $[0, 1]$ onto itself. Consider the composition of T and C, first for $x \leq 1/2$. We have,

$$\begin{aligned} C(T(x)) &= \sin^2(\pi x) = \left(2\sin(\tfrac{\pi x}{2})\cos(\tfrac{\pi x}{2})\right)^2 \\ &= 4\sin^2(\tfrac{\pi x}{2})\left(1 - \sin^2 \tfrac{\pi x}{2}\right) = 4C(x)(1 - C(x)) \\ &= L(C(x)). \end{aligned} \tag{1.30}$$

By proceeding in similar fashion one can prove that relation (1.30) holds for $x > 1/2$. Thus, we get that for all x,

$$C(T(x)) = L(C(x)). \tag{1.31}$$

Taking into account that C is a one-to-one transformation, equation (1.31) can be rewritten as

$$T(x) = C^{-1}(L(C(x))), \quad \text{for } x \in [0, 1]. \tag{1.32}$$

Two maps T and L are called **conjugate**, if there exists a transformation C that is one-to-one and onto and such that both C and C^{-1} are continuous so that (1.31) or (1.32) is satisfied.

Let $\bar{x}$ be the equilibrium point of L, i.e., $L(\bar{x}) = \bar{x}$. Then

$$\bar{x} = L(\bar{x}) = C(T(C^{-1}(\bar{x})),$$

and

$$C^{-1}(\bar{x}) = T(C^{-1}(\bar{x}),$$

which shows that $C^{-1}(\bar{x})$ is the equilibrium point of T. Likewise, one can show that if $\bar{y}$ is the equilibrium point of T, then $C^{-1}(\bar{y})$ is the equilibrium point of L. Let us check the second iterates of two maps:

$$T^2 = (C^{-1} \circ L \circ C) \circ (C^{-1} \circ L \circ C) = C^{-1} \circ L^2 \circ C.$$

Thus, the second iterates of two maps are conjugate as well. Likewise, we conclude that m-th iterates of two maps are conjugate. Consequently, if $\{x_0, x_1, \ldots, x_{k-1}\}$ is a period-k orbit of L, then $\{C^{-1}(x_0), C^{-1}(x_1), \ldots, C^{-1}(x_{k-1})\}$ is a period-k orbit of T. This indicates that the dynamical behavior of L and T is similar. The fact that m-th iterates of two maps are conjugate can be expressed as

$$C(T^m(x)) = L^m(C(x)), \quad \text{for } x \in [0, 1]. \tag{1.33}$$

Now, we can use this formula to find derivatives, and hence Lyapunov exponents, for the logistic map at its periodic points. Using the chain rule formula (1.33) gives:

$$C'(T^m(x))(T^m(x))' = (L^m(C(x)))'C'(x), \quad m = 1, 2, \ldots. \tag{1.34}$$

If we consider a period k orbit of the tent map T starting at x_0, we have that $T^k(x_0) = x_0$. Thus, the derivative of the C terms cancels in (1.34) and we obtain that

$$(T^m(x_0))' = (L^m(C(x_0)))'.$$

Since the Lyapunov exponent is obtained by taking the arithmetic mean of the logarithms of the absolute values of derivatives, it follows that for periodic orbits of the logistic map L it has the same value as for the corresponding orbits of the tent map T, namely, $\ln 2$. More generally, one can show that for an arbitrary orbit of the logistic map starting at $y_0 = \sin^2(\pi x_0/2)$,

$$\lambda(y_0) = \ln 2 + \lim_{m\to\infty} (\ln C'(x_m) - \ln C'(x_0)).$$

It can be shown, see [ASY], that there is an uncountably infinite subset of $y_0 \in [0, 1]$ such that $\lambda(y_0) = \ln 2$. One can find other maps that are conjugate to the tent map T and use the same technique to calculate exactly the corresponding Lyapunov exponents. In general, there are very few examples where the Lyapunov exponents can be computed exactly. In most cases one can just estimate them or compute them numerically. ∎

The `LyapunovNumbers` function with 50 points applied to the Logistic Map with $p = 1.38$ gives the following graph.

```
In[65]:= logispmap[x] = 3.8x(1 - x);

In[66]:= L1 = LyapunovNumbers[
              logispmap, {0.2}, 50];

         ListPlot[L1, PlotJoined → True];
```

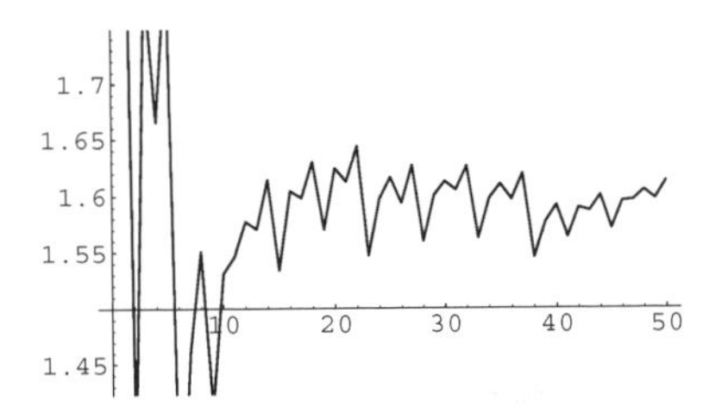

By taking more points we obtain a more precise graph. Both graphs indicate that the orbit starting at x_0 is chaotic, as its Lyapunov number is clearly greater than 1.

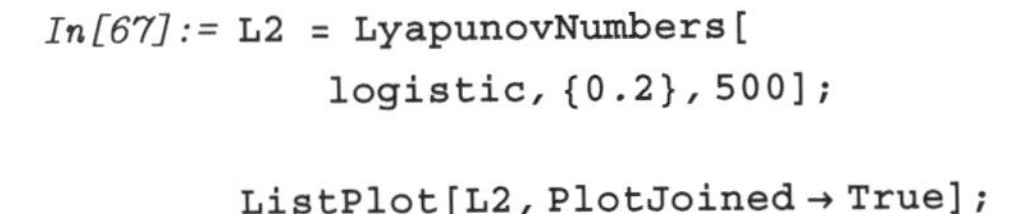

```
In[67]:= L2 = LyapunovNumbers[
              logistic, {0.2}, 500];

         ListPlot[L2, PlotJoined → True];
```

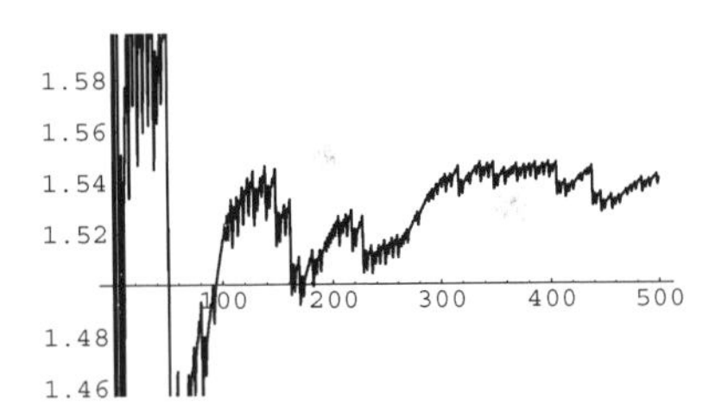

1.7.5 Box Dimension

One way to measure the complexity of a set (an orbit of the map, an attracting set of the map) is to compute its dimension over different scales of magnification. If after many magnifications the set looks like a simple geometric object such as line (polygon), then its dimension is simply the positive integer 1. However, there are many sets that can be effectively constructed, such as Cantors sets, that have a level of complication that does not simplify upon magnification. To explore this idea imagine the set lying on a grid of equal spacing, and then determine the number of grid boxes necessary to cover it. Then we see how this number varies as the grid size is made smaller.

Consider a grid of step size $(b - a)/n$ on the interval $[a, b]$. This means that the grid points are a, $a + (b - a)/n$, ..., b. Obviously, there are n subintervals (boxes) of length $(b-a)/n$. If we use the boxes of length $1/n$, then their number is $(b-a)n$. This fact is expressed by saying that the number of boxes of size r scales as $1/r$, meaning that the number of boxes of size r, $N(r)$ is proportional to $1/r$, i.e.,

$$N(r) \sim \frac{C}{r}.$$

Based on this example, it is natural to ask the following question. Given an object in one-dimensional space, how many intervals of length r does it take to cover the object? In the case of simple sets such as intervals, it is easy to see that this number is exactly C/r. Our goal is to extend this idea to more complicated sets, and to define the dimension d of the object in cases where this dimension is a positive noninteger number.

Let S be a bounded set in R. We would like to define S as a d-dimensional set when it can be covered by

$$N(r) = C/r^d \tag{1.35}$$

boxes of side-length r. Solving for d in (1.35) we get

$$d = \frac{\ln N(r) - \ln C}{\ln(1/r)} \tag{1.36}$$

Letting r go to 0 in (1.36) and assuming that the scaling constant C remains unchanged, we can neglect $\ln C$ for small r. This motivates the following definition.

DEFINITION 1.13 (Box Dimension)

A bounded set S in R has **box dimension** *(box-counting dimension)*

$$\text{BoxDimension}(S) = \lim_{r\to 0} \frac{\ln N(r)}{\ln(1/r)} \tag{1.37}$$

when the limit exists.

It is easy to check that (1.37) gives the expected values for all "simple" sets such as line, polygon, circle (BoxDimension = 1). See Chapter 6 for the definition of a box dimension in higher dimensional spaces and for its computation. In most cases, the only practical way of calculating the box dimension of an orbit of a dynamical system is through numerical approximations.

Here is an approximation to the box dimension of the logistic map with $p = 3.8$.

```
In[68]:= logismap[x]
Out[68]= 3.8x(1 - x)

In[69]:= BoxDimension[logismap,
              1000, 0.001, {0.4}]
Out[69]= 1.00014
```

Taking more iterates produces a more accurate approximation.

```
In[70]:= BoxDimension[logismap,
            8000,0.001,{0.4}]
Out[70]= 1.30105
```

The function `BoxDimension` may be applied to attractors of difference equations of any order, as well as to any time series.

1.8 Symbolic Dynamics for One-Dimensional Maps

In this section we explore some aspects of symbolic dynamics and Cantor sets. The computer-aided part of this section is based on [Mc].

1.8.1 The Cantor Set

We begin by exploring the standard **middle thirds Cantor set**, C. It is constructed from a closed interval I_0 by successive deletion of a family of subintervals. Typically, the starting set is $I_0 = [0, 1]$. The next set I_1 is obtained from I_0 by deleting the open middle third $(1/3, 2/3)$, hence

$$I_1 = I_0 \setminus \left(\frac{1}{3}, \frac{2}{3}\right) = \left[0, \frac{1}{3}\right] \cup \left[\frac{2}{3}, 1\right].$$

The next set I_2 is obtained from I_1 by deleting the open middle third of each of its two component intervals $\left[0, \frac{1}{3}\right]$ and $\left[\frac{2}{3}, 1\right]$. Thus,

$$I_2 = \left[0, \frac{1}{9}\right] \cup \left[\frac{2}{9}, \frac{3}{9}\right] \cup \left[\frac{6}{9}, \frac{7}{9}\right] \cup \left[\frac{8}{9}, 1\right].$$

We see that I_1 is the union of two copies obtained from I_0 by linear scaling by a $1/3$ factor. Similarly, I_2 is the union of two copies obtained from I_1 by linear scaling by a $1/3$ factor.

Continuing this process, at the n-th step I_n is the union of 2^n closed intervals produced from I_{n-1} by deleting the open middle thirds of each of its 2^{n-1} intervals. Again, I_n is the union of two copies of I_{n-1}, each obtained by linear scaling by a $1/3$ factor. Denoting by Sc the scaling by a $1/3$ factor and by Tr the scaling by a $1/3$ factor and translation to the point where the distance from the left end point of the interval is equal to $2/3$ of the length of this interval, that is, $Sc(x) = \frac{1}{3}x + \frac{2}{3}$, we have that

$$I_1 = Sc(I_0) \cup Tr(I_0),$$

$$I_2 = Sc(I_1) \cup Tr(I_1),$$

hence

$$I_n = Sc(I_{n-1}) \cup Tr(I_{n-1}).$$

Clearly I_n is an exact self-similar set, meaning that it is equal to the scaled copy of itself.

The classical Cantor set C is defined as

$$C = \cap_{n=0}^{\infty} I_n. \tag{1.38}$$

There are many points that belong to the Cantor set such as the endpoints of all deleted intervals: $1/3, 2/3, 2/9, 4/9, \ldots$. The length (or size) of the Cantor set is obtained through a simple limiting process. At the n-th stage there are 2^n intervals each of length $1/3^n$, so the length $L(I_n)$ is

$$L(I_n) = \left(\frac{2}{3}\right)^n.$$

Therefore,

$$L(C) = \lim_{n\to\infty} L(I_n) = 0.$$

1.8.1.1 The Cantor Set and the Tent Map

Now we show that there is a very close connection between the classical Cantor set and the following tent map

$$T_p(x) = \begin{cases} 3x, & x \leq \frac{1}{2} \\ 3(1-x), & x > \frac{1}{2}. \end{cases}$$

Notice that, for values of x to the left of $1/2$, the map T_p is exactly the scaling with factor 3. Set $I_0 = [0, 1]$. If $x_0 > 1$, then $x_1 = 3(1 - x_0) < 0$ and $x_2 = 3x_1 < x_1 < 0$. Likewise, we conclude that, for $n > 0$, the sequence $\{x_n\}$ is a strictly decreasing sequence that diverges to $-\infty$. Similarly, if $x_0 < 0$, we obtain the same conclusion. Thus, the only initial points whose corresponding sequence may not diverge to $-\infty$ are the points in I_0. Furthermore, if $x_0 \in (1/3, 2/3)$, then $x_1 = T_3(x_0) > 1$, hence the corresponding orbit diverges to $-\infty$. Thus, the set

$$I_1 = \left[0, \frac{1}{3}\right] \cup \left[\frac{2}{3}, 1\right]$$

is exactly the set of initial points x_0 for which $T_3(x_0)$ is still in I_0. Each of the intervals $\left[0, \frac{1}{3}\right]$ and $\left[\frac{2}{3}, 1\right]$ is mapped to the whole interval I_0 under the map T_3. Now, the middle thirds of the intervals $\left[0, \frac{1}{3}\right]$ and $\left[\frac{2}{3}, 1\right]$ are mapped to values larger than 1 by the second iterate of the map T_3. In other words,

$$T_3^2(x) > 1, \quad \text{for every } x \in \left(\frac{1}{9}, \frac{2}{9}\right) \cup \left(\frac{7}{9}, \frac{8}{9}\right).$$

Consequently, such initial points generate orbits which diverge to $-\infty$. Thus, the set I_2 in the above mentioned construction of the Cantor set can be described as the set of all initial points for which the first iterate is in I_1 and the second iterate is in I_0, that is,

$$I_2 = \{x_0 \in [0, 1] : T_3(x_0) \in I_1, T_3^2(x_0) \in I_0\}.$$

In general,

$$I_n = \{x_0 \in [0, 1] : T_3(x_0) \in I_{n-1}\}, \quad n = 0, 1, \ldots.$$

An interesting fact is that $T_3^2(1/3) = T_3^2(2/3) = 0$; this means that $1/3$ and $2/3$ are eventually equilibrium points. Similarly, $T_3^3(1/3^2) = T_3^3(2/3^2) = T_3^3(7/3^2) = T_3^3(8/3^2) = 0$, which means that $1/3^2, 2/3^2, 7/3^2$ and $8/3^2$ are eventually equilibrium points. Continuing in this fashion, one can show that all the points in the Cantor set are eventually equilibrium points.

Now, we show that the Cantor set is uncountable. Consider the points from the interval $I_0 = [0, 1]$ represented with respect to base 3 (ternary representation) as

$$0.a_1a_2\ldots = \frac{a_1}{3} + \frac{a_2}{3^2} + \ldots + \frac{a_n}{3^n} + \ldots$$

This representation of numbers is not unique; for example, $1/3$ can be represented as

$$.1 \quad \text{and} \quad .0222\ldots2\ldots.$$

Thus, we have representation by the terminating sequence and by the nonterminating sequence. If we make the convention that whenever this happens the nonterminating sequence is chosen, then we have the unique ternary representation for all numbers in I_0. Now it is easy to see that the tent map T_3 acts on numbers in I_0 by shifting the decimal point one position to the right in their ternary representation, i.e.,

$$T_3(.a_1a_2a_3\ldots) = \begin{cases} a_1.a_2a_3\ldots, & x \leq \frac{1}{2} \\ b_1.b_2b_3\ldots, & x > \frac{1}{2}, \end{cases}$$

where $b_j = 2 - a_j$, $j = 1, 2, \ldots$. This number is in I_0 provided that $a_1 \neq 1$. Thus, it follows that the middle third Cantor set is precisely the set of all points in I_0 whose ternary representation uses only the digits 0 and 2. Obviously, there is a one-to-one correspondence between such a set and the set of numbers whose binary (base 2) representation uses the digits 0 and 1 (this correspondence is obtained by replacing each 2 by 1 in the ternary representation of numbers from C), that is, the complete interval $I_0 = [0, 1]$.

1.8.2 Symbolic Dynamics and Chaos in the Sense of Devaney

Now we proceed to formalize some of the properties of the Cantor set. For this, we need to introduce the set Σ_2 of all one-sided binary sequences, that is,

$$\Sigma_2 = \{a = \{a_n\}_{n=0}^{\infty} : a_n = 0 \quad \text{or} \quad 1\}.$$

This set is known as the **code space**, see [B]. As usual, a bar over a group of digits indicates that the group is repeated indefinitely. For example,

$$\{110\overline{01}\ldots\}$$

denotes the sequence

$$\{110010101\ldots\}.$$

Let us define the **shift map** $\sigma : \Sigma_2 \to \Sigma_2$ by

$$\sigma(\{x_0x_1x_2x_3\ldots\}) = \{x_1x_2x_3\ldots\}.$$

This map has two equilibrium points $\{00\overline{0}\ldots\}$ and $\{11\overline{1}\ldots\}$ and infinity of eventually equilibrium points such as $\{x_0x\ldots x_m\overline{0}\ldots\}$ and $\{x_0x\ldots x_m\overline{1}\ldots\}$ for all $m > 0$. There are also 2^k periodic points of minimal period k, which are the points of the form $\{\overline{x_0\ldots x_{k-1}}\ldots\}$. There are many more eventually periodic points, all of which are of the form: $\{x_0\ldots x_m\overline{x_{m+1}\ldots x_{m+k}}\ldots\}$.

We would like to introduce the concept of continuity and density on the set Σ_2, and for this we need the concept of distance between two sequences. It takes no extra effort to introduce these notions in the general setting of metric spaces, which is what we do next.

DEFINITION 1.14 *A* **metric** *(or* **distance function***) on a set $X \neq \emptyset$ is a function $d : X \times X \to R^+$ with the following properties:*

(1) $d(x, y) \geq 0$; $\quad d(x, y) = 0 \quad$ *if and only if* $\quad x = y$

(2) $d(x, y) = d(y, x)$

(3) $d(x, y) \leq d(x, z) + d(z, y)$

The pair (X, d) is said to be a **metric space***.*

Familiar examples of metric spaces are the following.

(1) (R, d) where $d(x, y) = |y - x|$.

(2) (R^2, d) where $d(x, y) = d\left((x_1, y_1), (x_2, y_2)\right) = \sqrt{(x_2 - x_1)^2 + (y_2 - y_1)^2}$.

(3) (R^2, d) where $d(x, y) = d\left((x_1, y_1), (x_2, y_2)\right) = |x_2 - x_1| + |y_2 - y_1|$.

To simplify notation, we shall use X rather than (X, d) throughout the subsection to represent a metric space with metric d.

Let $x \in X$. The ϵ-open ball centered at x is the set

$$B_\epsilon(x) = \{y \in X : d(x, y) < \epsilon\}.$$

A subset B of X is said to be **bounded** if there exists $r > 0$ such that $B \subset B_r$. A subset G of X is said to be **open** if for each $x \in G$ there exists $\delta = \delta(x) > 0$ such that $B_\delta(x) \subset G$. A subset F of X is said to be **closed** if its complement $X \setminus F$ is open. We say that the sequence $\{a_n\}_{n=0}^{\infty}$ from X **converges** to a point $a \in X$, denoted as $\lim_{n\to\infty} a_n = a$, if for every $\epsilon > 0$ there exists a positive integer N such that $d(a_n, a) < \epsilon$, for $n > N$.

A point x is said to be a **limit point** of a subset A of X if every ϵ-open ball $B_\epsilon(x)$ contains a point of A other than x. The union of a set A and the set of its limit points is called the **closure** of A and is denoted by $CL(A)$ or $\bar{A}$. It can be shown that a set is closed if and only if it equals its closure. The set A is said to be **dense** in X if $CL(A) = X$. In other words, A is dense in X if every ϵ-open ball contains a point of A. For example, the set of rational numbers Q is dense in R because every open interval in R must contain a rational number.

DEFINITION 1.15 *A sequence $\{x_n\}$ in a metric space X is said to be* **convergent** *to a point $x_* \in X$ if for every $\epsilon > 0$ there exists a positive integer N such that $d(x_n, x_*) < \epsilon$ for $n \geq N$. In this case we write*

$$\lim_{n\to\infty} x_n = x_*$$

The following characterization of density will be helpful in the sequel.

THEOREM 1.14
Let A be a subset of X. The following statements are equivalent:

(1) The set A is dense in X.

(2) For each $x \in X$ and $\epsilon > 0$, there exists $a \in A$ such that $d(x, a) < \epsilon$.

(3) For each $x \in X$, there exists a sequence $\{a_n\}_{n=0}^{\infty}$ in A that converges to x.

Our next definition is that of a continuous function on metric space.

DEFINITION 1.16 *Let $f : X \to Y$ be a map from a metric space (X, d_1) into a metric space (Y, d_2).*

(1) f is **continuous** *at $x \in X$ if for every sequence $\{x_n\}$ in X that converges to x, the sequence $\{f(x_n)\}$ converges to $\{f(x)\}$ in Y.*

(2) f is **homeomorphism** *if it is one-to-one and onto, and if both f and f^{-1} (inverse map) are continuous.*

Now we define a distance function on the code space Σ_2 as follows:

$$d(x, y) = \sum_{i=0}^{\infty} \frac{|x_i - y_i|}{2^i}. \tag{1.39}$$

It is easy to show that the code space Σ_2 is a metric space and that the distance function d is a metric on Σ_2. In addition, one can show that the shift map is continuous on Σ_2.

An interesting fact is that Σ_2 is indeed a Cantor set. To show this we define the mapping $h : C \to \Sigma_2$ as follows:

$$h(0.x_0x_1x_2\ldots) = y = \left\{\frac{x_0}{2}, \frac{x_1}{2}, \frac{x_2}{2}, \ldots\right\}, \quad x_i \in 0, 2, \quad \text{for every } i = 0, 1, \ldots. \tag{1.40}$$

It is not difficult to see that h is a homeomorphism, that is, h is one-to-one, onto, and continuous and its inverse h^{-1} has the same properties. Thus Σ_2 has all the topological properties of C, and, in this sense, can be identified with C.

Let us define now chaos (in the sense of Devaney [D1] and [E2]). First, we define the notion of topological transitivity.

DEFINITION 1.17 *A map $f : X \to X$, is said to be (topologically)* **transitive** *if for any pair of nonempty open sets U and V, there exists a positive integer k such that $f^k(U) \cap V \neq \emptyset$.*

Intuitively, the orbit of a transitive map of a point wanders all over the space coming as close as we wish from every point in X. Technically, it is not easy to check this property. The following result gives a simple test for verifying transitivity.

THEOREM 1.15
The map $f : X \to X$, is transitive if it has a dense orbit. Furthermore, if X is a closed interval in R, then f is transitive if and only if it has a dense orbit.

Second, we define the notion of sensitive dependence on initial conditions.

DEFINITION 1.18 *A map $f : X \to X$, is said to possess sensitive dependence on initial conditions if there exists $\epsilon > 0$ such that for any open set $U \subset X$ containing x_0 there exists $y_0 \in U$ and a positive integer k such that $d(f^k(x_0), f^k(y_0)) > \epsilon$.*

The following result establishes that a transitive map with a dense set of periodic points possesses sensitive dependence on initial conditions.

THEOREM 1.16
Let $f : X \to X$, is transitive and its set of periodic points P_f is dense in X, i.e., $\overline{P_f} = X$. Then, f possesses sensitive dependence on initial conditions.

A chaotic map is defined next.

DEFINITION 1.19 *A map $f : X \to X$, is said to be chaotic if:*

(1) f is transitive.

(2) The set of periodic points P_f of f is dense in X.

REMARK 1.4 It has been shown, see [VB], that for continuous maps on intervals in R, transitivity implies that the set of periodic points is dense. This means that in this case transitivity implies chaos. ▮

Now we are ready for our main result.

THEOREM 1.17
The shift map $\sigma : \Sigma_2 \to \Sigma_2$ is chaotic on Σ_2.

It is easy to see that $\overline{P_{\Sigma_2}\sigma} = \Sigma_2$. A dense orbit on Σ_2 is

$$\{01\ 00011011\ 000001010100011101110111\ldots\}.$$

Recall that two maps $T : A \to A$ and $L : B \to B$ are called conjugate if there exists a homeomorphism $C : A \to B$ such that

$$T(x) = C^{-1}(L(C(x))), \quad \text{for every } x \in A. \tag{1.41}$$

One can also show that if two maps are conjugate, they must have identical topological properties. In particular, if a map is conjugate to the chaotic map, then it is chaotic. So far, we have an example of a chaotic map: this is a shift map on the space Σ_2. Now, we show that there exists a whole family of maps that are conjugate to the shift map, hence they are chaotic.

Example 1.19 Consider the logistic map

$$f_p(x) = px(1-x) \quad \text{on} \quad A_0 = [0, 1],$$

where $p > 4$. It is easy to see that the orbit of every point greater than 1 diverges to $-\infty$. Clearly, $f_p(1/2) = p/4 > 1$. By continuity and monotonicity of $f_p(x)$ as well as symmetry of its graph with respect to the line $x = 1/2$, there exist points a_0 and a_1 such that $f_p(a_0) = f_p(a_1) = 1$, and $f_p(x) > 1$ for every $x \in (a_0, a_1)$. The points a_0 and a_1 are the solutions to the quadratic equation $px(1-x) = 1$, that is,

$$a_0 = \frac{1}{2} - \frac{\sqrt{p^2 - 4p}}{2p}, \quad a_1 = \frac{1}{2} + \frac{\sqrt{p^2 - 4p}}{2p}.$$

Consequently, the orbit of every point in (a_0, a_1) diverges to $-\infty$. The orbits of all points from $I_0 = [0, a_0]$ or $I_1 = [a_1, 1]$ stay in $[0, 1]$ after one iteration. Denote

$$A_1 = I_0 \cup I_1.$$

Then we have

$$A_1 = I_0 \cup I_1 = \{x \in A_0 : f_p(x) \in A_0\}.$$

Notice the striking similarity to the first step of the construction of the Cantor middle third set. Denote

$$A_2 = \{x \in A_0 : f_p(x) \in A_1\} = \{x \in A_0 : f_p^2(x) \in A_0\}.$$

It is easy to see that $A_2 \subset A_1$ and that A_2 consists of the four closed intervals

$$A_2 = I_{00} \cup I_{01} \cup I_{11} \cup I_{10},$$

where

$$I_{ij} = \{x \in I_i : f_p(x) \in I_j\}, \quad i, j \in \{0, 1\}.$$

For example,

$$I_{00} = \{x \in I_0 : f_p(x) \in I_0\}, \quad I_{10} = \{x \in I_1 : f_p(x) \in I_0\}.$$

There is a great similarity to the second step of the construction of the Cantor middle third set.

Continuing with this process, we denote

$$A_n = \{x \in A_0 : f_p^n(x) \in A_0\} = \cup I_{s_0 s_1 \ldots s_{n-1}},$$

where $s_i \in \{0, 1\}$ and

$$I_{s_0 s_1 \ldots s_{n-1}} = \{x \in A_0 : x \in I_{s_0}, f_p(x) \in I_{s_1}, \ldots, f_p(x)^j \in I_{s_j}\}.$$

Obviously,

$$I_{s_0 s_1 \ldots s_n} \subset I_{s_0 s_1 \ldots s_{n-1}},$$

hence $A_{n+1} \subset A_n$. Now define

$$\Lambda = \cap_{n=1}^{\infty} A_n.$$

Let us consider the map $h : \Lambda \to \Sigma_2$ given by

$$h(x) = \{a_0 a_1 a_2 \ldots\}, \quad \text{where} \quad a_n = \begin{cases} 0 & \text{if } f_p^n(x) \in I_0 \\ 1 & \text{if } f_p^n(x) \in I_1. \end{cases}$$

Then $h(x) = \{a_0 a_1 a_2 \ldots\}$ if and only if $f_p^n(x) \in I_{a_n}$ for each positive integer n. The sequence $h(x)$ is called the **itinerary** of x. Similarly, if we prescribe the sequence $h(x)$, we can determine x; it is called the **inverse itinerary**.

Now, it is possible to show that h is homeomorphism, see [D1] and [E2], which implies that the logistic map f_p on Λ for $p > 4$ is conjugate to the shift map on Σ_2, hence is chaotic. ∎

1.8.3 *Dynamica* Session

In this section we present *Dynamica* functions for producing ternary expansions with respect to the specified base b, visualization of the Cantor sets and Cantor functions, calculation of itineraries, etc.

Here we find the base 3 expansion of π. The first answer is the base 3 expansion of 3 (the integer part of π) and the second is the base 3 expansion of 0.14159... (the fractional part of π).

```
In[1]:= << Dynamica`

In[2]:= BDigits[Pi, 3]
Out[2]= {1, 0}

In[3]:= BDecimals[Pi, 3]
Out[3]= {0, 1, 0, 2, 1, 1, 0, 1, 2, 2, 2, 2, 0, 1,
         0, 2, 1, 1, 0, 0, 2}
```

By default, one gets 20 base b decimal places, but one can get more by specifying the optional argument.

```
In[4]:= BDecimals[Pi, 3, 30]
Out[4]= {0, 1, 0, 2, 1, 1, 0, 1, 2, 2, 2, 2, 0, 1,
         0, 2, 1, 1, 0, 0, 2, 1, 1, 1, 1, 1, 0, 2,
         2, 1, 2}
```

One can also enter an expansion and find the corresponding number. The last answer is the value of the periodic base 3 decimal 0.102010201020...

```
In[5]:= BDigitValue[{1, 0, 2, 0}, 3]
Out[5]= 33

In[6]:= BDecimalValue[{1, 0, 2, 0}, 3]
Out[6]= 11/27

In[7]:= BDecimalValue[{1, 0, 2, 0}, 3,
                    periodic]
Out[7]= 33/80
```

Recall that we obtain elements of the Cantor set by specifying ternary expansions with only 0s and 2s. Any finite ternary decimal base of this form is one of the endpoints of the intervals used in the construction of C. Any other such decimal is an element of C which is not an endpoint.

In particular, any periodic expansion not ending in all 0s or all 2s is not an endpoint. Here, the first example is an endpoint, while the second example is not.

```
In[8]:= BDecimalValue[
          {0, 2, 2, 0, 2, 2, 0, 0, 0}, 3]
Out[8]= 224/729

In[9]:= BDecimalValue[
          {0, 2, 2, 0, 2, 2, 0, 0, 0}, 3,
          periodic]
Out[9]= 3024/9841
```

Now we take a look at the first few steps of the construction of the Cantor set in several ways. The first is just to compute a list of all of the intervals obtained in the first n stages of the construction of C. These intervals were computed using the fact that the standard Cantor set is an invariant set of a tent map T_3.

```
In[10]:= CIntervals[1]
```

Out[10]= $\left\{\left\{0, \frac{1}{3}\right\}, \left\{\frac{2}{3}, 1\right\}\right\}$

```
In[11]:= CIntervals[2]
```

Out[11]= $\left\{\left\{0, \frac{1}{9}\right\}, \left\{\frac{2}{9}, \frac{1}{3}\right\}, \left\{\frac{2}{3}, \frac{7}{9}\right\}, \left\{\frac{8}{9}, 1\right\}\right\}$

```
In[12]:= CIntervals[4]
```

Out[12]= $\left\{\left\{0, \frac{1}{81}\right\}, \left\{\frac{2}{81}, \frac{1}{27}\right\}, \left\{\frac{2}{27}, \frac{7}{81}\right\}, \left\{\frac{8}{81}, \frac{1}{9}\right\}, \left\{\frac{2}{9}, \frac{19}{81}\right\}, \left\{\frac{20}{81}, \frac{7}{27}\right\}, \left\{\frac{8}{27}, \frac{25}{81}\right\}, \left\{\frac{26}{81}, \frac{1}{3}\right\}, \left\{\frac{2}{3}, \frac{55}{81}\right\}, \left\{\frac{56}{81}, \frac{19}{27}\right\}, \left\{\frac{20}{27}, \frac{61}{81}\right\}, \left\{\frac{62}{81}, \frac{7}{9}\right\}, \left\{\frac{8}{9}, \frac{73}{81}\right\}, \left\{\frac{74}{81}, \frac{25}{27}\right\}, \left\{\frac{26}{27}, \frac{79}{81}\right\}, \left\{\frac{80}{81}, 1\right\}\right\}$

Cmap is the tent map

$$C_{map}(x) = \begin{cases} \frac{2x}{1-m} & \text{if } x \leq \frac{1}{2} \\ \frac{2(1-x)}{1-m} & \text{if } x > \frac{1}{2} \end{cases}$$

where one can specify the value of the parameter m.

```
In[13]:= m = 1/3;
In[14]:= Plot[{x, Cmap[x]}, {x, 0, 1},
             AspectRatio → 1,
             PlotRange → 0, 2]
```

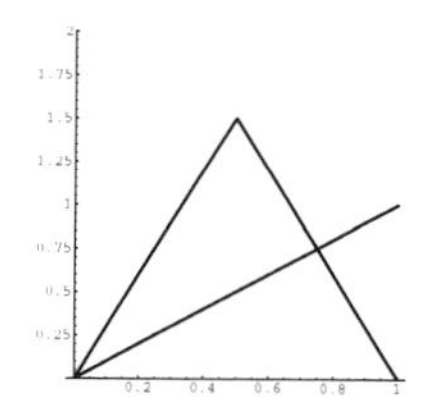

Cmap is not invertible but has, in a sense, two inverse maps, one taking [0, 1] onto [0, 1/3] and the other taking [0, 1] onto [2/3, 1].

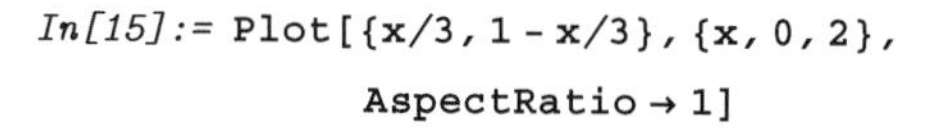

```
In[15]:= Plot[{x/3, 1 - x/3}, {x, 0, 2},
             AspectRatio → 1]
```

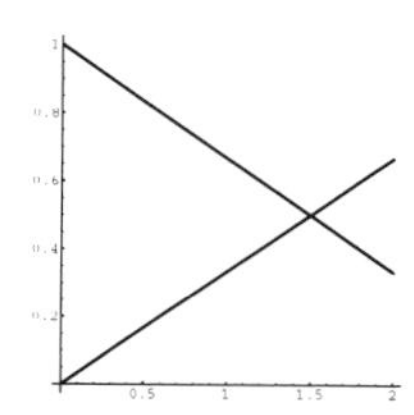

The intervals [0, 1/3] and [2/3, 1] are the first two intervals in the Cantor construction–they are components of I_1. Applying each of the two inverse functions to these two intervals we get the four intervals of the second stage of the Cantor construction I_2, and so on. At the nth stage we get the 2^n intervals we want.

An interesting generalization of the middle thirds Cantor set is the middle m-th Cantor set where m is some fraction between 0 and 1. Here we look at the first steps in the construction of middle 1/10-th Cantor intervals.

```
In[16]:= CIntervals[1, 1/10]
```

$$\text{Out[16]}= \left\{\left\{0, \frac{9}{20}\right\}, \left\{\frac{11}{20}, 1\right\}\right\}$$

```
In[17]:= CIntervals[3, 1/10]
```

$$\text{Out[17]}= \left\{\left\{0, \frac{729}{8000}\right\}, \left\{\frac{891}{8000}, \frac{81}{400}\right\}, \left\{\frac{99}{400}, \frac{2709}{8000}\right\}, \left\{\frac{2871}{8000}, \frac{9}{20}\right\}, \left\{\frac{11}{20}, \frac{5129}{8000}\right\}, \left\{\frac{5291}{8000}, \frac{301}{400}\right\}, \left\{\frac{319}{400}, \frac{7109}{8000}\right\}, \left\{\frac{7271}{8000}, 1\right\}\right\}$$

One nice way to visualize Cantor sets is through Cantor functions. These are nondecreasing and continuous functions defined on $[0, 1]$ that are constant on each of the deleted intervals in the construction of the corresponding Cantor set. In spite of the fact that they are constant almost everywhere, they are nondecreasing from 0 to 1 as x ranges over $[0, 1]$. All of the increase occurs on the Cantor set C itself. The formal definition is as follows:

Let f be a function defined as follows:

$$f(x) = f(0.c_1c_2\ldots) = 0.\frac{c_1}{2}\frac{c_2}{2}\ldots, \qquad x \in C,$$

where C is the middle third Cantor set. Here $0.c_1c_2\ldots$ is the ternary representation and $0.\frac{c_1}{2}\frac{c_2}{2}\ldots$ is the binary representation of x. Then $f(x)$ is defined everywhere on C. Extend the definition on the whole interval $[0, 1]$ by assuming that f is nondecreasing on $[0, 1] - C$. Now we have defined completely the Cantor function. It has the following properties:

f is nondecreasing and onto $[0, 1]$. Consequently, f must be continuous.

The function f is not one-to-one. For example, two points of C:

$$x_1 = 0.c_1c_2\ldots c_n2000\ldots \quad \text{and} \quad x_2 = 0.c_1c_2\ldots c_n0222\ldots$$

have the same image $f(x_1) = f(x_2) = 0.\frac{c_1}{2}\frac{c_2}{2}\ldots\frac{c_n}{2}0111\ldots$ in binary representation.

In the plot of the graph, the infinitely many flat spots form the so-called "Devil's staircase".

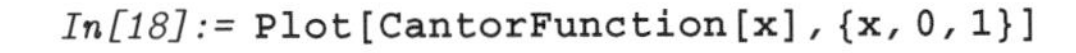

```
In[18]:= Plot[CantorFunction[x], {x, 0, 1}]
```

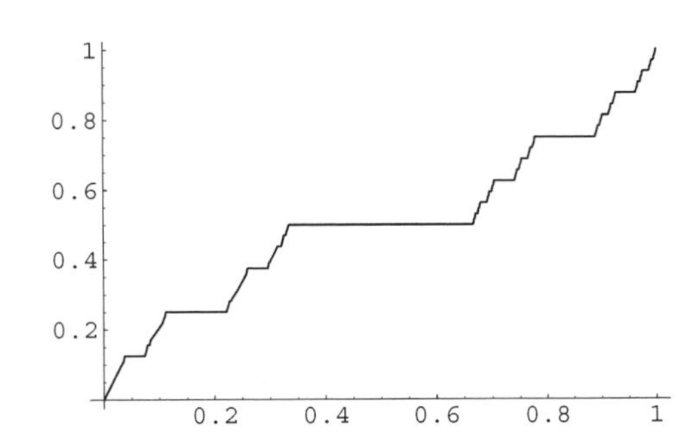

Zooming in on a small interval shows the smaller deleted intervals.

```
In[19]:= Plot[CantorFunction[x], {x, 0, 1/100}]
```

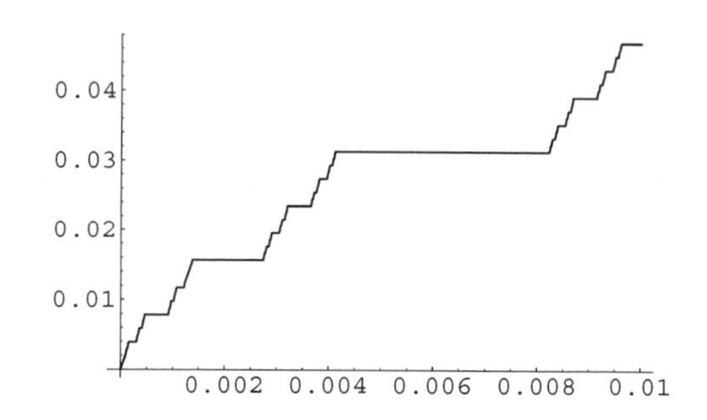

There are Cantor functions for the middle m-th Cantor sets as well.

This plot shows the intervals in the successive stages of the construction as well as the itineraries of the points in the various intervals for *Cmap*. Itineraries are read from bottom to top with dark representing 0 and light representing 1. Thus, for example, the itinerary of 1/3 is $0, 1, 0, 0, 0$ and the itinerary of 8/9 is $1, 0, 1, 0, 0$.

```
In[20]:= CantorPlot[5]
```

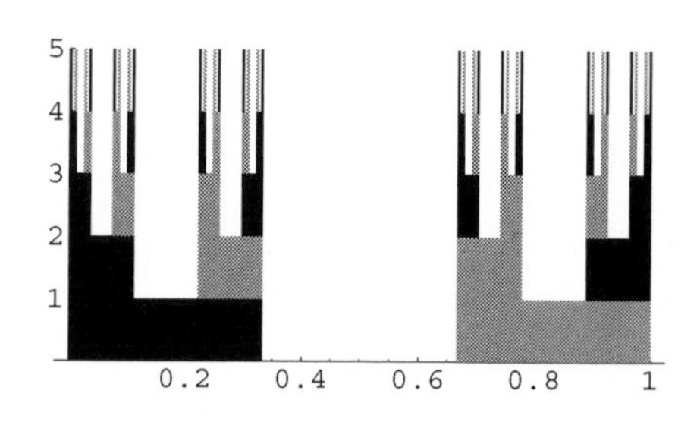

Other Cantor sets (called Cantor-like sets) can be obtained as follows. The function CantorPlot has the second argument which gives the size of the middle interval that is removed in the construction of this set. For example, in this case the middle half of the interval is removed.

```
In[21]:= CantorPlot[5, 1/2]
```

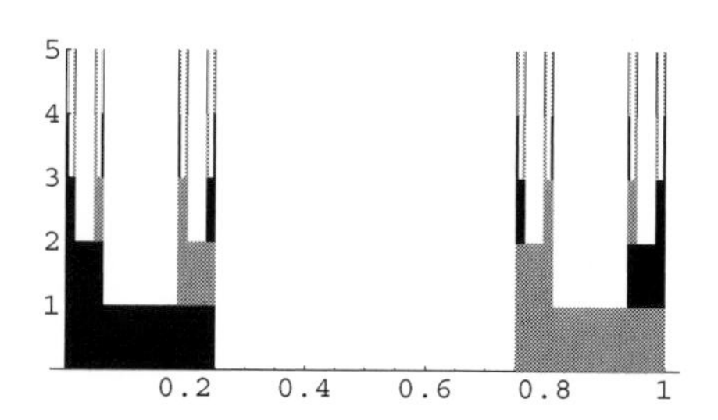

Now we use similar ideas to study the quadratic $Qmap[x] = x^2 + c$ with $c \leq -2$. We constructed an invariant Cantor-like set, C_L, on which the dynamics is conjugate to the shift. The conjugacy is given by the itinerary map. The following functions compute itineraries and inverse itineraries.

This is an itinerary of length 10 for initial condition x_0 = 1/3 and c=-2.1.

```
In[22]:= QItinerary[1/3, 10, -2.1]
Out[22]= {1, 0, 1, 1, 0, 0, 1, 1, 1, 0, -1}
```

Recall that a 0 in the i-th position means that the point x_i is in the interval I_0 and a 1 means that it is in I_1. We have used a -1 to indicate that x_i is in neither of the two intervals. This means that the chosen initial condition $x_0 \neq 1/3$ is not in the invariant set, L, although it stays around for 9 iterations.

We can visualize the result by plotting the corresponding staircase diagram.

```
In[23]:= PlotQOrbits[1/3, 10, -2.1]
```

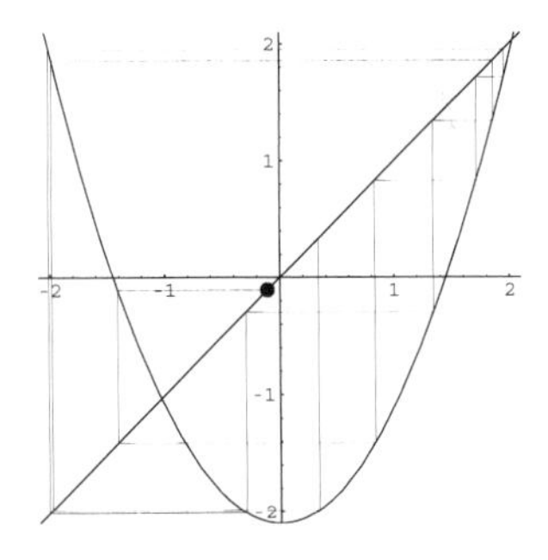

Next we turn to inverse itineraries. The inverse itinerary function returns the interval of all initial conditions that have the given finite itinerary.

The interval here is the one we called $I_{1011001110}$. Recall that we found points with infinite itineraries by intersecting intervals like this.

```
In[24]:= QInverseItinerary[
             {1, 0, 1, 1, 0, 0, 1, 1, 1, 0}, -2.1]
Out[24]= {0.332723, 0.33404}
```

A list of all of the intervals in the n-th stage of the construction of C_L can be found as shown here.

```
In[25]:= QIntervals[1, -2.1]
Out[25]= {{-2.03297, -0.2589},
             {0.2589, 2.03297}}

In[26]:= QIntervals[2, -2.1]
Out[26]= {{-2.03297, -1.53587},
             {-1.35687, -0.2589},
             {0.2589, 1.35687},
             {1.53587, 2.03297}}
```

```
In[27]:= QIntervals[4,-2.1]
Out[27]= {{-2.03297,-2.0017},
          {-1.98979,-1.9068},
          {-1.85927,-1.72106},
          {-1.68852,-1.53587},
          {-1.35687,-1.16143},
          {-1.11263,-0.862049},
          {-0.751085,-0.490646},
          {-0.43955,-0.2589},
          {0.2589,0.43955},
          {0.490646,0.751085},
          {0.862049,1.11263},
          {1.16143,1.35687},
          {1.53587,1.68852},
          {1.72106,1.85927},
          {1.9068,1.98979},
          {2.0017,2.03297}}
```

The method for computing the intervals is the same as for the Cantor set. The quadratic map has two inverse functions, $\pm\sqrt{x-c}$ which can be applied repeatedly to generate the intervals of our construction.

To illustrate the construction of C_L we plot the first few stages of the construction, coded according to itineraries as for the Cantor set.

```
In[28]:= LambdaPlot[5,-2.1]
```

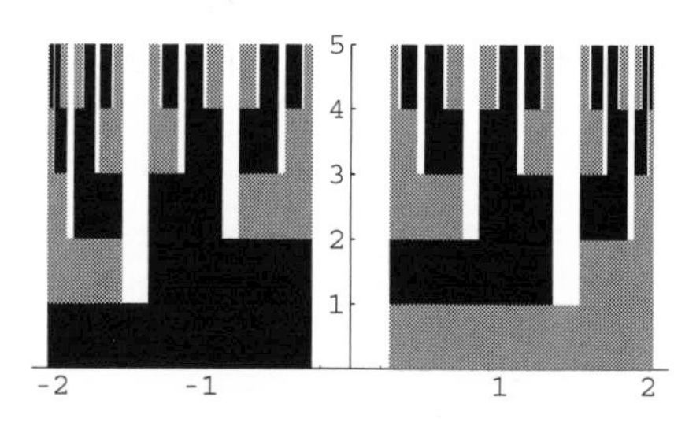

As before, dark means 0 and light means 1. Note that unlike the standard Cantor sets the intervals and the gaps between them have different sizes.

```
In[29]:= LambdaPlot[7,-2.01]
```

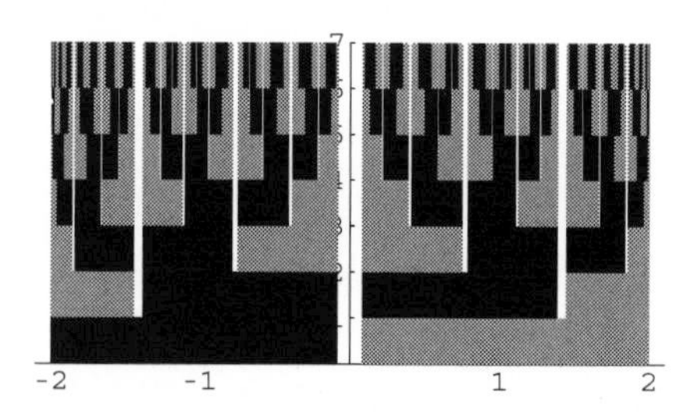

Finally, we investigate sensitive dependence on initial conditions. The following function follows two initial conditions under iteration of the quadratic map. Sensitive dependence means that very close initial conditions diverge under iteration. First, we consider a non-chaotic parameter value such as $c = -1.4$.

```
In[30]:= PlotQOrbits[0.4, 0.41, 20, -1.4]
```

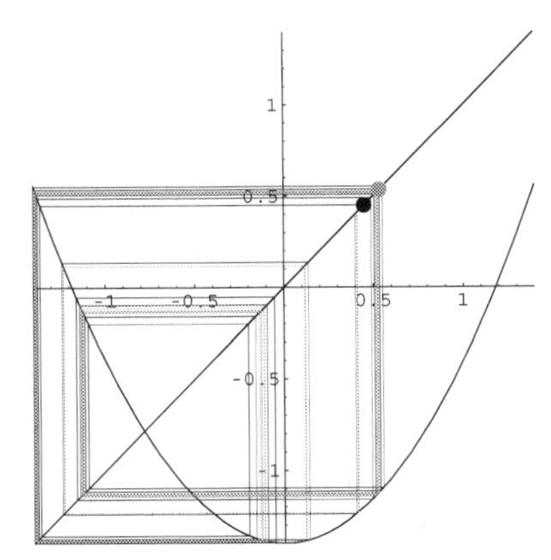

The orbits remain close. To see the dynamics one may animate a sequence of plots produced with the command

```
Do[PlotQOrbits[0.4,0.41,i,-1.4],{i,0,8}]
```

1.9 Dissipative Maps and Global Attractivity

In this section we want to emphasize a special class of difference equations

$$x_{n+1} = F(x_n), \quad n = 0, 1, \dots \tag{1.42}$$

where every orbit has the property to eventually enters an interval $[a, b]$. The interval $[a, b]$ is called an **absorbing interval**, or an **attracting interval**, see [SH]. An interval is called an **invariant interval** if every orbit that enters interval stays in this interval forever.

A map (difference equation) with an absorbing interval is called **dissipative map** (**dissipative difference equation**). If the function F is continuous and monotonic on the invariant interval $[a, b]$, then the dynamics of solutions of this equation that enter eventually $[a, b]$ is very simple as the following two results show, see [KL]. If the difference equation is dissipative we obtain global results.

THEOREM 1.18
Let $[a, b]$ be an interval of real numbers and assume that F is a continuous, nondecreasing function such that

$$F([a, b]) \subset [a, b]$$

and (1.42) has a unique equilibrium $\bar{x}$. Then every solution of (1.42) with initial point in $[a, b]$ converges to $\bar{x}$. If (1.42) is dissipative then every solution of (1.42) converges to $\bar{x}$.

Let us illustrate this result with some examples.

Example 1.20 The difference equation

$$x_{n+1} = \frac{x_n}{1 + x_n}, \qquad n = 0, 1, \ldots,$$

satisfies all conditions of Theorem 1.18 on the interval $[0, 1]$. Thus, every solution of this equation that starts in the invariant $[0, 1]$ converges to the unique equilibrium 0. In addition, every solution that starts in $[1, \infty)$ in one step enters the invariant interval and converges to the zero equilibrium. The situation for negative initial conditions is much more complicated as we face the problem of the existence of the solution (that is, x_n may become undefined for some n.) Fortunately, this equation can be solved in exact form, hence the negative initial values that generate a solution existing for all $n \geq 0$ can be found effectively, see Section 2.4.

The difference equation

$$x_{n+1} = \frac{x_n}{1 + x_n^p}, \qquad n = 0, 1, \ldots,$$

where $0 < p \leq 1$ is another example of this kind. ▮

Example 1.21 The difference equation

$$x_{n+1} = \frac{x_n^2}{1 + x_n^2}, \qquad n = 0, 1, \ldots,$$

satisfies all conditions of Theorem 1.18 on the interval $[0, 1]$, which implies that every solution of this equation that starts in the invariant $[0, 1]$ converges to the unique equilibrium 0. Similar to Example 1.20, every solution with an initial value in $[0, \infty)$ enters the invariant interval and converges to the zero equilibrium. The situation for negative initial conditions is very simple as the solution in one step becomes positive, thus it converges to zero. Thus, $[0, 1]$ is an absorbing interval and zero is a global attractor. ▮

The next result, first obtained in [DKL], deals with the case when the function F is decreasing on an invariant interval.

THEOREM 1.19

Let $[a, b]$ be an interval of real numbers and let F be a continuous, non-increasing function, such that $F([a, b]) \subset [a, b]$ and (1.42) has no prime period-

two solution. Then every solution of (1.42) that enters eventually $[a, b]$ converges to $\bar{x}$. In addition, the subsequences $\{x_{2n}\}$ and $\{x_{2n+1}\}$ of every solution of (1.42) converge monotonically to $\bar{x}$ by oscillating about the equilibrium such that

$$(x_{n+1} - \bar{x})(x_n - \bar{x}) < 0, \quad forn = 0, 1, \ldots$$

Let us illustrate this result with some examples.

Example 1.22 The difference equation

$$x_{n+1} = e^{-x_n}, \quad n = 0, 1, \ldots,$$

has $[0, 1]$ as an invariant interval. The corresponding map is decreasing everywhere and it does not possess a prime period-two solution. Indeed, if such a solution would exist it would satisfy $f(f(p)) = p$ for some $p \neq \bar{x}$, that is,

$$e^{-e^{-p}} = p.$$

Consider the function $G(u) = e^{-e^{-u}} - u$. Clearly, $G(0) = \frac{1}{e} > 0$, $G(1) = 1/e^{e^{-1}} - 1 < 0$ and $G'(u) = e^{-u-e^{-u}} - 1 < 0$ for $u \geq 0$. This shows that this function has a unique zero on interval $[0, 1]$. Thus, all conditions of Theorem 1.19 on the interval $[0, 1]$ are satisfied and every solution of the equation that starts in the invariant interval $[0, 1]$ converges to the unique equilibrium. In addition, every solution that starts in R enters the invariant interval in at most two steps and so converges to the equilibrium. ▮

The next example has been studied in [DKL].

Example 1.23 Consider the difference equation

$$x_{n+1} = 1 + \frac{A}{x_n^p}, \quad n = 0, 1, \ldots, \tag{1.43}$$

where A, p, and x_0 are positive numbers. This equation has an invariant interval $[1, 1 + A]$ and the corresponding map is decreasing for all positive values. Equation (1.43) has a unique positive equilibrium $\bar{x}$, where $\bar{x}$ is the unique positive root of the equation

$$u^{p+1} - u^p - A = 0.$$

Using linearized stability one can show that if $p \leq 1$, the equilibrium $\bar{x}$ is locally asymptotically stable. Similarly, if $p > 1$, $\bar{x}$ is locally asymptotically stable provided that $\bar{x} < \frac{p}{p-1}$. One can show that this condition is equivalent to the condition

$$A < \frac{p^p}{(p-1)^{p+1}}. \tag{1.44}$$

When (1.44) holds, the Schwarzian derivative (see Theorem 1.8), can be used to show that $\bar{x}$ is locally asymptotically stable. Thus, in both cases, (a) $p \leq 1$, and, (b) $p > 1$ and $A \leq \frac{p^p}{(p-1)^{p+1}}$, the unique equilibrium $\bar{x}$ is locally asymptotically stable. To show that (1.43) has no prime period-two solution we consider the second iterate

$$G(u) = F(F(u)) = 1 + \frac{A}{\left(1 + \frac{A}{u^p}\right)^p},$$

and the equation $H(u) = 0$, where $H(u) = G(u) - u$. Clearly, the equilibrium $\bar{x}$ is a solution of this equation. Next, we show that $\bar{x}$ is the only solution by checking that $H(u)$ is strictly decreasing. To complete this, observe that

$$H'(u) = \frac{(Ap)^2}{D(u)^{p+1}} - 1$$

where $D(u) = u + \frac{A}{u^{p-1}}$. Since the function $D(u)$ has a minimum at $u = u_0 = ((p-1)A)^{\frac{1}{p}}$,

$$D(u) \geq D(u_0) = \frac{p}{p-1}((p-1)A)^{\frac{1}{p}}.$$

Thus, we have that

$$H'(u) \leq \frac{(Ap)^2}{D(u_0)^{p+1}} - 1 \leq \left(\frac{A(p-1)^{p+1}}{p^p}\right)^{\frac{p-1}{p}} - 1 \leq 0,$$

if

$$A \leq \frac{p^p}{(p-1)^{p+1}},$$

where equality holds only if $u = u_0$. Therefore, $H(u)$ has a unique zero and this is the unique positive equilibrium. Hence, (1.43) does not possess a prime period-two solution and by Theorem 1.19 every solution of this equation that starts in the invariant interval $[1, 1 + A]$ converges to the unique equilibrium $\bar{x}$. In addition, every solution that starts in $[0, \infty)$ in at most two steps enters the invariant interval, hence converges to the equilibrium. Thus the invariant interval $[1, 1 + A]$ is an absorbing interval, and the difference equation is dissipative.

The remaining case

$$p > 1 \quad \text{and} \quad A > \frac{p^p}{(p-1)^{p+1}}$$

is qualitatively different. By using a fine analysis of the problem one can show that in this case (1.43) has two periodic solutions, each of period two, of the form $P_1 = \{p, q\}$ and $P_2 = \{q, p\}$ which are asymptotically stable and global attractors with a basin of attraction $\mathcal{B}(P_1) = (0, \bar{x})$ and $\mathcal{B}(P_2) = (\bar{x}, \infty)$, respectively. See [DKL]. ∎

The dynamics of equation (1.42) can become very complex in the case where the function F has a minimum point or a maximum point at the given interval, as we have seen from the example of the logistic equation

$$x_{n+1} = px_n(1 - x_n),$$

and all equations conjugate to this equation.

1.10 Parametrization and the Poincaré Functional Equation

Consider again the difference equation

$$x_{n+1} = f(x_n) \tag{1.45}$$

and assume that 0 is a source. If $f(x)$ is an **analytic function** at $x = 0$ (continuously differentiable as a function of complex variable z in some open disk around $z = 0$), then any orbit of equation (1.45) can be parametrized by the formula

$$x_n = G(t^n c), \tag{1.46}$$

where G is an analytic function of its complex argument z in some open disk around $z = 0$ and where t is the multiplier of the equilibrium point. The constant c is given by $c = G^{-1}(x_0)$. The existence of such a parametrizing function function is guaranteed by the following result discovered by Poincaré, [La1].

THEOREM 1.20 (Poincaré Theorem)
The functional equation

$$G(az) = f(G(z)), \tag{1.47}$$

where $f(z)$ is analytic at $z = 0$ with $f(0) = 0$ and $|f'(0)| > 1$, has a solution $G(z)$ analytic at $z = 0$ with $G(0) = 0$. If, in addition, we assume that $G'(0) = 1$, then the solution is unique. If $f(z)$ is an entire function, then $G(z)$ is an entire function as well.

Equation (1.47) is called the **Poincaré functional equation**, and the unique solution of this equation that satisfies the normalization condition $G'(0) = 1$ is called the **Poincaré function**.

There are very few known cases where the Poincaré function is an elementary function. Some of them are listed in the following table [La1] and [My4].

Difference equation	Poincare function	Corresponding solution
$x_{n+1} = ax_n$	$G(z) = z$	$x_n = t^n c$
$x_{n+1} = 2x_n(1 - x_n)$	$G(z) = \dfrac{e^{2z} - 1}{2}$	$x_n = \frac{e^{2^n c} - 1}{2}$
$x_{n+1} = 4x_n(1 - x_n)$	$G(z) = \sin^2 \sqrt{z}$	$x_n = \sin^2(2^n c)$
$x_{n+1} = \dfrac{ax_n}{1 + x_n}$	$G(z) = \frac{(a-1)z}{a-1+z}$	$x_n = \frac{(a-1)a^n}{a-1+a^n}$
$x_{n+1} = \dfrac{2x_n}{1 + x_n^2}$	$G(z) = \tanh(z)$	$x_n = \tanh(2^n c)$

In the general case, if one cannot express the Poincaré function as an elementary function, then one may attempt to solve the equation (1.47) by using the method of successive approximations or a power-series expansion with undetermined coefficients. Let us illustrate the second method.

Example 1.24 Consider the logistic equation

$$x_{n+1} = p\,x_n\,(1 - x_n), \quad p > 1.$$

With $t = p$ in (1.46), the Poincaré functional equation becomes

$$G(pz) = p\,G(z)\,(1 - G(z)).$$

Taking into account the normalization condition $G'(0) = 1$, the power-series expansion of solution of this equation is

$$G(z) = z + c_2 z^2 + c_3 z^3 + \ldots,$$

where the c_i's are the coefficients that should be determined from the above equation. We obtain the following relations:

$$c_2 = \frac{1}{1-p}, \quad c_3 = \frac{2}{(p-1)(p^2-1)}, \quad c_4 = \frac{p+5}{(1-p)(p^2-1)(p^3-1)}, \quad \ldots$$

∎

1.11 Exercises

Exercise 1.1 Use `RSolve` to find the general solutions of the following difference equations:

1. $x_{n+1} - 2x_n = 0, \quad n = 0, 1, \dots,$
2. $x_{n+1} - \frac{1}{4}x_n = 0, \quad n = 0, 1, \dots,$
3. $x_{n+1} + x_n = n, \quad n = 0, 1, \dots,$
4. $x_{n+1} - x_n = 2n + n^4, \quad n = 0, 1, \dots,$
5. $x_{n+1} - ax_n = bn^2, \quad n = 0, 1, \dots.$

Exercise 1.2 Use `RSolve` to find the solutions of the following IVPs:

1. $x_{n+1} - 2x_n = 0, \quad x_0 = 1,$
2. $x_{n+1} - \frac{1}{4}x_n = 0, \quad x_0 = 2,$
3. $x_{n+1} + x_n = n, \quad x_0 = 4,$
4. $x_{n+1} - x_n = 2n + n^4, \quad x_0 = A,$
5. $x_{n+1} - ax_n = bn^2, \quad x_0 = 2.$

Exercise 1.3 Use `RSolve` to find the general solutions of the following difference equations:

1. $x_{n+1} - \frac{n+2}{n+1}x_n = 0, \quad n = 0, 1, \dots,$
2. $x_{n+1} - \frac{n+1}{n+2}x_n = 0, \quad n = 0, 1, \dots,$
3. $x_{n+1} + e^{-n}x_n = 1, \quad n = 0, 1, \dots,$
4. $x_{n+1} - \dfrac{\ln(n+1)}{\ln(n+2)}x_n = 2n, \quad n = 1, 2, \dots.$

Exercise 1.4 Find the equilibrium points and the period-two solutions of the following difference equations. Analyze their stability and visualize it using both `StairCase` and `TimeSeriesPlot` functions.

1. $x_{n+1} = x_n - 2x_n^3, \quad n = 0, 1, \dots,$
2. $x_{n+1} = x_n - e^{-x_n}, \quad n = 0, 1, \dots,$

3. $x_{n+1} = x_n + e^{-x_n^2}, \quad n = 0, 1, \dots,$

4. $x_{n+1} = \ln(1 + x_n), \quad n = 0, 1, \dots,$

5. $x_{n+1} = \tan^{-1}(x_n), \quad n = 0, 1, \dots.$

Exercise 1.5 Find the equilibrium points and analyze their stability for the following difference equations. Use appropriate tests if the equilibrium point is nonhyperbolic.

1. $x_{n+1} = -\sin(x_n), \quad n = 0, 1, \dots,$

2. $x_{n+1} = e^{-x_n} - 1, \quad n = 0, 1, \dots,$

3. $x_{n+1} = e^{x_n} - 1 - \frac{x_n^2}{2}, \quad n = 0, 1, \dots,$

4. $x_{n+1} = x_n + \cos(x_n) - 1, \quad n = 0, 1, \dots,$

5. $x_{n+1} = \frac{1+x_n}{-1+2x_n}, \quad n = 0, 1, \dots.$

Use *Dynamica* to visualize your findings and/or to predict the results.

Exercise 1.6
Consider the difference equation

$$x_{n+1} = \frac{1}{2}\left(x_n + \frac{A}{x_n}\right), \quad n = 0, 1, \dots, \quad x_0 \neq 0, \quad A > 0. \tag{1.48}$$

This difference equation is known as Heron's iterative procedure for finding the square root of A by an iterative method. It is a special case of **Newton's iterative method** for numerically solving an equation $f(x) = 0$, and it has the form

$$x_{n+1} = x_n - \frac{f(x_n)}{f'(x_n)}, \quad n = 0, 1, \dots, \tag{1.49}$$

where x_0 is chosen in an appropriate way for the corresponding orbit to converge to the equilibrium solution of this equation. In the case of equation (1.48) we have $f(x) = x^2 - A$.

1. Find the equilibrium points and the period-two solutions of equation (1.48).

2. Investigate the stability of the equilibrium points.

3. Show that the equilibrium points are the global attractors with the following basins of attraction: $\mathcal{B}(\sqrt{A}) = (0, \infty)$ and $\mathcal{B}(-\sqrt{A}) = (-\infty, 0)$.

4. Use *Dynamica* to study the rate of convergence of (1.48).

Exercise 1.7 Consider the difference equation

$$x_{n+1} = \frac{1}{k}\left((k-1)x_n + \frac{B}{x_n^{k-1}}\right), \quad n = 0, 1, \ldots, \quad x_0 \neq 0, \quad B > 0. \tag{1.50}$$

This difference equation is an analogue of Heron's iterative procedure for finding the k-th root of a positive number B by an iterative method. Perform an analysis of this equation similar to the analysis in Exercise 1.6.

Exercise 1.8 Consider the difference equation

$$x_{n+1} = \frac{1}{2}\left(x_n - \frac{A}{x_n}\right), \quad n = 0, 1, \ldots, \quad x_0 \neq 0, \quad A > 0. \tag{1.51}$$

This difference equation is an analogue of the iterative process given with (1.48). It represents Newton's method for numerically solving the equation $x^2 + A = 0$, where $A > 0$.

1. Find the equilibrium points and the period-two solutions of equation (1.48).
2. Investigate the stability of the period-two solutions.
3. Try to use the complex numbers as the initial points.
4. Show that for every initial value $x_0 = \alpha + i\beta$, $\alpha > 0$ the corresponding orbit converges to $\sqrt{A}$, while if $\alpha < 0$ the corresponding orbit converges to $-\sqrt{A}$. (This can be quite challenging.)
5. Use *Dynamica* to study the rate of convergence of (1.51).

Exercise 1.9 Consider the difference equation

$$x_{n+1} = \frac{2x_n^3 - 1}{3x_n^2 + p}, \quad n = 0, 1, \ldots, \quad x_0 \neq 0, \quad A > 0. \tag{1.52}$$

This difference equation represents Newton's method for numerically solving the cubic equation $x^3 + px + 1 = 0$.

1. Find the equilibrium points and the period-two solutions of equation (1.52).
2. Investigate the stability of the equilibrium solutions.
3. Use *Dynamica* to plot the bifurcation diagram for this equation. Locate visually the parameter values where the period doubling takes place.

4. Try to use Theorem 1.13 to find the bifurcation values of the parameter where the period doubling takes place.

   See [HuW], pp. 283–285.

Exercise 1.10 Consider the difference equation

$$x_{n+1} = \frac{pe^{x_n}(1 - x_n)}{1 - pe^{x_n}}, \quad n = 0, 1, \dots, \quad p > 0. \tag{1.53}$$

This difference equation represents Newton's method for numerically solving the equation $x = pe^x$.

1. Find the equilibrium solutions of equation (1.53).
2. Investigate the stability of the equilibrium solutions.
3. Use *Dynamica* to plot the bifurcation diagram for this equation. Locate the parameter values where the period doubling takes place.
4. Try to use Theorem 1.13 to find the bifurcation values of the parameter where the period doubling takes place.

Exercise 1.11 Consider the difference equation

$$x_{n+1} = px_n{}^k(1 - x_n^r), \quad n = 0, 1, \dots, \quad p > 0, k > 0, r > 0. \tag{1.54}$$

1. Find the equilibrium solutions of equation (1.54).
2. Investigate the stability of the equilibrium solutions.
3. Investigate the oscillatory character of solutions of (1.54).
4. Use *Dynamica* to plot the bifurcation diagram for this equation.

See [BZ].

Exercise 1.12 Show that the map F is conjugate to the map G with the homeomorphism h that makes this conjugacy possible.

1. $F(x) = 2x - x^2$, $G(x) = x^2$ and $h(x) = 1 - x$.
2. $F(x) = 2x$, $G(x) = 2^{1/3}x$ and $h(x) = x^3$. Show that the multipliers of F and G are different.
3. If $F(x) = ax^2 + bx + c$, $a \neq 0$, $G(x) = x^2 + d$ find h in the form $h(x) = \alpha + \beta x$, that is, find h that satisfies $h \circ F = G \circ h$. This can also be posed as, find α and β such that

$$\alpha + \beta(ax^2 + bx + c) = (\alpha + \beta x)^2 + d.$$

Exercise 1.13 Show that if a conjugate mapping h is a diffeomorphism (i.e., both h and h^{-1} are differentiable), then the multipliers at the corresponding fixed points are equal.

Exercise 1.14 Consider the **baker map**

$$B(x) = \begin{cases} 2x & \text{if } 0 \leq x < \frac{1}{2} \\ 2x - 1 & \text{if } \frac{1}{2} \leq x < 1, \end{cases}$$

see [E2] and [KJ]. Clearly, $B : [0, 1) \to [0, 1)$.

1. Show that the baker map may be defined as $B(x) = 2x(mod 1)$ on $[0, 1)$. Recall that $a = b(mod p)$ if $b = mp + b$ where $0 \leq a < p$.

2. Show that every periodic point of the baker map B has the form $k/(2^n - 1), k = 0, 1, ..., 2^n - 2$.

3. Determine the stability of the period-two point.

4. Prove that B is transitive on $[0, 1]$, hence it is chaotic in the sense of Devaney.

5. Is the baker map conjugate to the logistic map or the tent map? Justify your answer and find the homeomorphism h that makes this conjugacy possible.

6. Describe the effect of B on the numbers from $[0, 1]$ in the binary representation. Does this representation of the baker map help in understanding its chaotic character? (Remember the shift map is a canonical example of the chaotic map.)

Exercise 1.15 Consider the map $S_d : [0, 1) \to [0, 1)$ defined as

$$S_d(x) = 10x \quad (mod\, 1).$$

Clearly, $S_d(.a_1a_2a_3...) = .a_2a_3...$ and so S_d is another shift map on $[0, 1)$ that can be called a decimal shift.

1. Find all periodic points of this map.

2. Determine the stability of the period-two point.

3. Prove that S_d is transitive on $[0, 1]$, and so is chaotic in the sense of Devaney.

4. Is the decimal shift map conjugate to the logistic map or the tent map? Justify your answer and find the homeomorphism h that makes this conjugacy possible.

5. Exhibit explicitly a dense orbit of this map.

Exercise 1.16 Consider the map S_p defined as

$$S_p(x) = px \quad (mod\, 1),$$

where $p > 2$ is a positive integer. Clearly, S_p is another shift map on $[0, 1)$ in the p-base representation and can be called p-base shift.

1. Find all periodic points of this map.
2. Is S_p transitive on $[0, 1]$ for every p?
3. Is S_p conjugate to any chaotic map for every value of p? If your answer is affirmative find the homeomorphism h that makes this conjugacy possible.
4. Is there a dense orbit of this map for every p? Exhibit explicitly a dense orbit if it exists.

Exercise 1.17 Consider the map $R_2 : S^1 \to S^1$ where S^1 is the unit circle defined as

$$R_2((\cos\theta, \sin\theta)) = (\cos(2\theta), \sin(2\theta)),$$

where $\theta \in [0, 2\pi]$. This map is known as the **double angle map**. Using complex numbers and the well-known Euler's formula $e^{i\theta} = \cos\theta + i\sin\theta$, we can represent this map in the form $R_2(e^{i\theta}) = e^{2i\theta}$.

1. Show that all periodic points of this map are given by $\theta = \frac{2n\pi}{2^k-1}$.
2. Show that R_2 is transitive on S^1.
3. Find a dense orbit of this map.
4. Define the map $h : S^1 \to [0, 1]$ as

   $$h(e^{i\theta}) = \sin^2\left(\frac{\theta}{2}\right).$$

   Show that R_2 satisfies

   $$h(R_2(e^{i\theta})) = f_4(h(e^{i\theta})), \quad \theta \in [0, 2\pi]$$

   where f_4 is the logistic map with parameter $p = 4$, i.e., $f_4(x) = 4x(1-x)$. Is this enough to establish the conjugacy between these two maps? See [E2], pp. 131–132.

Exercise 1.18 Perform the following analysis for each of the given equations:

a. Find equilibrium point(s).

b. Check the linearized stability of the equilibrium points.

c. Find periodic solutions of period-two (if they exist).

d. Plot the bifurcation diagram by fixing all parameters except one.

e. Plot the Lyapunov exponents for some orbits.

f. Find the values of parameters (if any) for which the corresponding map is conjugate to some known chaotic maps (can be quite challenging).

The equations are the following.

(1) $x_{n+1} = \frac{Ax_n}{1+x_n}$. (Riccati difference equation; also known as Pielou's difference equation, see [KoL] and [KL].)

(2) $x_{n+1} = \frac{Ax_n}{1+x_n{}^2}$. (Rational difference equation, see [KoL].)

(3) $x_{n+1} = p\sin(\pi x_n)$. (Unimodal difference equation)

(4) $x_{n+1} = px_n e^{-x_n}$.

(5) $x_{n+1} = (1-A)x_n + Ax_n{}^3$. (Cubic map, see [K].)

(6) $x_{n+1} = x_n exp(A\frac{1-x_n}{1+x_n})$. (Simple genotype selection model, see [KoL].)

(7) $x_{n+1} = Ax_n + \frac{B}{1+x_n{}^p}$. (Discrete analogue of a model of haematopoiesis, see [KoL].)

(8) $x_{n+1} = B + e^{-Ax_n^2}$. (The Gaussian map, see [D].)

Chapter 2

Dynamics of Two-Dimensional Dynamical Systems

2.1 Introduction

In this chapter we study the dynamics of systems of difference equations of the form

$$\begin{aligned} x_{n+1} &= f(x_n, y_n) \\ y_{n+1} &= g(x_n, y_n) \end{aligned} \qquad n = 1, 2, \ldots \tag{2.1}$$

where f and g are given functions. We also analyze here the dynamics of the second order difference equation

$$x_{n+1} = H(x_n, x_{n-1}), \tag{2.2}$$

where H is a given function.

The chapter begins with a presentation of the linear theory in Section 2.2. Equilibria and periodic points are introduced in Section 2.3. The Riccati difference equation is studied with *Dynamica* in Section 2.4. The main definitions and results from the theory of Linearized Stability Analysis are presented in Section 2.5 and illustrated in *Dynamica* sessions throughout the chapter, such as the one in Section 2.6, but especially in the session in Section 2.17. Period-doubling bifurcations, Lyapunov numbers, and box dimension are discussed in Sections 2.7, 2.8, and 2.9, respectively. Section 2.10 presents recent developments concerning semicycle analysis. Stable and unstable manifolds are discussed in Section 2.11, and illustrated in the *Dynamica* session on Henon's equation in Section 2.12. Invariants and Lyapunov functions are discussed in Sections 2.13 and 2.14, and their use is illustrated in the *Dynamica* session on Lyness' equation in Section 2.15. Dissipative maps and systems are presented in Section 2.16. An analysis of a dissipative system is performed with *Dynamica* in Section 2.17. Section 2.18 is a brief discussion on area-preserving maps and systems. The chapter ends with a collection of research projects in Sections 2.19 and 2.20. These are applications in biology and economics.

2.2 Linear Theory

In this section we present notions and results from the theory of linear systems and equations. The results are fairly standard, and the corresponding proofs can be found in introductory textbooks on difference equations such as [E1], [KP], and [LT].

2.2.1 Two-Dimensional Linear Systems

Consider the system

$$\begin{aligned} x_{n+1} &= ax_n + by_n \\ y_{n+1} &= cx_n + dy_n. \end{aligned} \tag{2.3}$$

System (2.3) may be written as

$$\begin{pmatrix} x_{n+1} \\ y_{n+1} \end{pmatrix} = A \begin{pmatrix} x_n \\ y_n \end{pmatrix}, \tag{2.4}$$

where

$$A = \begin{pmatrix} a & b \\ c & d \end{pmatrix}.$$

By setting

$$\mathbf{Z}_n = \begin{pmatrix} x_n \\ y_n \end{pmatrix}$$

we rewrite (2.3) as

$$\mathbf{Z}_{n+1} = A\mathbf{Z}_n$$

A **solution** of system (2.3) is an expression that satisfies this system for all the values of $n = 0, 1, \ldots$. The **general solution** is a solution that contains all solutions of the system. A **particular solution** of (2.3) is a solution that satisfies an **initial condition** of the form

$$x_0 = c, \quad y_0 = d, \tag{2.5}$$

where c and d are given real numbers. The problem of finding a particular solution to system (2.3) with specified initial conditions (2.5) is called an **initial value problem**, which is abbreviated as **IVP**.

It is easy to see that the solution of (2.3) has the form

$$\mathbf{Z}_n = A^n \mathbf{Z}_0. \tag{2.6}$$

Then, it is clear that efficient computation of A^n is a practical issue when solving linear difference equations.

Two solutions, $\mathbf{z}^1$ and $\mathbf{z}^2$, of (2.3) are said to be **linearly independent** if one is not a scalar multiple of the other for all n, or equivalently, if

$$c_1\mathbf{z}_n^1 + c_2\mathbf{z}^2 = \mathbf{0}, \quad \text{for all} \quad n,$$

implies $c_1 = c_2 = 0$. Otherwise, two solutions are said to be **linearly dependent**. A set $\{\mathbf{z}^1, \mathbf{z}^2\}$ of any two linearly independent solutions is called a **fundamental set** of solutions, and in this case the matrix $F = (\mathbf{z}^1, \mathbf{z}^2)$ is a **fundamental matrix** of (2.3). It can be shown that system (2.3) always has a fundamental set of solutions.

THEOREM 2.1

a. *There exists a fundamental set of solutions for system (2.3).*

b. *If $\{\mathbf{z}^1, \mathbf{z}^2\}$ is a fundamental set of solutions of system (2.3), then a general solution is*

$$\mathbf{z} = c_1\mathbf{z}^1 + c_2\mathbf{z}^2 = \mathbf{F}\begin{pmatrix} c_1 \\ c_2 \end{pmatrix}, \quad c_1, c_2 \in R.$$

The matrix

$$\mathbf{F} = \begin{pmatrix} \mathbf{z}^1 & \mathbf{z}^2 \end{pmatrix}$$

is called a **fundamental matrix** *of (2.3).*

System (2.3) is called a **homogeneous system**. Along with this system we can consider the **nonhomogeneous system** of the form

$$\mathbf{Z}_{n+1} = A\mathbf{Z}_n + B_n, \quad \mathbf{Z}_0 = \mathbf{d}, \quad n = 0, 1, \ldots \tag{2.7}$$

where $B_n \in R^2$ and $d \in R^2$ are given. The IVP (2.7) has the (unique) solution given by

$$\mathbf{Z}_n = A^n\mathbf{d} + \sum_{i=0}^{n} A^{n-i}B_i, \quad n = 0, 1, \ldots. \tag{2.8}$$

The basic notions, stated here for systems (2.3) and (2.7), can be extended to more general linear systems

$$\mathbf{Z}_{n+1} = A_n\mathbf{Z}_n, \quad n = 0, 1, \ldots \tag{2.9}$$

and

$$\mathbf{Z}_{n+1} = A_n\mathbf{Z}_n + B_n, \quad \mathbf{Z}_0 = \mathbf{d}, \tag{2.10}$$

where A_n is a real 2×2 matrix for every $n = 1, 2, \ldots$.

See Chapter 3 for an extension of the results presented here to linear systems of any finite dimension.

2.2.2 Powers of a Matrix and Jordan Normal Form

Let us recall some basic notions in linear algebra. If A is the two-by-two matrix

$$A = (a_{ij}) = \begin{pmatrix} a_{11} & a_{12} \\ a_{21} & a_{22} \end{pmatrix},$$

the determinant of A is $\det(A) = a_{11}a_{22} - a_{21}a_{12}$, and the trace of A is $tr(A) = a_{11} + a_{22}$. The polynomial

$$p(\lambda) = det(\lambda I - A) = \lambda^2 - (a_{11} + a_{22})\lambda + a_{11}a_{22} - a_{21}a_{12}$$

is called the **characteristic polynomial** of A. The zeros of $p(\lambda)$ are called the **eigenvalues** of A. Here I denotes the identity matrix

$$I = \begin{pmatrix} 1 & 0 \\ 0 & 1 \end{pmatrix},$$

If λ is an eigenvalue of A, a nonzero vector $\mathbf{v}$ such that $A\mathbf{v} = \lambda\mathbf{v}$ is called an **eigenvector** of A associated to λ. The matrix A is nonsingular if 0 is not an eigenvalue. Two matrices A and B are said to be **similar** if there exists a **nonsingular matrix** P such that $B = P^{-1}AP$. Here P^{-1} denotes the inverse matrix, i.e., a matrix that satisfies $P^{-1}P = I$. Similar matrices A and B have the same eigenvalues, and there is a correspondence between eigenvectors $P^{-1}\mathbf{v}$ of B with eigenvectors $\mathbf{v}$ of A. See [FB].

The next result is relevant for the computation of A^n.

THEOREM 2.2 (Jordan Normal Form)

Every 2×2 real matrix is similar to a matrix J given by one of the following cases.

1. *If A has real and different eigenvalues $\lambda_1 \neq \lambda_2$, then*

$$J = \begin{pmatrix} \lambda_1 & 0 \\ 0 & \lambda_2 \end{pmatrix},$$

2. *If A has real and equal eigenvalues $\lambda_1 = \lambda_2 = \lambda$, then*

$$J = \begin{pmatrix} \lambda & 1 \\ 0 & \lambda \end{pmatrix},$$

3. *If A has complex eigenvalues $\lambda = \alpha \pm i\beta$, then*

$$J = \begin{pmatrix} \alpha & \beta \\ -\beta & \alpha \end{pmatrix}$$

Theorem 2.2 gives an efficient tool for computing A^n, since

$$A^n = (PJP^{-1})^n = PJ^nP^{-1}$$

where J is as in one of the three cases of Theorem 2.2.

2.2.3 Second Order Linear Equations

In this section we present theory and methods of finding solutions of second order linear difference equations.

Consider the second order linear difference equations

$$x_{n+2} + a_n x_{n+1} + b_n x_n = 0, \quad n = 0, 1, \ldots \tag{2.11}$$

and

$$x_{n+2} + a_n x_{n+1} + b_n x_n = c_n, \quad n = 0, 1, \ldots \tag{2.12}$$

where $\{a_n\}_{n=0}^{\infty}$, $\{b_n\}_{n=0}^{\infty}$, and $\{c_n\}_{n=0}^{\infty}$ are given sequences of real numbers, and where

$$b_n \neq 0 \quad \text{for} \quad n = 0, 1, \ldots. \tag{2.13}$$

Equation (2.11) is called a **linear homogeneous difference equation of second order**, while equation (2.12) is called a **linear nonhomogeneous difference equation** of second order.

In what follows, we shall use the abbreviations $x_n^{(h)}$, $x_n^{(gen)}$, and $x_n^{(p)}$ to denote, respectively, the general solution of the homogeneous difference equation (2.12), the general solution of the nonhomogeneous difference equation (2.12), and any particular solution of the non homogeneous difference equation (2.12). The followings result won't be a surprise to the reader familiar with differential equations. See [E1], [KP], [LT], and [LL].

The next concept that we introduce is an analogue of the Wronski determinant or Wronskian for differential equations. In the case of difference equations it is called the **Casorati determinant** or **Casoratian**, and its main function is to help determine whether or not two solutions of (2.11) are linearly independent.

DEFINITION 2.1 *Let $x = \{x_n\}_{n=0}^{\infty}$ and $y = \{y_n\}_{n=0}^{\infty}$ be two solutions of (2.12). The Casoratian of x and y, denoted by $\{C(x, y; n)\}_{n=0}^{\infty}$, is given by*

$$C(x, y; n) = \begin{vmatrix} x_n & y_n \\ x_{n+1} & y_{n+1} \end{vmatrix} = x_n y_{n+1} - x_{n+1} y_n, \quad \textit{for} \quad n = 0, 1, \ldots.$$

The following results are analogue to results for linear differential equations, and can be found in [MT], p. 354.

THEOREM 2.3 (Heymann's Theorem)
Let $x = \{x_n\}_{n=0}^{\infty}$ and $y = \{y_n\}_{n=0}^{\infty}$ be two solutions of (2.11).

(i) The Casoratian $\{C_n\}_{n=0}^{\infty}$ satisfies the first order linear homogeneous difference equation

$$C_{n+1} - b_n C_n = 0, \quad n = 0, 1, \ldots,$$

and consequently,

$$C_n = \left(\prod_{k=0}^{n-1} q_k \right) C_0.$$

(ii) $x = \{x_n\}_{n=0}^{\infty}$ *and* $y = \{y_n\}_{n=0}^{\infty}$ *are linearly independent solutions of (2.11) if and only if* $C(x, y; n) \neq 0$ *for all* $n = 0, 1, \ldots$.

The next result gives the form of a general solution of (2.11) and (2.12).

THEOREM 2.4

Let $\{x_n\}_{n=0}^{\infty}$ *and* $\{y_n\}_{n=0}^{\infty}$ *be two linearly independent solutions of (2.11) and* $\{x_n^{(p)}\}_{n=0}^{\infty}$ *be a particular solution of (2.12). Then*

(1) A general solution of (2.11) is given by

$$x_n^{(h)} = c_1 x_n + c_2 y_n \quad \text{for} \quad n = 0, 1, \ldots$$

where c_1 *and* c_2 *are arbitrary constants.*

(2) the general solution of (2.12) is given by

$$x_n^{(gen)} = c_1 x_n + c_2 y_n + x_n^{(p)} \quad \text{for} \quad n = 0, 1, \ldots$$

where c_1 *and* c_2 *are arbitrary constants.*

In the case of linear equations with constant coefficients

$$x_{n+2} + p x_{n+1} + q x_n = 0, \quad n = 0, 1, \ldots \tag{2.14}$$

with $p, q \in \mathbf{R}$ and $q \neq 0$, there is a simple formula for the solution of (2.11).

If we look for a solution of (2.14) of the form

$$x_n = \lambda^n \quad \text{for} \quad n = 0, 1, \ldots$$

with $\lambda \neq 0$, then (2.14) leads to the quadratic equation

$$\lambda^2 + p\lambda + q = 0. \tag{2.15}$$

Equation (2.15) is called the **characteristic equation** of (2.14), and its roots are called the **characteristic roots** associated with (2.14). The roots λ_+ and λ_- of (2.15) are given by the formula

$$\lambda_{\pm} = \frac{-p \pm \sqrt{p^2 - 4q}}{2}.$$

There are two real and distinct roots if the **discriminant** $D = p^2 - 4q$ of the characteristic equation (2.15) is positive. If $D = 0$, then (2.15) has equal roots

$\lambda_+ = \lambda_- = -\frac{p}{2}$. In this case we say that (2.15) has a *double root.* Finally, when $D < 0$, the characteristic roots are *complex conjugate* numbers of the form $a \pm ib$. A complex root $a + ib$ can always be written in the form

$$r(\cos\theta + i\sin\theta)$$

where

$$r = \sqrt{a^2 + b^2}$$

is the *modulus* of the complex number $a + ib$, and θ is the *argument* of the complex number $a + ib$. The angle θ is given by

$$\cos\theta = \frac{a}{r}, \quad \sin\theta = \frac{b}{r}, \quad \text{and} \quad -\pi < \theta \leq \pi.$$

We summarize the discussion about equation (2.14) in the following result.

THEOREM 2.5

Let λ_+ and λ_- be the roots of the characteristic equation of

$$x_{n+2} + px_{n+1} + qx_n = 0, \qquad n = 0, 1, \ldots$$

with $p, q \in \mathbf{R}$ and $q \neq 0$,.

Then the general solution $\{x_n\}_{n=0}^{\infty}$ of the equation is given by one of the following cases.

1. *If λ_+, and λ_- are both real and distinct, then*

$$x_n = c_1\lambda_+^n + c_2\lambda_-^n \quad \text{for} \quad n = 0, 1, \ldots$$

where c_1 and c_2 are arbitrary constants;

2. *If $\lambda_+ = \lambda_-$, then*

$$x_n = c_1\lambda_+^n + c_2 n\lambda_+^n \quad \text{for} \quad n = 0, 1, \ldots$$

where c_1 and c_2 are arbitrary constants;

3. *If $\lambda_\pm = a \pm ib = re^{\pm i\theta}$ are both complex numbers, then*

$$x_n = c_1 r^n \cos n\theta + c_2 r^n \sin n\theta \quad \text{for} \quad n = 0, 1, \ldots$$

where

$$r = \sqrt{a^2 + b^2}, \ \cos\theta = \frac{a}{r}, \ \sin\theta = \frac{b}{r}, \quad \textit{and} \quad \theta \in (-\pi, \pi]$$

and where c_1 and c_2 are arbitrary constants.

Let us illustrate this result with some examples.

Example 2.1

(a) The equation

$$y_{n+2} - 5y_{n+1} + 6y_n = 0, \quad n = 0, 1, \ldots$$

has the solution

$$y_n = C_1 3^n + C_2 2^n,$$

where C_1 and C_2 are arbitrary constants. The corresponding IVP has the solution

$$y_n = (y_1 - 2y_0)3^n + (3y_0 - y_1)2^n.$$

(b) The equation

$$y_{n+2} - 4y_{n+1} + 4y_n = 0, \quad n = 0, 1, \ldots$$

has the solution

$$y_n = C_1(n+1)2^n + C_2 2^n,$$

where C_1 and C_2 are arbitrary constants. The corresponding IVP has the solution

$$y_n = (\frac{1}{2}y_1 - y_0)(n+1)2^n + (3y_0 - y_1)2^n.$$

(c) The equation

$$y_{n+2} - 2y_{n+1} + 2y_n = 0, \quad n = 0, 1, \ldots$$

has the solution

$$y_n = C_1(1-i)^n + C_2(1+i)^n = D_1 \cos\left(n\frac{\pi}{4}\right) + D_2 \sin\left(n\frac{\pi}{4}\right),$$

where C_1, C_2, D_1 and D_2 are arbitrary constants. The corresponding IVP has the solution

$$y_n = y_0 \cos\left(n\frac{\pi}{4}\right) + (\sqrt{2}y_1 - y_0) \sin\left(n\frac{\pi}{4}\right).$$

∎

Example 2.2

(a) The IVP

$$y_{n+2} + 5y_{n+1} + 4y_n = 0, \quad y_0 = -2, \ y_1 = 1,$$

has the solution

$$y_n = -\frac{7}{3}(-1)^n + \frac{1}{3}(-4)^n.$$

(b) The IVP

$$y_{n+2} + 2y_{n+1} + y_n = 0, \quad y_0 = -2, y_1 = 1$$

has the solution

$$y_n = (n+1)(-1)^n - 3(-1)^n.$$

∎

There are several methods for finding a particular solution of a linear non-homogeneous difference equation such as the method of undetermined coefficients and in some cases the Z-transform method. We shall mention only the method of **variation of constants**, which is more general as it works for equations with variable coefficients as well. In the case where the linear difference equation has variable coefficients, we need to know two linearly independent solutions of the associated homogeneous equation.

Given a sequence $\{z_n\}_{n=0}^{\infty}$, the **forward difference** $\{\Delta z_n\}_{n=0}^{\infty}$ is defined by

$$\Delta z_n := z_{n+1} - z_n, \quad n \geq 0,$$

see [E1] and [KP].

THEOREM 2.6

Consider the nonhomogeneous difference equation

$$x_{n+2} + a_n x_{n+1} + b_n x_n = c_n, \quad n = 0, 1, \ldots \tag{2.16}$$

where $\{a_n\}_{n=0}^{\infty}$, $\{b_n\}_{n=0}^{\infty}$, *and* $\{c_n\}_{n=0}^{\infty}$ *are sequences of real numbers, and where* $b_n \neq 0$ *for an infinite number of indices* n. *Let* $\{x_n\}_{n=0}^{\infty}$ *and* $\{y_n\}_{n=0}^{\infty}$ *be two linearly independent solutions of the corresponding homogeneous difference equation*

$$y_{n+2} + a_n y_{n+1} + b_n y_n = 0, \quad n = 0, 1, \ldots. \tag{2.17}$$

Then a particular solution of equation (2.16) is given by

$$x_n^{(p)} = \sum_{k=0}^{n-1} c_k \frac{\begin{vmatrix} x_{k+1} & y_{k+1} \\ x_n & y_n \end{vmatrix}}{\begin{vmatrix} x_{k+1} & y_{k+1} \\ x_{k+2} & y_{k+2} \end{vmatrix}}, \quad n = 0, 1, \ldots. \tag{2.18}$$

Mathematica's built-in package `RSolve` computes the solution of a given difference equation or a system of difference equations using the method of generating functions. The equations and systems that can be solved are mainly linear. With `RSolve` one can find the general solution of the system of equations, as well as particular solutions with prescribed initial conditions.

This loads the RSolve package.

```
In[1]:= << DiscreteMath`RSolve`
```

This is the difference equation satisfied by the well-known Fibonacci numbers.

```
In[2]:= RSolve[{a[n] == a[n - 1] + a[n - 2],
             a[0] == a[1] == 1}, a[n], n]
```

$$\text{Out[2]= } \left\{\left\{a[n] \rightarrow \frac{2^{-1-n}\left(-\left(1-\sqrt{5}\right)^{1+n}+\left(1+\sqrt{5}\right)^{1+n}\right)}{\sqrt{5}}\right\}\right\}$$

Here are the first 14 terms in the solution.

```
In[3]:= Table[(a[n]/.%)[[1]], {n, 0, 14}]
             //Expand
Out[3]= {1, 1, 2, 3, 5, 8, 13, 21, 34, 55,
             89, 144, 233, 377, 610}
```

2.3 Equilibrium Solutions

We shall assume that the functions f and g in (2.1) are continuously differentiable in their domain.

DEFINITION 2.2 *Let D_f and D_g be the domains of the functions f and g in (2.1), and set $D = D_f \cap D_g$. The* **forbidden set** *of system (2.1) is the set*

$$F := \left\{ (c,d) \in R^2 \ : \ \begin{array}{l} \text{there exists } n \in N \text{ such that when } (x_0, y_0) = (c,d), \\ (x_k, y_k) \in D \text{ for } k = 0, \ldots n-1, \text{ and } (x_n, y_n) \notin D \end{array} \right\} \tag{2.19}$$

Thus, by definition, solutions $\{(x_n, y_n)\}$ of (2.1) cannot have $(x_0, y_0) \in F$. We shall also assume that the systems we consider from now on have a forbidden set that is not equal to all of R^2, i.e., we assume that solutions exist. It should be mentioned that the problem of the existence of solutions of system (2.1) does not have a well-developed theory; see [GLMT] for the existence of specific system of rational equations and [KL] for some open problems in this area.

DEFINITION 2.3

1. *An* **equilibrium solution** *of (2.1) or* **fixed point** *of the vector map*

$\mathbf{F} = (f, g)$ is a point $(\overline{x}, \overline{y})$ that satisfies

$$\overline{x} = f(\overline{x}, \overline{y})$$
$$\overline{y} = g(\overline{x}, \overline{y}).$$

2. *A* **periodic point** (x_p, y_p) **of period** *m is a fixed point of the m-th iterate $\mathbf{F}^m$ of the map $\mathbf{F}$. A periodic point of minimal period m or prime period m is defined as in Definition 1.1.*

3. *Let (x_0, y_0) be a given element of R^2. The pairs (x_1, y_1), (x_2, y_2), ... defined inductively by (2.1) are called the* **iterates** *of (x_0, y_0), and the sequence $\{(x_n, y_n)\}_{n=0}^{\infty}$ is called the* **positive orbit** *of (x_0, y_0) and is denoted by $\gamma^+((x_0, y_0))$; that is,*

$$\gamma^+((x_0, y_0) = \{(x_0, y_0), \mathbf{F}(x_0, y_0), \ldots, \mathbf{F}^k(x_0, y_0), \ldots\}.$$

4. *If the map $\mathbf{F}$ is invertible, we define the* **negative orbit** *of (x_0, y_0) to be*

$$\gamma^-((x_0, y_0) = \{(x_0, y_0), \mathbf{F}^{-1}(x_0, y_0), \ldots, \mathbf{F}^{-k}(x_0, y_0), \ldots\},$$

where $\mathbf{F}^{-n}$ denotes the n-th iterate of $\mathbf{F}^{-1}$.

5. *When both the positive and negative orbits exist, the* **complete orbit** *$\gamma((x_0, y_0))$ is the union of the positive and negative orbits*

$$\gamma((x_0, y_0)) = \gamma^+((x_0, y_0)) \cup \gamma^-((x_0, y_0)).$$

DEFINITION 2.4 *A point (a, b) is called an* **ω-limit point** *of the positive orbit $\gamma^+((x_0, y_0)$ if there is a sequence of positive integers n_i such that*

$$n_i \to \infty \quad \text{and} \quad \mathbf{F}^{n_i}(x_0, y_0) \to \infty \quad \text{as} \quad i \to \infty.$$

The **ω-limit set** *$\omega(x_0, y_0)$ of $\gamma^+((x_0, y_0))$ is the set of all its ω-limit points.*

In the case F is invertible, the **α-limit set** *of the negative orbit is defined similarly by taking n_i to be negative integers.*

2.3.1 Examples

We start with an observation on linear difference equations and linear maps. Consider a linear system in the plane given by

$$\mathbf{z}_{n+1} = \mathbf{A}\mathbf{z}_n \tag{2.20}$$

where $\mathbf{A}$ is a 2×2 matrix. The positive orbit of $\mathbf{z}_0$ is the following sequence:

$$\gamma^+(\mathbf{z}_0) = \{\mathbf{z}_0, \mathbf{A}\mathbf{z}_0, \ldots, \mathbf{A}^n\mathbf{z}_0, \ldots\}.$$

The Jordan Normal Form of the matrix $\mathbf{A}$ may be used to compute $\mathbf{A^n}$, to find orbits of a linear map.

Example 2.3 (A hyperbolic sink with monotonic convergence) Consider system (2.20) where

$$\mathbf{A} = \begin{pmatrix} a & 0 \\ 0 & d \end{pmatrix},$$

with $a, d \in (0, 1)$. The powers of $\mathbf{A}$ are given by

$$\mathbf{A}^n = \begin{pmatrix} a^n & 0 \\ 0 & d^n \end{pmatrix},$$

therefore $\mathbf{A}^n$ approaches the zero matrix as $n \to \infty$. Thus, the origin, which is also the only equilibrium point, is a limit point of each orbit. To find a positive orbit notice that the eigenvalues a and d of $\mathbf{A}$ have the corresponding eigenvectors $\mathbf{v}^1 = (1, 0)^T$ and $\mathbf{v}^2 = (0, 1)^T$, respectively. For any initial value $\mathbf{z}_0 = (x_0, y_0)^T$, we have

$$\mathbf{A}^n\mathbf{z}_0 = a^n x_0 \mathbf{v}^1 + d^n y_0 \mathbf{v}^2.$$

Consequently, the positive orbits approach the origin in a monotonic way in the directions of both eigenvectors $\mathbf{v}^1$ and $\mathbf{v}^2$. ∎

Example 2.4 (A hyperbolic sink with oscillatory convergence) Consider system (2.20) where

$$\mathbf{A} = \begin{pmatrix} a & 0 \\ 0 & -d \end{pmatrix},$$

with $a, d \in (0, 1)$. The powers of $\mathbf{A}$ are given by

$$\mathbf{A}^n = \begin{pmatrix} a^n & 0 \\ 0 & (-1)^n d^n \end{pmatrix},$$

hence $\mathbf{A}^n$ approaches the zero matrix as $n \to \infty$. Using the notations of the previous example, we conclude that the origin, which is also the only equilibrium point, is a limit point of each orbit. Orbits have the form

$$\mathbf{A}^n\mathbf{z}_0 = a^n x_0 \mathbf{v}^1 + (-1)^n d^n y_0 \mathbf{v}^2.$$

Consequently, the positive orbits approach the origin in a monotonic way in the direction of $\mathbf{v}^1$ and in an oscillatory way in the direction of $\mathbf{v}^2$.

Notice that in this example, if we took $a \in (-1, 0)$, then the convergence would be oscillatory in the direction of $\mathbf{v}^1$ as well. ∎

Example 2.5 (A hyperbolic source with monotonic divergence) Consider system (2.20) where

$$\mathbf{A} = \begin{pmatrix} a & 0 \\ 0 & d \end{pmatrix},$$

with $a, d \in (1, \infty)$. The positive orbit is given by

$$\mathbf{A}^n \mathbf{z}_0 = a^n x_0 \mathbf{v}^1 + d^n y_0 \mathbf{v}^2.$$

Consequently, the positive orbits are diverging to $\pm\infty$ in a monotonic way in the directions of both eigenvectors $\mathbf{v}^1$ and $\mathbf{v}^2$.

If we assume that $a \in (-\infty, -1)$, then the divergence is oscillatory in the direction of $\mathbf{v}^1$. A similar conclusion holds if $d \in (-\infty, -1)$. ∎

2.4 The Riccati Equation

The difference equation

$$x_{n+1} = \frac{\alpha + \beta x_n}{A + B x_n}, \qquad n = 0, 1, \ldots \tag{2.21}$$

where α, β, A, B are real numbers is called the **Riccati difference equation**. The Riccati difference equation is the most general linear fractional equation of the first order. In addition to its low order and to its generality, it is an important equation because it is one of the few nonlinear difference equations with fully explained dynamics [KL]. The study of (2.21) is also beneficial to the student of difference equations, as it provides a model to be used when more complicated equations are considered, see [CK] and [CKS].

The forbidden set of equation (2.21) is the set

$$F := \left\{ d \in R \; : \; \begin{array}{c} \text{there exists } n \in N \text{ such that if } x_0 = d, \\ A + Bx_\ell \neq 0 \text{ for } 0 \leq \ell \leq n-1, \text{ and } A + Bx_n = 0 \end{array} \right\}. \tag{2.22}$$

Thus, solutions $\{x_n\}$ of (2.21) cannot have $x_0 \in F$.

One of the goals of this section is to determine the forbidden set of the Riccati equation. Another goal is to describe both short and long term behavior of solutions of the Riccati equation. We shall derive a general solution, and use it to reach both goals. The exposition in this section is based on [KL].

Clearly, when $B = 0$, equation (2.21) reduces to a linear equation. Also, when $\alpha B - \beta A = 0$, equation (2.21) reduces to the trivial relation $x_n = C$, where C is a constant. To avoid these cases, we shall assume throughout this section that

$$B \neq 0 \quad \text{and} \quad \alpha B - \beta A \neq 0. \tag{2.23}$$

Note that the Riccati differential equation is of order one, so it appears that its proper place in this book is in Chapter 1. However, we discuss it in the present chapter since we use the theory of linear second order difference equations to study it.

Before we start the discussion of equation (2.21), we would like to mention a connection to the well-known Riccati differential equation

$$\frac{dz}{dt} = az^2 + bz + c, \tag{2.24}$$

where $a, b,$ and c are constants. The process of discretizing equation (2.24) consists of replacing the continuous function $z(t)$ by a set of discrete points $z_n = z(nh)$, $n = 0, 1, \ldots$, which represent a sampling of $z(t)$ at equally spaced intervals, and where h is a constant, the so-called step of a discretization. The derivative dz/dt is approximated by a difference quotient

$$\frac{z_{n+1} - z_n}{h}$$

and the quadratic term z^2 may be replaced by any of the following expressions

$$z_n^2, \quad z_n z_{n-1}, \quad z_n \frac{z_n + z_{n-1}}{2}, \quad z_n z_{n+1}, \ldots .$$

If we choose $z_n z_{n+1}$ and solve for z_{n+1}, then we obtain the discretization of (2.24) in the form

$$z_{n+1} = \frac{(hb+1)z_n + hc}{1 - haz_n},$$

which has the form of equation (2.21). See [Sa].

2.4.1 The Riccati Number and Changes of Variables

We begin by considering certain changes of variables that, as we shall see, under suitable conditions transform Riccati's equation into a linear equation of the second order.

We begin the session by defining the Riccati difference equation and the corresponding map.

```
In[1]:= << Dynamica`

In[2]:= ricc = x[n + 1] == (α + β x[n])/(A + B x[n]);

In[3]:= riccmap = DEToMap[ricc];
```

The fixed points will be needed later.

```
In[4]:= {fp1, fp2} = x /.
            Solve[riccmap[x] == x, x]
```

$$Out[4]= \left\{-\frac{A - \beta - \sqrt{4 B \alpha + (-A + \beta)^2}}{2 B}, -\frac{A - \beta + \sqrt{4 B \alpha + (-A + \beta)^2}}{2 B}\right\}$$

We now introduce a change of variable which is valid under the assumption $A+\beta \neq 0$

```
In[5]:= Roots[ricc /. {x[i_] → (β + A)/B w[i] - A/B},
            w[n + 1]]
Out[5]= w[1 + n] ==
   (B α - A β + A^2 w[n] + 2 A β w[n] + β^2 w[n]) / ((A + β)^2 w[n])
```

The substitution

$$R = \frac{-(B\alpha - A\beta)}{(A+\beta)^2} \qquad (2.25)$$

simplifies the expression for w_{n+1}. R is called the **Riccati number**.

```
In[6]:= % /. {B α - A β → -R (A + β)^2}//Simplify
Out[6]= R/w[n] + w[1 + n] == 1
In[7]:= riccw = Roots[%, w[n + 1]]
Out[7]= w[1 + n] == (-R + w[n])/w[n]
```

Now we introduce another change of variables:

$$w_i = \frac{y_{i+1}}{y_i}$$

The resulting difference equation is a second order linear equation.

```
In[8]:= riccy =
          Roots[riccw /. {w[i_] → y[i + 1]/y[i]},
            y[n + 2]]
Out[8]= y[2 + n] == -R y[n] + y[1 + n]
```

The characteristic equation of the second order equation has the roots shown on the right column.

```
In[9]:= Solve[% /. y[i_] → y^i, y]
Out[9]= {{y → 1/2 (1 - Sqrt[1 - 4 R])},
          {y → 1/2 (1 + Sqrt[1 - 4 R])}}
```

We conclude that, if $A+\beta \neq 0$, then Riccati's equation is transformed into a linear second order equation. Thus, a study of Riccati's equation may be done by analyzing cases, which we do with *Dynamica* in Subsections 2.4.2, 2.4.3, 2.4.4, and 2.4.5.

2.4.2 The Case when $A+\beta = 0$

First A is set in terms of β.

```
In[10]:= A = -β;
```

By displaying a few terms of an orbit of Riccati's map we see that, whenever $A+\beta = 0$, all solutions are periodic, with period 2

```
In[11]:= Orbit[riccmap, x, 6]//Simplify
Out[11]= {x, (α + x β)/(B x - β), x, (α + x β)/(B x - β), x, (α + x β)/(B x - β), x}
In[12]:= Clear[A];
```

Thus, $A + \beta =$ is *sufficient* for all solutions to be periodic with period two. Now we try to determine whether $A + \beta = 0$ is a *necessary* condition for solutions to be periodic with period two.

For a period-two solution, this difference must be zero. Then, either $A + \beta = 0$, or, $Ax + Bx^2 - \alpha - x\beta = 0$.

```
In[13]:= Simplify[riccmap[riccmap[x]] - x]
```

$$Out[13]= -\frac{(A+\beta)\ (A x + B x^2 - \alpha - x\beta)}{A^2 + A B x + B\ (\alpha + x\beta)}$$

But $Ax + Bx^2 - \alpha - x\beta = 0$ yields the equilibrium points of Riccati's equation.

$$In[14]:= \text{Solve}\left[A x + B x^2 - \alpha - x\beta == 0, x\right]$$

$$Out[14]= \left\{\left\{x \to \frac{-A - \sqrt{4 B\alpha + (A-\beta)^2} + \beta}{2 B}\right\}, \left\{x \to \frac{-A + \sqrt{4 B\alpha + (A-\beta)^2} + \beta}{2 B}\right\}\right\}$$

We conclude that if the solution is periodic with prime period two, then $A + \beta = 0$. The following result collects our conclusions.

THEOREM 2.7

Let α, β, A, and B be such that (2.23) holds. Equation (2.21) has a prime period-two solution if and only if $A + \beta = 0$. In either case, the forbidden set is $F = \{-A/B\}$, and every solution is periodic with period two.

2.4.3 The Case when $A + \beta \neq 0$ and $R = \frac{1}{4}$

When $R = \frac{1}{4}$ and $A + \beta \neq 0$, the characteristic equation of

$$y_{n+2} - y_{n+1} + R y_n = 0$$

has a root of multiplicity two, $\bar{y} = \frac{1}{2}$. According to Theorem 2.5, the general solution has the form

$$y_n = \left(c_1 + c_2 n\right)\left(\frac{1}{2}\right)^n$$

Here we may take $y_0 = 1$. Also, recall that $y_1 = w_0$.

We may use the `RSolve` package to produce a solution.

```
In[15]:= << DiscreteMath`RSolve`
```

Here is the solution with $y_0 = 1$, $y_1 = w_0$ produced by RSolve.

```
In[16]:= RSolve[{riccy /. R → 1/4,
                 y[0] == 1, y[1] == w[0]}, y[n], n]
```

$$Out[16]= \{\{y[n] \to 2^{-n}\ (1 + n\ (-1 + 2\, w[0]))\}\}$$

The forbidden set F is determined by points x_0 for which $y_n = 0$. Here we get the answer in terms of w_0.

```
In[17]:= solny[n_] = %[[1]];

In[18]:= Solve[(y[n] /.solny[n]) == 0, w[0]]//
           Simplify
```

$$Out[18]= \left\{\left\{w[0] \to \frac{-1+n}{2n}\right\}\right\}$$

We now substitute w_0 by its expression in terms of x_0. The resulting formula specifies which points belong to the forbidden set.

$$In[19]:= f[n\_] = \frac{\beta + A}{B} w[0] - \frac{A}{B} /. \%[[1]]$$

$$Out[19]= -\frac{A}{B} + \frac{(-1+n)\ (A+\beta)}{2 B n}$$

Here is a list of the first few terms in the forbidden set.

```
In[20]:= Table[f[n], {n, 1, 10}]//Simplify
```

$$Out[20]= \left\{-\frac{A}{B}, \frac{-3A+\beta}{4B}, \frac{-2A+\beta}{3B}, \frac{-5A+3\beta}{8B}, \frac{-3A+2\beta}{5B}, \frac{-7A+5\beta}{12B}, \frac{-4A+3\beta}{7B}, \frac{-9A+7\beta}{16B}, \frac{-5A+4\beta}{9B}, \frac{-11A+9\beta}{20B}\right\}$$

The limit point of the points in the forbidden set F is the following.

```
In[21]:= Limit[f[n], n → ∞]
```

$$Out[21]= \frac{-A+\beta}{2B}$$

Since $R = \frac{1}{4}$, one can solve for α in terms of A, B, and β.

$$In[22]:= \alpha rule = Solve\left[-\frac{B\alpha - A\beta}{(A+\beta)^2} == \frac{1}{4}, \alpha\right] // Simplify$$

$$Out[22]= \left\{\left\{\alpha \to -\frac{(A-\beta)^2}{4B}\right\}\right\}$$

We now verify that the limit of the sequence $\{f_n\}$ of elements of the forbidden set is precisely the only fixed point of the Riccati map.

```
In[23]:= {fp1 , fp2}/. αrule[[1]] //Simplify
```

$$Out[23]= \left\{\frac{-A+\beta}{2B}, \frac{-A+\beta}{2B}\right\}$$

To produce a general solution for the Riccati equation, the first step is to find a formula for w_n.

$$In[24]:= solnw[n\_] = w[n] \to \left(\frac{y[n+1] /.solny[n+1]}{y[n] /.solny[n]}\right)$$

$$Out[24]= w[n] \to \frac{1 + (1+n)\ (-1 + 2 w[0])}{2\ (1 + n\ (-1 + 2 w[0]))}$$

The general solution of the Riccati equation is given by the replacement rule shown in the right column.

```
In[25]:= solnx[n_] = Simplify[x[n] →
           (w[n] /. solnw[n] /.
              w[0] → B/(β + A) (x[0] + A/B))]
Out[25]= x[n] →
   (A (2 + n) - n β + 2 B x[0] + 2 B n x[0]) / (2 (A + A n + β - n β + 2 B n x[0]))
```

The following are the first four terms of an orbit.

```
In[26]:= Table[x[n]/.solnx[n], {n, 1, 4}]
Out[26]= {(3 A - β + 4 B x[0])/(4 A + 4 B x[0]), (2 A - β + 3 B x[0])/(3 A - β + 4 B x[0]),
   (5 A - 3 β + 8 B x[0])/(8 A - 4 β + 12 B x[0]), (3 A - 2 β + 5 B x[0])/(5 A - 3 β + 8 B x[0])}
```

The conclusions of the preceding computer session can be summarized as a theorem.

THEOREM 2.8

Let α, β, A, and B be such that assumption (2.23) holds. If $A+\beta \neq 0$ and $R = \frac{1}{4}$, then the only equilibrium point of (2.21) is

$$\bar{x} = \frac{-A+\beta}{2B}.$$

The forbidden set of (2.21) is $\mathbf{F} = \{ f_n \ : \ n = 1, 2, \ldots \}$, where

$$f_n = \frac{-A}{B} + \frac{n-1}{n}\frac{A+\beta}{2B} = \frac{-A+\beta}{2B}\left(1+\frac{1}{n}\right) \quad \text{for} \quad n = 1, 2, \ldots \tag{2.26}$$

Furthermore, the sequence $\{f_n\}$ converges to the equilibrium.

For $x_0 \notin \mathbf{F}$, the solution of (2.21) is given by

$$x_n = \frac{A(n+2) - \beta n + 2B(n+1)x_0}{2\left(A(n+1) - \beta(1-n) + 2Bnx_0\right)} \quad \text{for} \quad n = 0, 1, \ldots . \tag{2.27}$$

2.4.4 The Case when $A+\beta \neq 0$ and $R < \frac{1}{4}$

When $R < \frac{1}{4}$ and $A+\beta \neq 0$, the characteristic equation of

$$y_{n+2} - y_{n+1} + Ry_n = 0$$

has two distinct real roots, $\bar{y} = \frac{1}{2}(1 \pm \sqrt{1-4R})$. According to Theorem 2.5, the general solution has the form

$$y_n = c_1\left(\frac{1}{2}(1+\sqrt{1-4R})\right)^n + c_2\left(\frac{1}{2}(1-\sqrt{1-4R})\right)^n.$$

As before, we may take $y_0 = 1$ and $y_1 = w_0$.

We may use `RSolve` to obtain the solution of

$$R y_n - y_{n+1} + y_{n+2} = 0$$

```
In[27]:= RSolve[{riccy,
              y[0] == 1, y[1] == w[0]}, y[n], n]
```

$$\text{Out[27]=} \left\{\left\{y[n] \to \frac{1}{2\sqrt{1-4R}} \left(\left(\frac{1}{2} - \frac{1}{2}\sqrt{1-4R}\right)^n \left(-1+\sqrt{1-4R}\right) + 2^{n} \left(1+\sqrt{1-4R}\right)^{1 n} + 2^{1 n}\left(-\left(1-\sqrt{1-4R}\right)^n + \left(1+\sqrt{1-4R}\right)^n\right)(-1+w[0])\right)\right\}\right\}$$

```
In[28]:= solny[n_] = %[[1]];
```

To characterize the forbidden set, begin by solving $y_n = 0$ for w_0.

```
In[29]:= Solve[(y[n] /. solny[n]) == 0,
              w[0]]//Simplify
```

$$\text{Out[29]=} \left\{\left\{w[0] \to \left(\left(1-\sqrt{1-4R}\right)^n - \left(1+\sqrt{1-4R}\right)^n + \left(1-\sqrt{1-4R}\right)^n \sqrt{1-4R} + \left(1+\sqrt{1-4R}\right)^n \sqrt{1-4R}\right) \Big/ \left(2\left(1-\sqrt{1-4R}\right)^n - 2\left(1+\sqrt{1-4R}\right)^n\right)\right\}\right\}$$

Now replace w_0 by its expression in terms of x_0. This gives a formula for the elements of the forbidden set.

```
In[30]:= f[n_] = (β + A/B w[0] - A/B /. %[[1]])
              //Simplify
```

$$\text{Out[30]=} \frac{1}{B}\left(-A + \left(\left(\left(1-\sqrt{1-4R}\right)^n - \left(1+\sqrt{1-4R}\right)^n + \left(1-\sqrt{1-4R}\right)^n \sqrt{1-4R} + \left(1+\sqrt{1-4R}\right)^n \sqrt{1-4R}\right)(A+\beta)\right) \Big/ \left(2\left(1-\sqrt{1-4R}\right)^n - 2\left(1+\sqrt{1-4R}\right)^n\right)\right)$$

The sequence $\{f_n\}$ has a limit. Indeed, since $R < \frac{1}{4}$, then $(1-\sqrt{1-4R})^n \to 0$ when $n \to \infty$, a fact we may use to find the limit of the y_n's.

```
In[31]:= Simplify[f[n_] /. (1 - √(1 - 4 R))^n → 0]
```

$$\text{Out[31]=} -\frac{A + A\sqrt{1-4R} - \beta + \sqrt{1-4R}\,\beta}{2B}$$

Recall that $w_n = y_{n+1}/y_n$. Thus, the solution in terms of w_0 is given by the following expression.

```
In[32]:= solnw[n_] =
           w[n] → (y[n + 1]/.solny[n + 1])/(y[n]/.solny[n])
```

$$\text{Out[32]= } w[n] \to \left(\left(\frac{1}{2}-\frac{1}{2}\sqrt{1-4R}\right)^{1+n}\left(-1+\sqrt{1-4R}\right)+2^{1+n}\left(1+\sqrt{1-4R}\right)^{2+n}+2^{n}\left(-\left(1-\sqrt{1-4R}\right)^{1+n}+\left(1+\sqrt{1-4R}\right)^{1+n}\right)(-1+w[0])\right)\Big/\left(\left(\frac{1}{2}-\frac{1}{2}\sqrt{1-4R}\right)^{n}\left(-1+\sqrt{1-4R}\right)+2^{n}\left(1+\sqrt{1-4R}\right)^{1+n}+2^{1+n}\left(-\left(1-\sqrt{1-4R}\right)^{n}+\left(1+\sqrt{1-4R}\right)^{n}\right)(-1+w[0])\right)$$

The terms of the solution of the Riccati equation are given by the following formula, in terms of the Riccati number R and the initial point x_0.

```
In[33]:= solnx[n_] = Simplify[
           x[n] → (w[n] /. solnw[n] /.
             w[0] → B/(β + A) (x[0] + A/B))]
```

$$\text{Out[33]= } x[n] \to \left(\left(\frac{1}{2}-\frac{1}{2}\sqrt{1-4R}\right)^{1+n}\left(-1+\sqrt{1-4R}\right)+2^{1+n}\left(1+\sqrt{1-4R}\right)^{2+n}+\frac{1}{A+\beta}\left(2^{n}\left(-\left(1-\sqrt{1-4R}\right)^{1+n}+\left(1+\sqrt{1-4R}\right)^{1+n}\right)(-\beta+B\,x[0])\right)\right)\Big/\left(\left(\frac{1}{2}-\frac{1}{2}\sqrt{1-4R}\right)^{n}\left(-1+\sqrt{1-4R}\right)+2^{n}\left(1+\sqrt{1-4R}\right)^{1+n}+\frac{1}{A+\beta}\left(2^{1+n}\left(\left(1-\sqrt{1-4R}\right)^{n}-\left(1+\sqrt{1-4R}\right)^{n}\right)(\beta-B\,x[0])\right)\right)$$

Here are the terms x_1 and x_2 of the general solution.

```
In[34]:= Table[x[n]/.
            solnx[n] /.R → -(B α - A β)/(A + β)^2,
         {n, 1, 2}]//Simplify

Out[34]= {(A^2 + A B x[0] + B (α + β x[0]))/((A + β) (A + B x[0])),
          (A^3 + A^2 B x[0] + A B (2 α + β x[0]) +
              B (β^2 x[0] + α (β + B x[0])))/
           ((A + β) (A^2 + A B x[0] + B (α + β x[0])))}
```

THEOREM 2.9
Let α, β, A, and B be such that assumption (2.23) holds. If $A+\beta \neq 0$ and $R<\frac{1}{4}$, then there are two distinct equilibrium points of (2.21),

$$\overline{x}_{\pm} = -\frac{-A-\beta \pm \sqrt{4B\alpha + (-A+\beta)^2}}{2B}.$$

The forbidden set of (2.21) is $\mathbf{F} = \{ f_n \ : \ n = 1, 2, \ldots \}$, where

$$f_n = \frac{\beta + A}{2B}\left(\frac{(1-\sqrt{1-4R})^n - (1+\sqrt{1-4R})^n + (1-\sqrt{1-4R})^n\sqrt{1-4R} + (1-\sqrt{1-4R})^n\sqrt{1-4R}}{(1-\sqrt{1-4R})^n - (1+\sqrt{1-4R})^n}\right) - \frac{A}{B} \quad (2.28)$$

Furthermore, the sequence $\{f_n\}$ converges to the equilibrium.
For $x_0 \notin \mathbf{F}$ and $n = 1, 2, \ldots$, the term x_n of the solution of (2.21) is given by

$$\frac{\left(\frac{1}{2} - \frac{\sqrt{1-4R}}{2}\right)^{1+n}\left(-1+\sqrt{1-4R}\right) + 2^{-1-n}\left(1+\sqrt{1-4R}\right)^{2+n} + \frac{\left(-\left(1-\sqrt{1-4R}\right)^{1+n} + \left(1+\sqrt{1-4R}\right)^{1+n}\right)(-\beta + B x_0)}{2^n (A+\beta)}}{\left(\frac{1}{2} - \frac{\sqrt{1-4R}}{2}\right)^{n}\left(-1+\sqrt{1-4R}\right) + \frac{\left(1+\sqrt{1-4R}\right)^{1+n}}{2^n} + \frac{2^{1-n}\left(\left(1-\sqrt{1-4R}\right)^{n} - \left(1+\sqrt{1-4R}\right)^{n}\right)(\beta - B x_0)}{A+\beta}} \quad (2.29)$$

2.4.5 The Case when $A+\beta \neq 0$ and $R > \frac{1}{4}$

When $R > \frac{1}{4}$ and $A+\beta \neq 0$, the characteristic equation of

$$y_{n+2} - y_{n+1} + R y_n = 0$$

has two complex conjugate roots, $y = \frac{1}{2}(1 \pm i\sqrt{4R-1})$. According to Theorem 2.5, the general solution has the form

$$y_n = r^n \left(c_1 \cos(n\phi) + c_2 n \sin(n\phi)\right)$$

where, in this case, $r^2 = R$ and $\phi \in (0, \frac{\pi}{2})$ is such that

$$\cos\phi = \frac{1}{2\sqrt{R}} \quad \text{and} \quad \sin\phi = \frac{\sqrt{4R-1}}{2\sqrt{R}}.$$

As before, we may take $y_0 = 1$ and $y_1 = w_0$.

Here we define y_n in terms of constants c_1 and c_2.

```
In[35]:= ruley[n_] = y[n] →
            R^(n/2) (c[1] Cos[n φ] + c[2] Sin[n φ])
Out[35]= y[n] →
            R^(n/2) (c[1] Cos[n φ] + c[2] Sin[n φ])
```

Since $y_0 = c_1$, we have that $c_1 = 1$.

```
In[36]:= ruley[0]
Out[36]= y[0] → c[1]

In[37]:= c[1] = 1;
```

To find c_2, one can use the fact that $w_0 = y_1$.

```
In[38]:= w[0] == y[1] /. (ruley[1])
Out[38]= w[0] == √R (Cos[φ] + c[2] Sin[φ])

In[39]:= Solve[% , c[2]]
Out[39]= {{c[2] → -

            Csc[φ] (√R Cos[φ] - w[0])
            -------------------------  }}
                       √R

In[40]:= c[2] = Simplify[c[2]/.%[[1]] /.
                          1                    2√R
            {Cos[φ] → -------, Csc[φ] → ----------}]
                        2√R                √(4R - 1)
          -1 + 2 w[0]
Out[40]= ------------
          √(-1 + 4 R)
```

These are the w_0's for which $y_n = 0$ for some n.

```
In[41]:= Solve[(y[n] /. ruley[n]) == 0,
            w[0]]//Simplify
                    1
Out[41]= {{w[0] → - (1 - √(-1 + 4 R) Cot[n φ])}}
                    2
```

This formula gives the elements of the forbidden set.

```
                  β + A          A
In[42]:= f[n_] = (----- w[0] - - /. %[[1]])//
                    B            B
            Simplify
            A - β + √(-1 + 4 R) (A + β) Cot[n φ]
Out[42]= - -------------------------------------
                           2 B
```

The terms of the solution of the Riccati equation are given by the following formula.

```
In[43]:= Simplify[x[n] →
            ((β + A)/B w[n] - A/B //. % /.
              w[0] → B/(β + A) (x[0] + A/B))]

Out[43]= x[n] → 1/B ( - A+
   (√R (A + β) (√(-1 + 4 R) (A + β) Cos[(1 + n) φ] +
        Sin[(1 + n) φ] (A - β + 2 B x[0])))/
   (√(-1 + 4 R) (A + β) Cos[n φ] +
        Sin[n φ] (A - β + 2 B x[0])))
```

The session just finished supports the following theorem and its corollary.

THEOREM 2.10

Let α, β, A, and B be such that assumption (2.23) holds, and let $\phi \in (0, \frac{\pi}{2})$ be such that $\cos(\phi) = (2R)^{-\frac{1}{2}}$. If $A+\beta \neq 0$ and $R > \frac{1}{4}$, then there are no equilibrium points of (2.21). The forbidden set of (2.21) is $\mathbf{F} = \{ f_n : n = 1, 2, \ldots \}$, where

$$f_n = \frac{A - \beta + (A + \beta)\sqrt{4R - 1}\,\cot(n\phi)}{2B}$$

For $x_0 \notin \mathbf{F}$ and $n = 1, 2, \ldots$, the solution of (2.21) is given by

$$x_n = \frac{B\sqrt{-1+4R}\,(A+\beta)\cos(n\phi)x_0 - \sin(n\phi)\left(2A^2R + \beta\left(2R\beta - Bx_0\right) + A\left(-2\beta + 4R\beta + Bx_0\right)\right)}{B\left(\sqrt{-1+4R}\,(A+\beta)\cos(n\phi) + \sin(n\phi)\left(A - \beta + 2Bx_0\right)\right)} \tag{2.30}$$

COROLLARY 2.1

If there exist positive coprime natural numbers k and p such that $\phi = \frac{k}{p}\pi$ for the number ϕ in Theorem 2.10, then the forbidden set is given by $\mathbf{F} = \{ f_1, f_2, \ldots, f_{p-1} \}$, and every solution is periodic with period p.

It can be shown that if ϕ is not a rational multiple of π, then no solution is periodic. See [KL].

2.5 Linearized Stability Analysis

In this section we present basic notions and definitions concerning the stability of equilibria and periodic solutions of system (2.1).

By $\|\cdot\|$ we denote the euclidean norm in R^2 given by

$$\|(x, y)\| = \sqrt{x^2 + y^2}.$$

Next, we have definitions of equilibrium and periodic points and their stability for system (2.1).

DEFINITION 2.5

1. *An equilibrium solution* $(\bar{x}, \bar{y})$ *of (2.1) (fixed point of map* $\mathbf{F}$*) is said to be* **stable** *if for any* $\epsilon > 0$ *there is* $\delta > 0$ *such that for every initial point* (x_0, y_0) *for which* $\|(x_0, y_0) - (\bar{x}, \bar{y})\| < \delta$*, the iterates* (x_n, y_n) *of* (x_0, y_0) *satisfy* $\|(x_n, y_n) - (\bar{x}, \bar{y})\| < \epsilon$ *for all* $n > 0$*. An equilibrium (fixed) point* $(\bar{x}, \bar{y})$ *of (2.1) is said to be* **unstable** *if it is not stable.*

2. *An equilibrium (fixed) point* $(\bar{x}, \bar{y})$ *of (2.1) is said to be* **asymptotically stable** *if there exists* $r > 0$ *such that* $(x_n, y_n) \to (\bar{x}, \bar{y})$ *as* $n \to \infty$ *for all* (x_0, y_0) *that satisfy* $\|(x_0, y_0) - (\bar{x}, \bar{x})\| < r$.

3. *A periodic point* (x_p, y_p) *of period* m *is stable (respectively, unstable, asymptotically stable) if* (x_p, y_p) *is a stable (resp. unstable, asymptotically stable) fixed point of* F^m.

DEFINITION 2.6 *A set* M *is said to be* **invariant under the map** $F = (f, g)$ *if* $F(M) \subset M$.

Therefore, every orbit that enters an invariant set M stays in M "forever".

DEFINITION 2.7

1. *Let* $(\bar{x}, \bar{y})$ *be a fixed point of a map* $\mathbf{F} = (f, g)$*, where* f *and* g *are continuously differentiable functions at* $(\bar{x}, \bar{y})$*. The* **Jacobian matrix of F at** $(\bar{x}, \bar{y})$ *is the matrix*

$$J_{\mathbf{F}}(\bar{x}, \bar{y}) = \begin{pmatrix} \frac{\partial f}{\partial x}(\bar{x}, \bar{y}) & \frac{\partial f}{\partial y}(\bar{x}, \bar{y}) \\ \frac{\partial g}{\partial x}(\bar{x}, \bar{y}) & \frac{\partial g}{\partial y}(\bar{x}, \bar{y}) \end{pmatrix}.$$

The linear map $J_{\mathbf{F}}(\bar{x}, \bar{y}) : R^2 \to R^2$ *given by*

$$J_{\mathbf{F}}(p, q)(x, y) = \begin{pmatrix} \frac{\partial f}{\partial x}(\bar{x}, \bar{y})x + \frac{\partial f}{\partial y}(\bar{x}, \bar{y})y \\ \frac{\partial g}{\partial x}(\bar{x}, \bar{y})x + \frac{\partial g}{\partial y}(\bar{x}, \bar{y})y \end{pmatrix} \quad (2.31)$$

is called the **linearization of the map** F **at the fixed point** $(\bar{x}, \bar{y})$.

2. *An equilibrium point $(\bar{x}, \bar{y})$ of the map $\mathbf{F}$ is said to be* **hyperbolic** *if the linearization (2.31) of $\mathbf{F}$ is hyperbolic, that is, if the Jacobian matrix $J_{\mathbf{F}}(\bar{x}, \bar{y})$ at $(\bar{x}, \bar{y})$ has no eigenvalues on the unit circle. If $J_{\mathbf{F}}(\bar{x}, \bar{y})$ has at least one eigenvalue on the unit circle, then it is a* **nonhyperbolic** *equilibrium point.*

The main result in the linearized stability analysis is the following result.

THEOREM 2.11 (Linearized Stability Theorem)
Let $\mathbf{F} = (f, g)$ be a continuously differentiable function defined on an open set W in R^2, and let $(\bar{x}, \bar{y})$ in W be a fixed point of $\mathbf{F}$.

(a) If all the eigenvalues of the Jacobian matrix $J_F(\bar{x}, \bar{y})$ have modulus less than one, then the equilibrium point $(\bar{x}, \bar{y})$ is asymptotically stable.

(b) If at least one of the eigenvalues of the Jacobian matrix $J_{\mathbf{F}}(\bar{x}, \bar{y})$ has modulus greater than one, then the equilibrium point $(\bar{x}, \bar{y})$ is unstable.

From a direct analysis of the characteristic equation of the Jacobian matrix $J_{\mathbf{F}}(\bar{x}, \bar{y})$, which is a quadratic equation, we may obtain explicit conditions for the equilibrium point to be locally a sink, source, saddle point, and a nonhyperbolic equilibrium point. One can also used a general method called the Schur-Cohn criterion, see Section 3.4.

THEOREM 2.12

1. *An equilibrium point $(\bar{x}, \bar{y})$ of (2.1) is locally asymptotically stable if and only if every solution of the characteristic equation*

$$\lambda^2 - \operatorname{tr} J_{\mathbf{F}}(\bar{x}, \bar{y})\lambda + \det J_{\mathbf{F}}(\bar{x}, \bar{y}) = 0 \tag{2.32}$$

lies inside the unit circle, that is, if and only if

$$|\operatorname{tr} J_{\mathbf{F}}(\bar{x}, \bar{y})| < 1 + \det J_{\mathbf{F}}(\bar{x}, \bar{y}) < 2.$$

2. *Similarly, an equilibrium point $(\bar{x}, \bar{y})$ of (2.1) is locally a repeller if and only if every solution of the characteristic equation (2.32) lies outside the unit circle, that is, if and only if*

$$|\operatorname{tr} J_{\mathbf{F}}(\bar{x}, \bar{y})| < |1 + \det J_{\mathbf{F}}(\bar{x}, \bar{y})| \quad \textit{and} \quad |\det J_{\mathbf{F}}(\bar{x}, \bar{y})| > 1.$$

3. *An equilibrium point $(\bar{x}, \bar{y})$ of (2.1) is locally a saddle point if and only if the characteristic equation (2.32) has one root that lies inside the unit circle and one root that lies outside the unit circle, that is, if and only if*

$$|\operatorname{tr} J_{\mathbf{F}}(\bar{x}, \bar{y})| > |1 + \det J_{\mathbf{F}}(\bar{x}, \bar{y})| \quad \textit{and} \quad \operatorname{tr} J_{\mathbf{F}}(\bar{x}, \bar{y})^2 - 4 \det J_{\mathbf{F}}(\bar{x}, \bar{y}) > 0.$$

4. *An equilibrium point* $(\bar{x}, \bar{y})$ *of (2.1) is nonhyperbolic if and only if the characteristic equation (2.32) has at least one root that lies on the unit circle, that is, if and only if*

$$|\operatorname{tr} J_{\mathbf{F}}(\bar{x}, \bar{y})| = |1 + \det J_{\mathbf{F}}(\bar{x}, \bar{y})| \quad or \quad \det J_{\mathbf{F}}(\bar{x}, \bar{y}) = 1 \ \ and \ \ |\operatorname{tr} J_{\mathbf{F}}(\bar{x}, \bar{y})| \le 2.$$

DEFINITION 2.8

1. *If a hyperbolic equilibrium point* $(\bar{x}, \bar{y})$ *of system (2.1) is asymptotically stable, then there is an open neighborhood* O *of* $(\bar{x}, \bar{y})$ *in which all points converge to the equilibrium point under forward iterations, that is,*

$$F^n(a, b) \to (\bar{x}, \bar{y}) \quad for\ every \quad (a, b) \in O.$$

Such an equilibrium point is called a **sink**. *When an equilibrium point is unstable, there are two qualitatively distinct cases.*

2. *If both eigenvalues of the linearization of* F *at the equilibrium point are outside the unit circle, then there is an open neighborhood of the equilibrium such that backward iterations of points in the neighborhood converge to the equilibrium point; such a fixed point is called a* **source** *or a* **repeller**.

3. *If one eigenvalue is inside the unit circle and the other eigenvalue is outside the unit circle, then in any open neighborhood of the fixed point, the forward iterates of some points in the neighborhood converge to the fixed point, while the backward iterations of other points in the neighborhood converge to the fixed point under backward iterations; such a fixed point is called a* **saddle**.

In the case of nonhyperbolic equilibrium point the situation is more complicated, as the linear stability analysis is inconclusive. In such cases, stability of the equilibrium is determined by a higher order term in Taylor's expansion of the map **F**.

For convenience of the reader, we also present analogous results for the general second order difference equation

$$x_{n+1} = f(x_n, x_{n-1}), \tag{2.33}$$

where f is a continuously differentiable function.

It is very convenient to transform difference equations into corresponding maps and vice versa. The standard transformation used in literature to convert the difference equation

$$x_{n+1} = f(x_n, x_{n-1}) \tag{2.34}$$

into the corresponding system

$$\begin{aligned} u_{n+1} &= v_n \\ v_{n+1} &= f(v_n, u_n) \end{aligned} \tag{2.35}$$

is

$$\begin{aligned} u_n &= x_{n-1} \\ v_n &= x_n. \end{aligned} \tag{2.36}$$

The corresponding map is

$$F(x, y) = (y, f(x, y)).$$

THEOREM 2.13

1. *An equilibrium point* $\bar{x}$ *of (2.33) is locally asymptotically stable if and only if every solution of the characteristic equation*

$$\lambda^2 - P\lambda - Q = 0, \tag{2.37}$$

where

$$P = \frac{\partial f}{\partial x}(\bar{x}, \bar{x}) \quad \text{and} \quad Q = \frac{\partial f}{\partial y}(\bar{x}, \bar{x}),$$

lies inside the unit circle, that is, if and only if

$$|P| < 1 - Q < 2.$$

2. *An equilibrium point* $\bar{x}$ *of (2.33) is a local repeller if and only if every solution of the characteristic equation (2.37) lies outside the unit circle, that is, if and only if*

$$|P| < |1 - Q| \quad \text{and} \quad |Q| > 1.$$

3. *An equilibrium point* $\bar{x}$ *of (2.33) is a saddle point if and only if the characteristic equation (2.37) has one root that lies inside the unit circle and one root that lies outside the unit circle, that is, if and only if*

$$|P| > |1 - Q| \quad \text{and} \quad P^2 + 4Q > 0.$$

4. *An equilibrium point* $\bar{x}$ *of (2.33) is nonhyperbolic if and only if the characteristic equation (2.37) has at least one root that lies on the unit circle, that is, if and only if*

$$|P| = |1 - Q| \quad \text{or} \quad Q = -1 \text{ and } |P| \le 2.$$

2.6 *Dynamica* Session

Let us consider the problem of finding and analyzing the equilibrium (fixed) points of the discrete predator prey difference equation

$$\begin{aligned} x_{n+1} &= A(1-x_n)x_n - x_n y_n \\ y_{n+1} &= \frac{x_n y_n}{B}, \end{aligned} \tag{2.38}$$

see [La2]. The system of difference equations (2.38) is a discrete model of interactions between two species. Here x_n and y_n represent the population of prey and of predator, respectively. The model is given in terms of the parameters A, and B.

The map is defined in terms of the parameters A and B.

```
In[1]:= << Dynamica`

In[2]:= DPPMap[{x_, y_}] :=
          {A (1 - x) x - x y, x y/B}

In[3]:= DPPMap[{x, y}]
Out[3]= {A (1 - x) x - x y, x y/B}
```

The *Mathematica* `Solve` command may be used to find fixed points.

```
In[4]:= Solve[DPPMap[{x, y}] == {x, y},
          {x, y}]
Out[4]= {{x -> 0, y -> 0}, {x -> -(1 - A)/A, y -> 0},
          {y -> -1 + A - A B, x -> B}}
```

Here we name the fixed points `fp1`, `fp2`, and `fp3`.

```
In[5]:= {fp1, fp2, fp3} = {x, y} /. %
Out[5]= {{0, 0}, {-(1 - A)/A, 0}, {B, -1 + A - A B}}
```

The Jacobian of the map is needed to classify the fixed points.

```
In[6]:= jacmat = Jac[DPPMap, {x, y}];

In[7]:= MatrixForm[jacmat]
Out[7]= ( A (1 - x) - A x - y   -x  )
        (          y/B          x/B )
```

As an example, we now choose specific values for A and B.

```
In[8]:= A = 3.5; B = 0.3;
```

As both eigenvalues clearly have modulus greater than 1, the equilibrium point fp3 is a source.

```
In[9]:= {x, y} = fp3;

In[10]:= Eigenvalues[jacmat]
Out[10]= {0.475 + 1.08369 i, 0.475 - 1.08369 i}

In[11]:= Abs[%]
Out[11]= {1.18322, 1.18322}
```

The equilibrium point fp2 is a source as well.

```
In[12]:= {x, y} = fp2;

In[13]:= Eigenvalues[jacmat]
Out[13]= {2.38095, -1.5}
```

However, fp1 is a saddle point.

```
In[14]:= {x, y} = fp1;

In[15]:= Eigenvalues[jacmat]
Out[15]= {3.5, 0.}
```

We now go back to the case where A and B are unspecified and calculate the eigenvalues of Jacobian at each of the three equilibrium points.

The fixed point fp1 cannot be a source. However it can be a sink (for $|A| < 1$), or a saddle (for $|A| > 1$) or non-hyperbolic point.

```
In[16]:= Clear[A, B, x, y];

In[17]:= {x, y} = fp1;

In[18]:= Eigenvalues[jacmat]//Simplify
Out[18]= {0, A}
```

The fixed point fp2 is a sink if $A < 3, B > 2/3$, a source if $A > 3, B < 2/3$ or $0 < A < 1, B > A/(1 - A)$, a nonhyperbolic point if $A = 1$ or $A = 3$ or $B = \pm(A - 1)/A$ and saddle in all other cases.

```
In[19]:= {x, y} = fp2;

In[20]:= Eigenvalues[jacmat]//Simplify
```

$$\text{Out[20]= } \left\{2 - A, \frac{-1 + A}{A\,B}\right\}$$

Just as before we may find conditions under which this equilibrium point has different characters.

```
In[21]:= {x, y} = fp3;

In[22]:= Eigenvalues[jacmat]//Simplify
```

$$\text{Out[22]= } \left\{\frac{1}{2}\left(2 - A\,B - \sqrt{4 - 4\,A + 4\,A\,B + A^2\,B^2}\right), \frac{1}{2}\left(2 - A\,B + \sqrt{4 - 4\,A + 4\,A\,B + A^2\,B^2}\right)\right\}$$

Next, we shall do some numerical explorations. An orbit of a two-dimensional map is a list of ordered pairs determined by successive applications of the map to the initial point.

These are the first 10 elements of the orbit $\gamma^+((1,2))$ for a particular choice of A and B in the Predator–Prey model.

```
In[23]:= A = 3.4; B = 0.3;

In[24]:= orb = Orbit[DPPMap, {0.1, 0.2}, 2000];

In[25]:= Take[orb, {1, 10}]
Out[25]= {{0.1, 0.2}, {0.286, 0.0666667},
          {0.675227, 0.0635556},
          {0.70269, 0.143048},
          {0.609798, 0.335062},
          {0.604691, 0.681066},
          {0.400901, 1.37278},
          {0.26626, 1.8345},
          {0.17579, 1.62818},
          {0.206401, 0.954057}}
```

A total of 2,000 points are shown here. The phase portrait suggests that the orbit might be eventually periodic.

```
In[26]:= PhasePortrait[orb,
           PlotRange → {{0, 1}, {0, 2.3}}]
```

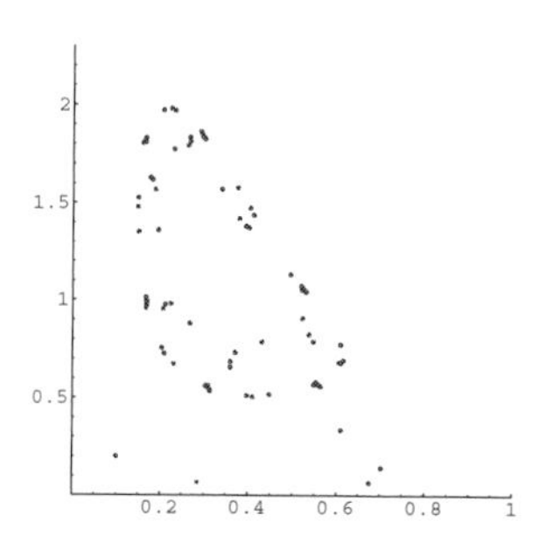

A small change in the parameters yields a more interesting plot.

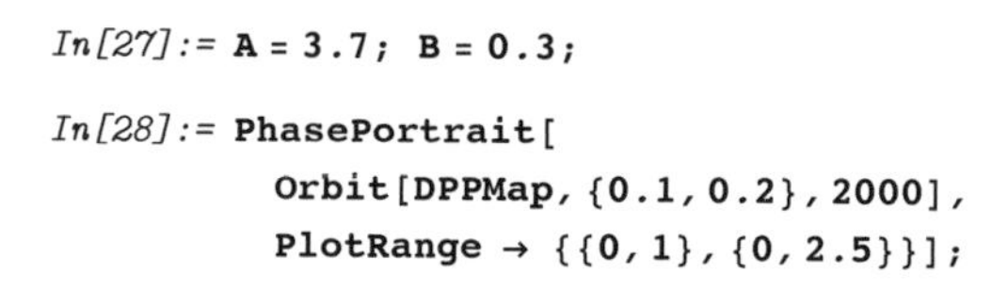

```
In[27]:= A = 3.7;  B = 0.3;

In[28]:= PhasePortrait[
           Orbit[DPPMap, {0.1, 0.2}, 2000],
           PlotRange → {{0, 1}, {0, 2.5}}];
```

The plot shows what in the literature is sometimes called *strange attractor*.

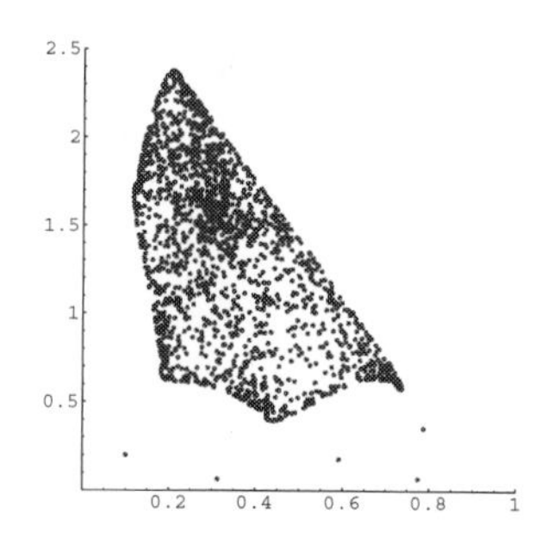

2.7 Period-Doubling Bifurcation

As the parameter ρ passes through 1, the linear system

$$\mathbf{z}_{n+1} = \rho \begin{pmatrix} \cos\omega & \sin\omega \\ -\sin\omega & \cos\omega \end{pmatrix} \mathbf{z}_n \tag{2.39}$$

undergoes a specific bifurcation (change in global behavior) of the following type: for $\rho \neq 1$, there are no invariants (except the trivial one, satisfied by the equilibrium point), and for $\rho = 1$, there is a one-parameter family of invariants that are closed curves. In the nonlinear case, the counterpart of this phenomenon is the so-called period-doubling bifurcation, also known as the Andronov-Hopf and Naimark-Sacker bifurcation, see [ASY], [E1], [GH], and [HK]. Let us illustrate the corresponding nonlinear phenomenon by one example.

Example 2.6 Consider the system with a real parameter ρ,

$$\mathbf{x}_{n+1} = (\rho - x_n^2 - y_n^2) \begin{pmatrix} \cos\omega & \sin\omega \\ -\sin\omega & \cos\omega \end{pmatrix} \mathbf{x}_n. \tag{2.40}$$

The origin is an equilibrium point of this system for all values of ρ. The linearized system about the origin is system (2.39). To investigate the effects of the nonlinear terms, we transform the system to polar coordinates. We have,

$$r_{n+1} = \rho r_n - r_n^3$$

$$\theta_{n+1} = \theta_n + \omega.$$

Thus, we obtain a system with decoupled equations, from which it is evident that for $\rho > 1$, there is an invariant circle of radius $r = \sqrt{\rho - 1}$. The ω-limit set of every positive orbit, except the origin, is contained in this circle. The asymptotic behavior on the invariant circle is a rotation with rotation number ω. ∎

The birth of an invariant circle in Example 2.6 is a typical local bifurcation near an equilibrium point as a pair of complex conjugate eigenvalues move from the inside to the outside of the unit disk. In fact, there exists a general result for two-dimensional systems that is associated with the names of Andronov, Hopf, Naimark, Poincare, and Sacker, see [E1], [GH], and [HK].

THEOREM 2.14 (Period-Doubling Bifurcations for Two-Dimensional Systems)
Let $\lambda_0 \in R$, and let $\mathbf{G}(\lambda, \mathbf{x})$ be an R^2-valued function of $\lambda \in R$ and $\mathbf{x} \in R^2$.

Suppose that $\mathbf{G}$ *is four times continuously differentiable, and that it satisfies the following conditions.*

i. $\mathbf{G}(\lambda, \mathbf{0}) = \mathbf{0}$ *for all* λ *in a neighborhood* U_{λ_0} *of* λ_0.

ii. *For each* $\lambda \in U_{\lambda_0}$, *the Jacobian of* $\mathbf{G}$ *in* $\mathbf{x}$, $J_{\mathbf{G}}(\lambda, \mathbf{0})$, *has two complex conjugate eigenvalues* $\mu(\lambda)$ *and* $\overline{\mu(\lambda)}$ *such that* $|\mu(\lambda)| = 1$ *for* $\lambda \in U_{\lambda_0}$.

iii.

$$\left\{\frac{d}{d\lambda}\,|\mu(\lambda)|\right\}\Big|_{\lambda=\lambda_0} > 0.$$

iv. $\mu^k(\lambda_0) \neq 1$ *for* $k = 1, 2, 3, 4$.

Then, there exists a smooth function $\mathbf{H}(\lambda, \mathbf{x})$ *such that for* $\lambda \in U_{\lambda_0}$,

$$\mathbf{G}(\lambda, \mathbf{x}) = \mathbf{H}(\lambda, \mathbf{x}) + O(\|\mathbf{x}^5\|)$$

where the function $\mathbf{H}(\lambda, \mathbf{x})$ *in polar coordinates is given by*

$$r_{n+1} = |\mu(\lambda)|\, r_n - a(\lambda) r_n^3$$

$$\theta_{n+1} = \theta_n + \omega(\lambda) + b(\lambda) r_n^2.$$

Here $a(\lambda), b(\lambda)$, *and* $\omega(\lambda)$ *are smooth functions with the following properties.*

(a) *If* $a(\lambda_0) > 0$, *then there is a neighborhood* U *of the origin and a* $\delta > 0$ *such that for* $|\lambda - \lambda_0| < \delta$ *and* $\mathbf{x}_0 \in U$ *the* ω*-limit set of* $\mathbf{x}_0$ *is the origin if* $\lambda < \lambda_0$, *and it belongs to a closed invariant* C^1 *curve* $C(\lambda)$ *encircling the origin if* $\lambda > \lambda_0$. *In addition,* $C(\lambda_0) = 0$.

(b) *If* $a(\lambda_0) > 0$, *then there is a neighborhood* U *of the origin and a* $\delta > 0$ *such that for* $|\lambda - \lambda_0| < \delta$ *and* $\mathbf{x}_0 \in U$ *the* α*-limit set of* $\mathbf{x}_0$ *is the origin if* $\lambda > \lambda_0$, *and it belongs to a closed invariant* C^1 *curve* $C(\lambda)$ *encircling the origin if* $\lambda < \lambda_0$. *In addition,* $C(\lambda_0) = 0$.

2.8 Lyapunov Numbers

It is well-known that chaotic dynamics is characterized by an exponential divergence of initially close points. In the case of a one-dimensional discrete map of an interval (a, b) into itself

$$x_{n+1} = f(x_n)$$

the Lyapunov exponent is a measure of the divergence of two orbits starting with slightly different initial conditions. We have already discussed this in Section 1.7.4.

For a map on R^2, each orbit has two Lyapunov numbers that measure the rates of separation from the current orbit point along two orthogonal directions. These directions are determined by the dynamics of the map. The first direction is the direction along which the separation between nearby points is the greatest. The second is the direction of greatest separation, chosen perpendicular to the first. The stretching factors in each of these two directions are the **Lyapunov numbers of the orbit**.

To illustrate this concept, let us consider a circle S of a small radius centered on the first point of the orbit. If we examine the image $\mathbf{f}(S)$ under one iteration of the map, we see an approximately elliptical shape, with the long axes along the expanding direction of $\mathbf{f}$ and the short axis along the contracting direction of $\mathbf{f}$. After n iterates of the map $\mathbf{f}$, the small circle evolves into a longer and thinner elliptic-like object. The per-iterate changes of the axes of this approximately elliptical shape are the Lyapunov numbers. They quantify the amount of stretching and shrinking due to the dynamics near the orbit beginning at the first point of the orbit. The natural logarithm of each Lyapunov number is a **Lyapunov exponent**.

For a formal definition, replace the small circle about the first point of the orbit and the map $\mathbf{f}$ by the unit circle U and the Jacobian D_f. Because we are interested in the infinitesimal behavior near the equilibrium point, the map $\mathbf{f}$ can be approximated by its linearization $D_{\mathbf{f}}$. Let $D_{(n)} = D_{\mathbf{f}^n}(x_0, y_0)$ denote the Jacobian matrix of the n-th iterate of $\mathbf{f}$. Then $D_{(n)}(U)$ is an ellipse. This is a consequence of the simple fact that the linear transformation of a circle is an ellipse. See [ASY] and [E2]. The axis is longer than 1 in the expanding direction and shorter than 1 in the contracting direction. The two average multiplicative expansion or contraction rates of the two orthogonal axes are the Lyapunov numbers.

DEFINITION 2.9 *Let* $\mathbf{f}$ *be a smooth map on* R^2*, and let* $D_n = D_{\mathbf{f}^n}(x_0, y_0)$*. For* $k = 1, 2, \ldots,$*, let* $r^{k(n)}$ *be the length of the k-th longest orthogonal axis of the ellipse* $D_n(U)$ *for an orbit with initial point* (x_0, y_0)*. Then* $r^{k(n)}$ *measures the contraction or expansion near the orbit that starts at* $Z_0 = (x_0, y_0)$ *during the first n iterations.*

The k**-th Lyapunov number** *of this trajectory is*

$$L(k)(Z_0) = \lim_{n \to \infty} \sqrt[n]{r^{k(n)}}$$

if the limit exists. The k**-th Lyapunov exponent** *of the trajectory that starts at* $Z_0 = (x_0, y_0)$ *is*

$$h(k)(Z_0) = \ln \left(L(k)(Z_0) \right), \quad k = 1, 2.$$

We shall use the notation $L(k)$ and $h(k)$ when it is clear from the context what the initial point Z_0 is. Clearly, $L(1) \geq L(2)$ and $h(1) \geq h(2)$. Using the concept of a Lyapunov exponent, we can extend the definition of the chaotic orbit of the one-dimensional map in the sense of positive Lyapunov exponents from Chapter 1 to orbits of higher dimensional maps. For technical reasons which are described in [ASY], it will be required that no Lyapunov exponent is exactly zero for a chaotic orbit.

DEFINITION 2.10 *Let* **f** *be a smooth map on* R^2. *An orbit* $\{x_0, x_1, x_2, \ldots\}$ *is* **asymptotically periodic** *if it converges to a periodic orbit as* $n \to \infty$; *that is, there exists a periodic orbit* $\{y_0, y_1, y_2, y_3, \ldots,\}$ *such that* $\lim \|x_n - y_n\| = 0$, *as* $n \to \infty$.

DEFINITION 2.11 *Let* **f** *be a map of* R^2, *and let* $\{x_0, x_1, x_2, \ldots\}$ *be a bounded orbit of* **f**. *The orbit is said to be* **chaotic** *if*

a. $\{x_0, x_1, x_2, \ldots\}$ *is not asymptotically periodic,*

b. no Lyapunov number is exactly 1, *and,*

c. the Lyapunov exponent satisfies $h(1) > 0$ *(equivalently,* $L(1) > 1$.

Definitions 2.9 and 2.11 extend to the m-dimensional case R^m, with the word "circle" replaced by the word "sphere", and the word "ellipse" replaced by the word "ellipsoid". Similarly, as in the one-dimensional case, an important interpretation of the Lyapunov exponent is the measure information loss during the process of iteration.

Using Lyapunov exponents we can easily define the Lyapunov dimension, which gives an important piece of information about the structure of the attracting set. Here we give the definition in the most general case of R^m.

DEFINITION 2.12 *Let* **f** *be a map on* R^m. *Consider an orbit with Lyapunov exponents* $h(1) \geq h(2) \geq \ldots \geq h(m)$, *and let* p *denote the largest integer such that*

$$h(1) + h(2) + \ldots + h(p) \geq 0$$

if such p *exists. Define the* **Lyapunov dimension** $D(L)$ *of the orbit as*

$$D(L) = \begin{cases} 0 & \text{if no such } p \text{ exists} \\ p + \dfrac{h(1) + \ldots + h(p)}{|h(p+1)|} & \text{if } p < m \\ m & \text{if } p = m \end{cases}$$

In the case of a two-dimensional map with $h(1) > 0 > h(2)$ and $h(1)+h(2) < 0$ (for example, the Henon map), we have

$$D(L) = 1 + \frac{h(1)}{|h(2)|}.$$

In general, this dimension is close in value to the box dimension which is defined in the next subsection, see [ASY].

The *Dynamica* function `LyapunovNumbers` computes approximations to the Lyapunov numbers.

```
In[29]:= ?LyapunovNumbers
"LyapunovNumbers[f,vars,vars0,niterations]
finds approximate Lyapunov numbers of
niterations of the map f starting at the point
vars0.  Here vars is a list of the variables
of the expression f.  Example, with plot:
Clear[a,b];
fn = {a - x^2 + b*y, x} ;
{a,b} ={1.2,0.4};
Ly = LyapunovNumbers[fn , {x,y},{0.0, 0.0},50];
ListPlot[Ly, PlotJoined→ True];"
```

Here is an example of the computation of Lyapunov numbers for the Discrete Predator-Prey model (2.38) for specific values of the parameters.

```
In[30]:= A = 3.7; B = 0.3;

In[31]:= DPPMap[{x, y}]
```

$$\text{In[32]:= } \{A\,(1-x)\,x - x\,y,\ \frac{x}{B}\}$$

```
In[33]:= LN = LyapunovNumbers[
            DPPMap[{x, y}], {x, y}, {0.1, 0.2},
            50];
```

The plot indicates that the simulated orbit is chaotic, as it is clearly larger than one. However, one should be careful with the interpretation of these numerical results because of numerical error build-up, see [ASY].

```
In[34]:= ListPlot[LN, PlotJoined → True,
            PlotRange → All]
```

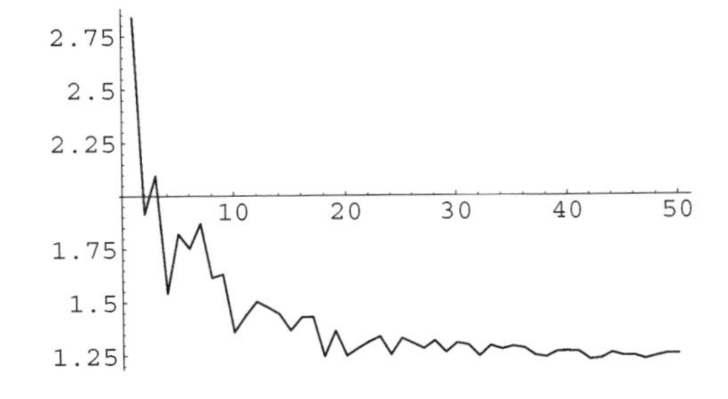

2.9 Box Dimension

As we have seen in Section 1.7.5 one way to measure the complexity of a set (an orbit of the map, an attracting set of the map) is to compute its dimension over different scales of magnification. Such a dimension is called the Box Dimension and gives the same information in two dimensions as in one dimension. In fact, we can easily define the Box Dimension for the set in d-dimensional space R^d.

In the case of dimension 2, notice that the square $[a, b]\times[a, b]$ can be covered by $N(r) = C(1/r)^2$ boxes of side length $(b - a)/n$. If we use the boxes of length $1/n$, then their number is $((b - a)n)^2$. This fact is expressed by saying that the number of boxes of side length r scales as $(1/r)^2$, meaning that $N(r)$, the number of boxes with areas r^2, is proportional to $(1/r)^2$, i.e.,

$$N(r) \sim (1/r)^2 \iff N(r) = C(1/r)^2.$$

Similarly, a m-dimensional cube requires $C(1/r)^m)$ boxes with volume r^m, that is,

$$N(r) \sim (1/r)^m \iff N(r) = C(1/r)^m. \tag{2.41}$$

Based on these examples, it is natural to ask the following question. Given an object in m-dimensional space, how many m-dimensional boxes of side-length r does it take to cover the object? In the case of simple sets such as squares, polygons, circles, etc. it is easy to see that this number is exactly $C(1/r)^m$. In the case of more complicated sets the definition follows.

Let S be a bounded set in R^m. We would like to define that S is a d-dimensional set when it can be covered by $N(r) = C(1/r)^d$ boxes of side-length r. Solving (2.41) for d we get

$$d = \frac{\ln N(r) - \ln C}{\ln(1/r)}.$$

Letting r approach 0 and assuming that the scaling constant C remains unchanged, we may neglect $\ln C$ in this formula for small r. This justifies the following:

DEFINITION 2.13 *A bounded set S in R^m has* **box dimension** *(box-counting dimension)*

$$BoxDimension(S) = \lim_{r\to 0} \frac{\ln(N(r))}{\ln\left(\frac{1}{r}\right)}$$

when the limit exists.

See Chapter 6 for visualization of the complicated sets and calculation of their box dimension which has noninteger values. In most cases when one has to compute BoxDimension of an orbit of a dynamical system, or time series, or a given complicated set obtained experimentally we can get only the approximate value. Let us see some examples.

Dynamica's `BoxDimension` function calculates numerically the Box Dimension of the attractor of a difference equation. In addition, this command can be used to calculate numerically the Box Dimension of any time series.

```
In[35]:= ?BoxDimension
"BoxDimension[h,niter,epsilon,vars0] produces
the box dimension of the data generated
by a dynamical system, specified by h.
BoxDimension[data,epsilon] produces the box
dimension of the list data.  epsilon is the
radius of the covering elements."
```

We may calculate the box dimension for the "chaotic orbits" of the discrete predator prey map (2.38). We expect to get a corresponding box dimension substantially higher than 1.

```
In[36]:= A = 3.7; B = 0.3;

In[37]:= BoxDimension[DPPMap, 1000,
           0.001, {0.1, 0.2}]
Out[37]= 1.00014
```

Taking more iterates produces a more accurate approximation.

```
In[38]:= BoxDimension[DPPMap, 4000,
           0.001, {0.1, 0.2}]
Out[38]= 1.20072
```

A list of `BoxDimension` calculations as a function of iteration numbers may be produced as shown.

```
In[39]:= bdim =
           Table[BoxDimension[DPPMap,
               1000 * i, 0.001, {0.1, 0.2}],
             {i, 1, 10}];

In[40]:= ListPlot[bdim, PlotJoined → True,
           Frame → True,
           Axes → None]
```

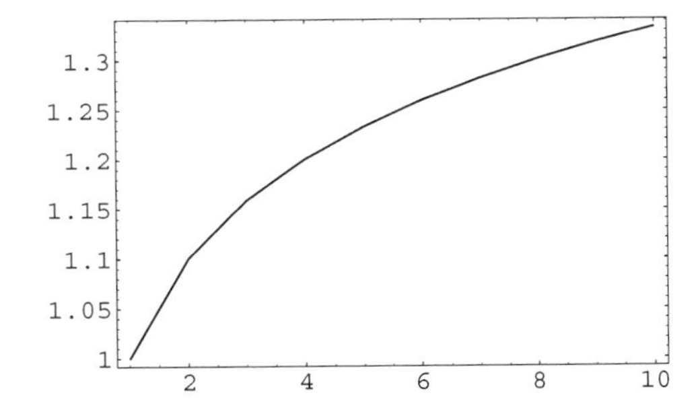

2.10 Semicycle Analysis

It has been shown in [KoL] and [KL] that a semicycle analysis of the solutions of a difference equation is a powerful tool for better understanding of the character of solutions, and often it leads to straightforward proofs of results about their long-term behavior.

In this section, we present some results about the semicycle character of difference equation solutions of the form

$$x_{n+1} = f(x_n, x_{n-1}), \quad n = 0, 1, \ldots \tag{2.42}$$

under appropriate hypotheses on the function f.

First, we give the definitions for the positive and negative semicycle of a solution of (2.42) relative to an equilibrium point $\overline{x}$.

DEFINITION 2.14

A **positive semicycle** *(with respect to an equilibrium $\overline{x}$) of a solution $\{x_n\}$ of (3.43) consists of an ordered set $\{x_l, x_{l+1}, \ldots, x_m\}$, with all terms greater than or equal to the equilibrium $\overline{x}$, with $l \geq 0$ and $m \leq \infty$ and such that*

$$\text{either } l = 0, \quad \text{or} \quad l > 0 \text{ and } x_{l-1} < \overline{x}$$

and

$$\text{either } m = \infty, \quad \text{or} \quad m < \infty \text{ and } x_{m+1} < \overline{x}.$$

A **negative semicycle** *(with respect to an equilibrium $\overline{x}$) of a solution $\{x_n\}$ of (2.42) consists of a an ordered set $\{x_l, x_{l+1}, \ldots, x_m\}$, with all terms less than the equilibrium $\overline{x}$, with $l \geq -1$ and $m \leq \infty$ and such that*

$$\text{either } l = 0, \quad \text{or} \quad l > 0 \quad \text{and } x_{l-1} \geq \overline{x}$$

and

$$\text{either } m = \infty, \quad \text{or} \quad m < \infty \text{ and } x_{m+1} \geq \overline{x}.$$

DEFINITION 2.15 (Oscillation)

a. *A sequence $\{x_n\}$ is said to* **oscillate about zero** *or simply to* **oscillate** *if the terms x_n are neither eventually all positive nor eventually all negative. Otherwise the sequence is said to be* **nonoscillatory**. *A sequence $\{x_n\}$ is called* **strictly oscillatory** *if for every $n_0 \geq 0$, there exist $n_1, n_2 \geq n_0$ such that $x_{n_1} x_{n_2} < 0$.*

b. *A sequence $\{x_n\}$ is said to* **oscillate about $\overline{x}$** *if the sequence $x_n - \overline{x}$ oscillates. The sequence $\{x_n\}$ is called* **strictly oscillatory about $\overline{x}$** *if the sequence $x_n - \overline{x}$ is strictly oscillatory.*

The first result was established in [GKL1] and in [KL].

THEOREM 2.15

Let $f : (0, \infty) \times (0, \infty) \to (0, \infty)$ be a continuous function such that $f(x, y)$ is decreasing in x for each fixed y, and $f(x, y)$ is increasing in y for each fixed x. Let $\bar{x}$ be a positive equilibrium of (2.42).

Then, except possibly for the first semicycle, every solution of (2.42) has semicycles of length 1.

The next result applies when the function f is decreasing in both arguments and was established in [KL].

THEOREM 2.16

Let $f : (0, \infty) \times (0, \infty) \to (0, \infty)$ be a continuous function such that $f(x, y)$ is decreasing in both arguments. Let $\bar{x}$ be a positive equilibrium of (2.42).

Then, every oscillatory solution of (2.42) has semicycles of length at most 2.

One can obtain similar results for the remaining two cases of the monotonic behavior of the function f.

Now we present an example which illustrates the use of semicycle analysis in studying the global behavior of second order difference equation, see [KL] and [KLP2].

Example 2.7 Consider the difference equation

$$y_{n+1} = \frac{py_n + y_{n-1}}{q + y_{n-1}}, \quad n = 0, 1, \ldots \tag{2.43}$$

where p and q are positive parameters and the initial conditions are nonnegative numbers.

The equilibrium points of (2.43) are the solutions of the equation

$$\bar{y} = \frac{p\bar{y} + \bar{y}}{q + \bar{y}}.$$

So $\bar{y} = 0$ is an equilibrium point of (2.43), and when

$$p + 1 > q,$$

equation (2.43) also possesses the unique positive equilibrium $\bar{y} = p + 1 - q$.

The linearized equation at the zero equilibrium is

$$z_{n+1} - \frac{p}{q} z_n - \frac{1}{q} z_{n-1} = 0,$$

with characteristic equation

$$\lambda^2 - \frac{p}{q}\lambda - \frac{1}{q} = 0.$$

The linearized equation at the positive equilibrium is

$$z_{n+1} - \frac{p}{p+1}z_n - \frac{q-p}{p+1}z_{n-1} = 0,$$

with characteristic equation

$$\lambda^2 - \frac{p}{p+1}\lambda - \frac{q-p}{p+1} = 0.$$

The following result follows from Theorem 2.11.

PROPOSITION 2.1

a. If

$$p + 1 \le q,$$

then the zero equilibrium of (2.43) is locally asymptotically stable.

b. If

$$p + 1 > q, \tag{2.44}$$

then the zero equilibrium of (2.43) is unstable and the positive equilibrium $\bar{y} = p+1-q$ of (2.43) is locally asymptotically stable. Furthermore, the zero equilibrium is a saddle point when

$$1 - p < q < 1 + p$$

and a repeller when

$$q < 1 - p.$$

Using a global asymptotic stability result (such as Theorem 1.3.1 in [KL]), one can prove that the condition $p + 1 \le q$ is sufficent for global asymptotic stability of the zero equilibrium.

In the remainder of this example we investigate the character of the positive equilibrium of (2.43), hence we assume from now on that (2.44) holds. Our goal is to show that when (2.44) holds and $y_{-1} + y_0 > 0$, the positive equilibrium $\bar{y} = p + 1 - q$ of (2.43) is globally asymptotically stable.

Now we find the invariant intervals and perform a semicycle analysis. Let $\{y_n\}_{n=-1}^{\infty}$ be a positive solution of (2.43). Then the following identities are easily established:

$$y_{n+1} - 1 = p\frac{y_n - \frac{q}{p}}{q + y_{n-1}} \quad \text{for } n \ge 0, \tag{2.45}$$

$$y_{n+1} - \frac{q}{p} = \frac{p^2[y_n - \left(\frac{q}{p}\right)^2] + (p-q)y_{n-1}}{p(q + y_{n-1})} \quad \text{for } n \geq 0, \tag{2.46}$$

and

$$y_{n+1} - y_n = \frac{(p-q)y_n + (1-y_n)y_{n-1}}{q + y_{n-1}} \quad \text{for } n \geq 0. \tag{2.47}$$

Note that the positive equilibrium

$$\bar{y} = p + 1 - q$$

of equation (2.43) satisfies

$$\bar{y} < 1 \quad \text{if} \quad p < q < p + 1$$

$$\bar{y} = 1 \quad \text{if} \quad p = q$$

$$\bar{y} > 1 \quad \text{if} \quad p > q.$$

When $p = q$, that is, for the difference equation

$$y_{n+1} = \frac{py_n + y_{n-1}}{p + y_{n-1}}, \quad n = 0, 1, \ldots \tag{2.48}$$

the identities (2.45)-(2.47) reduce to the following:

$$y_{n+1} - 1 = (y_n - 1)\frac{p}{p + y_{n-1}} \quad \text{for } n \geq 0 \tag{2.49}$$

and

$$y_{n+1} - y_n = (1 - y_n)\frac{y_{n-1}}{p + y_{n-1}} \quad \text{for } n \geq 0. \tag{2.50}$$

The following three lemmas are now direct consequences of identities 2.45 through 2.50.

LEMMA 2.1
Assume that

$$q - 1 < p < q \tag{2.51}$$

and let $\{y_n\}_{n=-1}^{\infty}$ *be a positive solution of equation (2.43). If* $N \geq 0$*, then*

i. $y_N < \frac{q}{p} \Rightarrow y_{N+1} < 1$

ii. $y_N = \frac{q}{p} \Rightarrow y_{N+1} = 1$

iii. $y_N > \frac{q}{p} \Rightarrow y_{N+1} > 1$

iv. $y_N > \left(\frac{q}{p}\right)^2 \Rightarrow y_{N+1} < \frac{q}{p}$

v. $y_N \leq 1 \Rightarrow y_{N+1} < 1$

vi. $y_N \geq 1 \Rightarrow y_{N+1} < y_N$

LEMMA 2.2
Assume that

$$p = q \tag{2.52}$$

and let $\{y_n\}_{n=-1}^{\infty}$ *be a positive solution of equation (2.43). Then the following statements are true:*

i. If $y_0 < 1$, *then* $y_n < 1$ *for all* $n \geq 0$, *and the solution is strictly increasing;*

ii. If $y_0 = 1$, *then* $y_n = 1$ *for* $n \geq 0$;

iii. If $y_0 > 1$, *then* $y_n > 1$ *for all* $n \geq 0$, *and the solution is strictly decreasing.*

LEMMA 2.3
Assume that

$$p > q \tag{2.53}$$

and let $\{y_n\}_{n=-1}^{\infty}$ *be a positive solution of equation (2.43). If* $N \geq 0$, *then,*

i. $y_N < \frac{q}{p} \Rightarrow y_{N+1} < 1$

ii. $y_N = \frac{q}{p} \Rightarrow y_{N+1} = 1$

iii. $y_N > \frac{q}{p} \Rightarrow y_{N+1} > 1$

iv. $y_N > \left(\frac{q}{p}\right)^2 \Rightarrow y_{N+1} > \frac{q}{p}$

v. $y_N \leq 1 \Rightarrow y_{N+1} > y_N$

vi. $y_N \geq 1 \Rightarrow y_{N+1} > \frac{q}{p}$ *and* $y_{N+2} > 1$

When (2.51) holds, it follows from Lemma 2.1 that if a solution $\{y_n\}_{n=-1}^{\infty}$ is such that

$$y_n \geq 1 \quad \text{for all} \quad n \geq 0,$$

then the solution decreases and its limit lies in the interval $[1, \infty)$. This is impossible because $\bar{y} < 1$. Hence, by Lemma 2.1, every positive solution of (2.43) eventually enters and remains in the interval $(0, 1)$. Now in the interval $(0, 1)$ the function

$$f(u, v) = \frac{pu + v}{q + v}$$

is increasing in both arguments and by Theorem 2.24, $\bar{y}$ is a global attractor.

When (2.52) holds, it follows from Lemma 2.2 that every solution of (2.43) converges to the equilibrium $\bar{y} = 1$.

Next, assume that (2.53) holds. Here it is clear from Lemma 2.3 that every solution of (2.43) eventually enters and remains in the interval $(1, \infty)$. Without loss of generality we assume that

$$y_n > 1 \quad \text{for} \quad n \geq -1.$$

Set

$$y_n = 1 + (q+1)u_n \quad \text{for} \quad n \geq -1.$$

Then we can see that $\{u_n\}_{n=-1}^{\infty}$ satisfies the difference equation

$$u_{n+1} = \frac{\frac{p-q}{(q+1)^2} + \frac{p}{q+1}u_n}{1 + u_{n-1}}, \quad n = 0, 1, \ldots \tag{2.54}$$

with positive parameters and positive initial conditions.

This equation was investigated in [KL] where it was established that every solution converges to the positive equilibrium

$$\bar{u} = \frac{p-q}{q+1}.$$

From the above observations and in view of Theorem 2.1 we have the following result:

THEOREM 2.17

Assume that $p + 1 > q$ and that $y_{-1} + y_0 > 0$. Then the positive equilibrium $\bar{y} = p + 1 - q$ of (2.43) is globally asymptotically stable.

■

2.11 Stable and Unstable Manifolds

In this section we study invariant sets near the equilibrium and periodic points of system (2.1). A major motivation for the study of invariant sets is to understand the behavior of system (2.1) near an equilibrium point or a periodic point which, in the linear approximation, is a saddle point or a nonhyperbolic fixed point. Consider the linear system

$$\mathbf{Z}_{n+1} = \mathbf{A}\mathbf{Z}_n, \tag{2.55}$$

where $\mathbf{A}$ is a 2×2 matrix. If the zero equilibrium of this system is a saddle point, then there are invariant subspaces on which the solution is forward or backward asymptotic to the zero equilibrium. The subspaces are spanned by eigenvectors associated to eigenvalues inside the unit circle or outside the unit circle, respectively. Let us define **stable**, **unstable**, and **center subspaces**.

$E^u = span\{\mathbf{v}^u :$	$\mathbf{v}^u$ is a generalized eigenvector for an eigenvalue λ_u of A with $\lvert\lambda_u\rvert > 1\}$
$E^s = span\{\mathbf{v}^s :$	$\mathbf{v}^s$ is a generalized eigenvector for an eigenvalue λ_s of A with $\lvert\lambda_s\rvert < 1\}$
$E^c = span\{\mathbf{v}^c :$	$\mathbf{v}^c$ is a generalized eigenvector for an eigenvalue λ_c of A with $\lvert\lambda_c\rvert = 1\}$

See [GH] and [R]. Our objective is to understand what happens to these subspaces if small nonlinear terms are added to the right-hand side of equation 2.55. Let us start with simple examples:

Example 2.8 Consider system (2.55) with

$$\mathbf{A} = \begin{pmatrix} a & 0 \\ 0 & d \end{pmatrix},$$

where $|a| < 1$ and $|d| > 1$. Let $\mathbf{Z}_n = (x_n, y_n)^T$. Then,

$$x_n = a^n x_0, \quad y_n = d^n y_0.$$

Observe that

i. if $x_0 = 0$, then $x_n = 0$ for all n, and $\mathbf{Z}_n \to (0,0)^T$ as $n \to -\infty$,

ii. if $y_0 = 0$, then $y_n = 0$ for all n, and $\mathbf{Z}_n \to (0,0)^T$ as $n \to \infty$.

If $\mathbf{Z} = (x, y)^T$, then the set $\{(x, y) : y = 0\}$ is the stable subspace of the equilibrium point $(0, 0)$ and the set $\{(x, y) : x = 0\}$ is the unstable subspace of $(0, 0)$. ∎

Here we give basic notions concerning stable and unstable manifolds of equilibrium and periodic points of system (2.1), see [E2], [GH], and [R].

DEFINITION 2.16 *Let U be a neighborhood of a fixed point $(\bar{x}, \bar{y})$ of (2.1). The local stable manifold W^s_{loc} and the local unstable manifold W^u_{loc} are the sets*

$$W^s_{loc} = \{(x, y) : \mathbf{F}^n(x, y) \in U \ for\ all\ n \geq 0,\ and\ \mathbf{F}^n(x, y) \to (\bar{x}, \bar{y})\ as\ n \to \infty\}$$

$$W^u_{loc} = \{(x, y) : \mathbf{F}^{-n}(x, y) \in U \ for\ all\ n \geq 0,\ and\ \mathbf{F}^{-n}(x, y) \to (\bar{x}, \bar{y})\ as\ n \to \infty\}$$

Some basic geometric properties of local invariant manifolds are described in the following theorem, see [E2], [GH], and [R].

THEOREM 2.18 (Local Stable Manifold)
Let $\mathbf{F} : R^2 \to R^2$ be a continuously differentiable map with a hyperbolic saddle point $(\overline{x}, \overline{y})$, with associated eigenvalues (respectively, eigenvectors) d_1 and d_2 (resp. v_1 and v_2). Suppose that $d_1 < 1$ and $d_2 > 1$. Then W^s_{loc} is a smooth curve tangent to v_1 at $(\overline{x}, \overline{y})$, and W^u_{loc} is a smooth curve tangent to v_2 at $(\overline{x}, \overline{y})$.

If we do not restrict our attention to a local neighborhood of $(\overline{x}, \overline{y})$, then we obtain the global analogs of the two invariant manifolds.

DEFINITION 2.17 *The global stable manifold W^s and the global unstable manifold W^u of an equilibrium point $(\overline{x}, \overline{y})$ are the sets*

$$W^s(\overline{x}, \overline{y}) = \{(x, y) : \ \mathbf{F}^n(x, y) \to (\overline{x}, \overline{y}) \ \text{as } n \to \infty\}$$

$$W^u(\overline{x}, \overline{y}) = \{(x, y) : \ \mathbf{F}^{-n}(x, y) \to (\overline{x}, \overline{y}) \ \text{as } n \to \infty\}$$

It is easy to see from the definition that

$$W^s(\overline{x}, \overline{y}) = \cup_{n \geq 0} \mathbf{F}^{-n}(W^s_{loc}(\overline{x}, \overline{y}))$$

$$W^u(\overline{x}, \overline{y}) = \cup_{n \geq 0} \mathbf{F}^{n}(W^u_{loc}(\overline{x}, \overline{y}))$$

It is possible for two invariant manifolds of an equilibrium point of (2.1) to cross each other at some point different from the equilibrium point without coinciding.

DEFINITION 2.18 *Let $(\overline{x}, \overline{y})$ be a saddle point of a map $\mathbf{F}$. A point (r, s) is a* **homoclinic point** *for $\mathbf{F}$ if $\mathbf{F}^n(r, s) \to (\overline{x}, \overline{y})$ as $n \to \infty$ and as $n \to -\infty$. A homoclinic point (r, s) is said to be* **transversal** *if the tangent vectors to W^s and W^u at (r, s) do not coincide.*

The presence of a transversal homoclinic point of planar maps is a clear sign of dynamical complexity (see [ASY]). We will explore this subject visually in Section 2.12.

In this case there is an equilibrium point, since we specified the period to search for to be 1.

```
In[41]:= A = 2; B = 0.3;

In[42]:= po = PeriodicOrbit[DPPMap, 1]
Out[42]= {{0., 0.}, {0.3, 0.4}, {0.5, 0.}}
```

`Multipliers` is now used to find the eigenvalues at the specified fixed point.

```
In[43]:= Multipliers[DPPMap, po[[1]]]
Out[43]= {2., 0.}
```

Therefore, `po[[1]]` is a saddle point with an oscillation in the unstable direction.

`UnstableManifold` approximates the unstable manifold W^u_{loc} by iterating forward points on the line determined by the unstable eigenvector that is close to the fixed point.

```
In[44]:= Wu = UnstableManifold[DPPMap,
              po[[1]], 1, 3];

In[45]:= PhasePortrait[{Wu, {po[[1]]}},
           PlotRange →
           {{-0.25, 0.25}, {-0.25, 0.25}},
           Size → {0.02, 0.04}]
```

The unstable manifold can be seen in this plot on the horizontal axis.

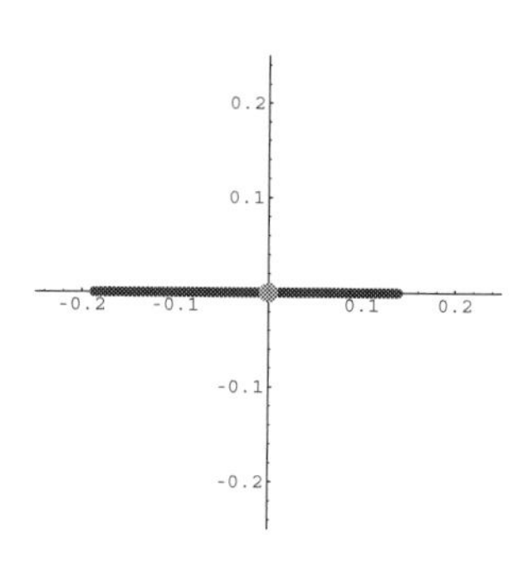

The Local Stable Manifold Theorem guarantees that a small piece of W^u_{loc} looks pretty much like a line segment. However, longer pieces can exhibit surprisingly complicated behavior.

Additional input parameters in `UnstableManifold` yield a larger piece of W^u_{loc}. Here we specify period one, with 10 iterations and 50 subdivisions. We see the zero equilibrium point and a piece of the local unstable manifold.

```
In[46]:= Wu = UnstableManifold[DPPMap,
              po[[1]], 1, 10, 50];

In[47]:= PhasePortrait[{Wu, {po[[1]]}},
           PlotRange → {{-2, 2}, {-2, 2}},
           Size → {0.02, 0.04}]
```

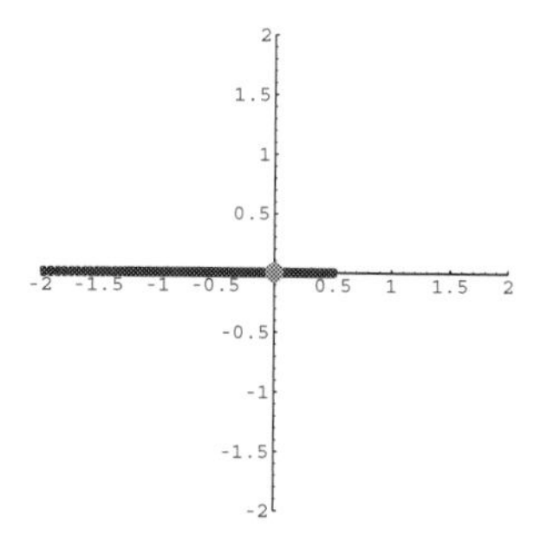

As another example ([La1], pp. 61), we consider Lauwerier's difference equation

$$x_{n+1} = 2x_n \qquad n = 0, 1, \ldots$$
$$y_{n+1} = \tfrac{y_n}{2} + 7x_n^2$$

```
In[48]:= LawerMap[{x_, y_}] :=
           {2 x, y/2 + 7 x^2}

In[49]:= LawerMap[{x, y}]
Out[49]= {2 x, 7 x^2 + y/2}
```

The only fixed point is (0, 0).

```
In[50]:= Solve[LawerMap[{x, y}] == {x, y},
           {x, y}]
Out[50]= {{y -> 0, x -> 0}}

In[51]:= fp = {x, y}/.%[[1]];
```

We see that the fixed point is a saddle.

```
In[52]:= Eigenvalues[Jac[LawerMap, fp]]
Out[52]= {2, 1/2}
```

The inverse map is calculated and given a name.

```
In[53]:= Solve[LawerMap[{u, v}] == {x, y},
           {u, v}]
Out[53]= {{v -> 1/2 (-7 x^2 + 4 y), u -> x/2}}

In[54]:= LawerInvMap[{x_, y_}] =
           {u, v} /. %[[1]]
Out[54]= {x/2, 1/2 (-7 x^2 + 4 y)};
```

We name Ws and Wu the stable and unstable manifolds. Note that the stable manifold is the unstable manifold of the inverse map.

```
In[55]:= Wu = UnstableManifold[LawerMap,
           {0., 0.}, 1, 10, 50];

In[56]:= Ws = UnstableManifold[LawerInvMap,
           {0., 0.}, 1, 10, 50];
```

The plot of the stable and unstable manifolds about the fixed point (0, 0) shows that the local stable manifold is the y-axis, and the local unstable manifold is a parabolic arc.

```
In[57]:= PhasePortrait[{Wu, Ws, {{0., 0.}}},
    Size → {0.015, 0.015, 0.03},
    PlotRange → {{-2, 2}, {-2, 2}}];
```

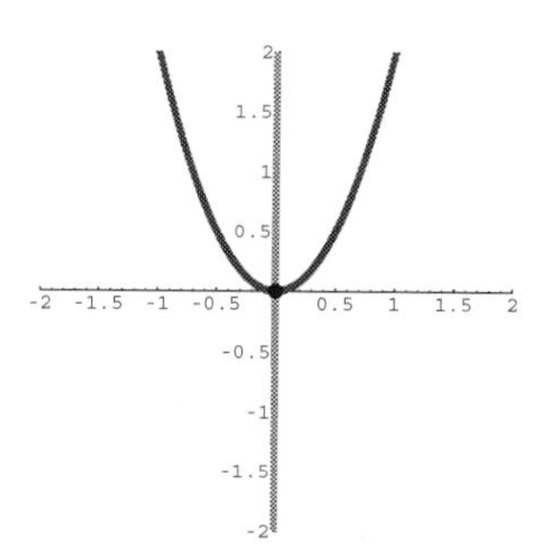

2.12 *Dynamica* Session on Henon's Equation

In this section we use *Dynamica* to study Henon's Equation

$$x_{n+1} = A - x_n^2 + Bx_{n-1} \tag{2.56}$$

where A and B are real numbers.

Equation (2.56) and the corresponding map are named after their discoverer, Michel Henon, an astronomer at the Nice Observatory in France, see [H]. Curious about the degradation of celestial orbits, he began to model the orbits of stars around the centers of their galaxies. Henon considered gravitational centers as three-dimensional objects and carefully studied the orbits of the stars. To simplify the task of trying to track a three-dimensional orbit, he considered, instead, the intersection of a plane with these orbits, the so-called Poincaré maps. Initially, the intersection points appeared to be completely random in their location, moving from one side of the plane to another. However, after a few dozen points had been plotted, a closed, egg-shaped curve began to appear, the celebrated Henon attractor. This mapping was apparently a cross section of a torus. Henon (along with one of his graduate students) continued to study this mapping and continued plotting the points for a system with increased energy levels. Once the newer mappings were made, though, the continuous curve began fading and random points began to appear proportionally to the energy. Over the years, Henon tried many ways to predict the upcoming points of his high-energy graph until 1976, when he decided to abandon the differential equations approach and use difference equations.

The global dynamics of Henon's equation and the corresponding map are quite complicated. However, it should be noted that for $A = 1.4$ and $B = 0.3$, Henon discovered, through numerical simulations, that in a region of the plane almost all solutions get attracted to a set, the Henon attractor. The Henon attractor is neither an equilibrium point nor a periodic orbit, but rather a "strange attractor." Although it is composed entirely of "lines of dust", orbits on this set do not flow continuously, but jump from one location in the attractor to another. Single points that are initially separated by any value eventually diverge (sensitive dependence on initial conditions). The Henon attractor also has a fine structure resembling a Cantor middle-third set in the sense that successive magnifications provide an ever-increasing degree of detail. Any cross section made through a branch of the Henon attractor forms a Cantor middle-thirds set. It is interesting to observe that this attractor appears to be the global unstable manifold of the saddle point. Some of these facts have been finally proved in [BC] and [PT].

Henon's equation and map are classical example of chaotic behavior in two dimensions and are part of all textbooks and major packages on nonlinear dynamics, such as [ASY], [GH], [HK], [K], [KJ], and [NY].

In this session we shall classify the fixed points, find numerically stable and unstable manifolds, find basic characteristics of semicycles such as the maxima and length of positive semicycles and the minima and length of negative semicycles, and plot bifurcation diagrams.

Let us consider the problem of finding and analyzing the equilibrium (fixed) points of the Henon map.

Henon's map has been defined in *Dynamica* in terms of parameters A and B.

```
In[58]:= Clear[A, B, x, y]

In[59]:= HenonMap[{x, y}]
```

Out[59]= $\{A - x^2 + B\,y,\ x\}$

The *Mathematica* `Solve` command is used here to find fixed points. There are two fixed points, named here `fp1` and `fp2`

```
In[60]:= Solve[HenonMap[{x, y}] == {x, y},
            {x, y}]
```

Out[60]= $\left\{\left\{y \to -\frac{1}{2} + \frac{B}{2} - \frac{1}{2}\sqrt{1 + 4A - 2B + B^2},\ x \to \frac{1}{2}\left(-1 + B - \sqrt{1 + 4A - 2B + B^2}\right)\right\},\ \left\{y \to -\frac{1}{2} + \frac{B}{2} + \frac{1}{2}\sqrt{1 + 4A - 2B + B^2},\ x \to \frac{1}{2}\left(-1 + B + \sqrt{1 + 4A - 2B + B^2}\right)\right\}\right\}$

```
In[61]:= {fp1, fp2} = {x, y} /. %;
```

As an example, we choose specific values for A and B. To classify the fixed points, the eigenvalues of the Jacobian of the map must be calculated. We see here that this fixed point is a sink.

```
In[62]:= A = 7/16; B = -1/2;

In[63]:= Eigenvalues[Jac[HenonMap, fp2]]
```

$$Out[63]= \left\{\frac{1}{4}\left(-1-\mathrm{i}\sqrt{7}\right), \frac{1}{4}\left(-1+\mathrm{i}\sqrt{7}\right)\right\}$$

```
In[64]:= Abs[%]
```

$$Out[64]= \left\{\frac{1}{\sqrt{2}}, \frac{1}{\sqrt{2}}\right\}$$

This is an orbit of Henon's equation with the parameter values originally studied by Henon in [H].

```
In[65]:= A = 1.4; B = 0.3;

In[66]:= orb1 = Orbit[HenonMap, {0., 0.},
              100];
```

A `PhasePortrait` plot shows that the orbit is concentrated on the famous "Henon attractor," which is discussed in detail in [BC], [MV], and in [PT].

```
In[67]:= PhasePortrait[orb1,
           PlotRange → {{-2, 2}, {-2, 2}}]
```

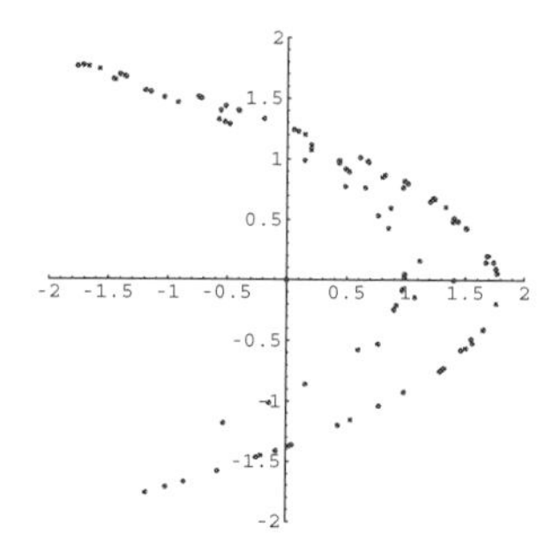

To get a better look at Henon's attractor, we first plot a long list of points of the orbit.

```
In[68]:= orb2 = Orbit[HenonMap, {0., 0.},
            10000];

In[69]:= pp = PhasePortrait[{orb2},
          {Black},
           PlotRange → {{-2, 2}, {-2, 2}}];
```

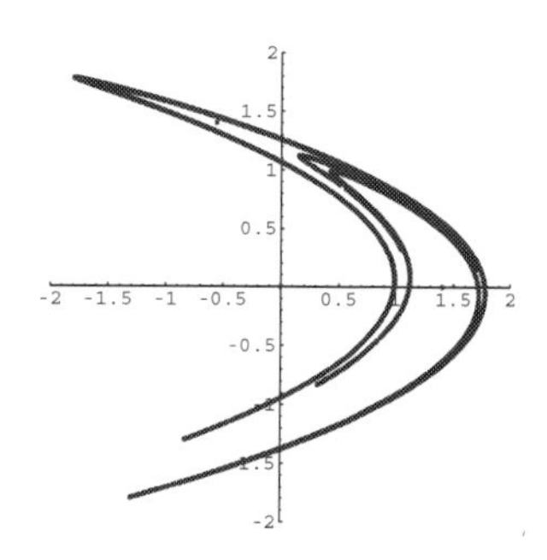

Next, we zoom in to get an idea of the complex structure of the attractor.

```
In[70]:= Show[pp, PlotRange →
          {{0.2, 0.6}, {0.8, 1.2}}]
```

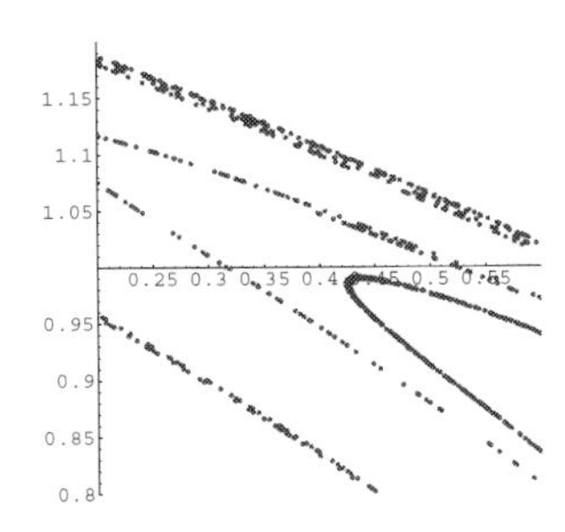

Visualization of maps can be done by using `PoincarePlot2D`, which is a function that takes the default circle of radius 1/2 centered at (1, 1) and applies the map generated by the difference equation or the system of difference equations a prescribed number of times, showing the evolution of an initial circle. The default values for the center and the radius can be changed.

Here we apply 4 times `PoincarePlot2D` on an initial circle to show the evolution of this circle under Henon's map. One can use different initial shapes to produce similar plots. Again we see that the initial circle is streched along the Henon's attractor.

```
In[71]:= Henon[{x_, y_}] := {y, A - y^2 + Bx};

In[72]:= PoincarePlot2D[Henon, 4,
          Center → (1, 1)];
```

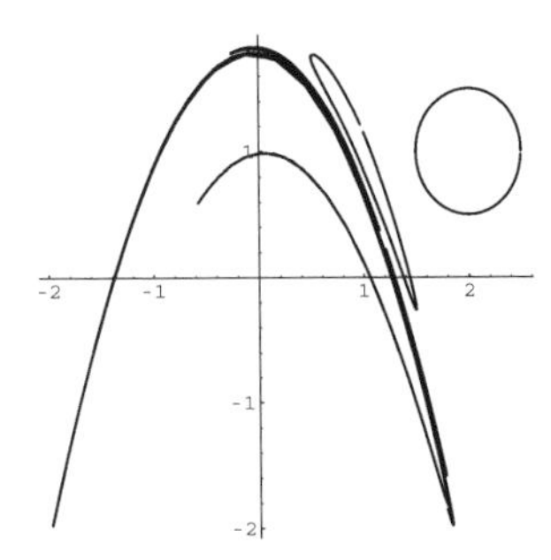

Now, we explore another aspect of the Henon map that is key to understanding the shape of the attractor.

Here are the fixed points for $A = 1.4$ and $B = 0.3$.

```
In[73]:= {fp1, fp2}
Out[73]= {{-1.5839, -1.5839},
          {0.883896, 0.883896}}
```

`Multipliers` may be used to find the eigenvalues at the fixed point.

```
In[74]:= Multipliers[HenonMap, fp1]
Out[74]= {-4.98193, -0.0180653}
```

The next plot shows the fixed point and a small piece of the unstable manifold as a curve through the fixed point.

```
In[75]:= Wu = UnstableManifold[HenonMap,
              fp1, 1, 3, 10];

In[76]:= PhasePortrait[{Wu, {fp1}},
           Size → {0.01, 0.03},
           PlotRange → {{-3, 0}, {-3, 0}}];
```

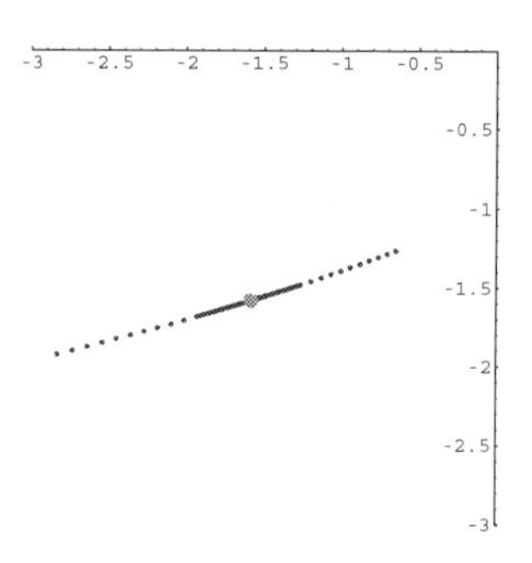

To extend the manifold, the last input parameter in the `UnstableManifold` function is increased. This should be done with care, as problems with round off errors and numerical stability may arise.

```
In[77]:= Wu = UnstableManifold[HenonMap,
            fp1, 1, 10, 50];

In[78]:= PhasePortrait[{Wu, {fp1}},
           Size → {0.01, 0.03},
           PlotRange → {{-2, 2}, {-2, 2}}];
```

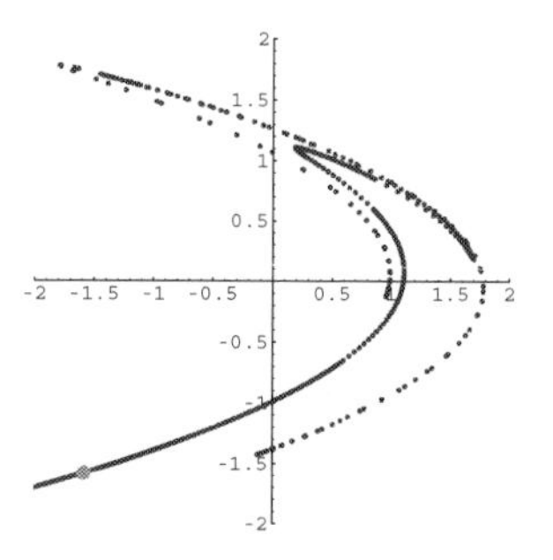

Here is another orbit.

```
In[79]:= orb3 = Orbit[HenonMap,
              {-1.5, -1.5}, 500];

In[80]:= PhasePortrait[orb3,
           Size → {0.01},
           PlotRange → {{-2, 2}, {-2, 2}}];
```

This orbit also approaches the Henon attractor.

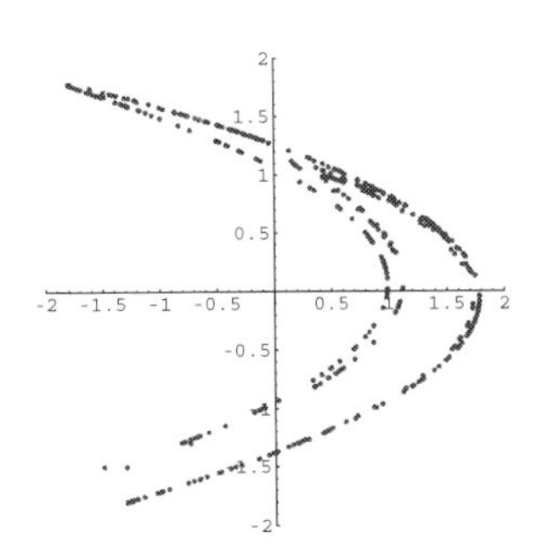

The relationship of orbits with Henon's attractor become clearer when an orbit and the attractor are plotted together. Clearly, the unstable manifold is part of the attractor.

```
In[81]:= PhasePortrait[Wu, {orb3},
           Size → {0.01, 0.01},
           PlotRange → {{-2, 2}, {-2, 2}}];
```

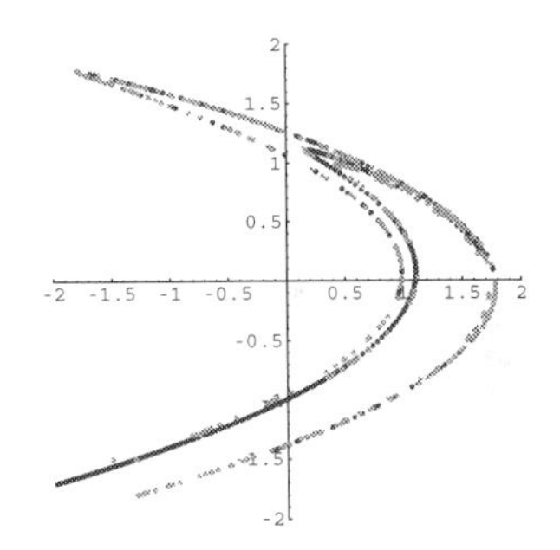

PeriodicOrbit may also be used to search for fixed points. In this case we calculated the fixed points previously, and the output of PeriodicOrbit merely confirms the previous values.

```
In[82]:= A = 0.7; B = -0.5;

In[83]:= {fp1, fp2}
Out[83]= {{-1.87361, -1.87361},
           {0.37361, 0.37361}}

In[84]:= PeriodicOrbit[HenonMap, 1]
Out[84]= {{-1.87361, -1.87361},
           {0.37361, 0.37361}}
```

The fixed point fp1 is a saddle.

```
In[85]:= Multipliers[HenonMap, fp1]
Out[85]= {3.60867, 0.138555}
```

The stable manifold is the unstable manifold for the inverse map. Here is a calculation of the inverse map. Note that the inverse map is already available in *Dynamica*.

```
In[86]:= Solve[HenonMap[{u, v}] == {x, y},
            {u, v}]
```

$$Out[86]= \{\{v \to 0.2\,(7. - 10.\,x - 10.\,y^2),\ u \to 1.\,y\}\}$$

```
In[87]:= HenonInverseMap[{x, y}]
```

$$Out[87]= \{y,\ -2.\,(-0.7 + x + y^2)\}$$

We name `Wu` and `Ws` the unstable and stable manifolds.

```
In[88]:= Wu = UnstableManifold[HenonMap,
            fp1, 1, 3, 200];

In[89]:= Ws = UnstableManifold[
            HenonInverseMap, fp1, 1,
            3, 200];
```

Here we plot both local stable and unstable manifolds and the orbit. Note that the plot is much simpler than the previous one where we had a chaotic attractor. The values of the parameters are in the region where the behavior of Henon's map is less complex.

```
In[90]:= PhasePortrait[{Wu, Ws, {fp1}},
          Size → {0.01, 0.01, 0.03},
          PlotRange → {{-3, 3}, {-3, 3}}];
```

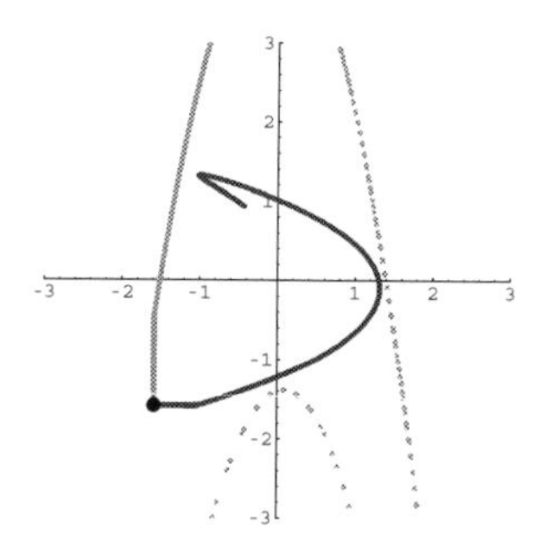

We now study bifurcation diagrams.

We fix $A = 1.25$, and use B as a parameter in bifurcation plots. Our plots reproduce the ones in [NY], p. 79.

```
In[91]:= Clear[A, B]

In[92]:= A = 1.25;
```

First, we start with the bifurcation diagram where $0.1 \leq p \leq 0.35$, with 300 steps, 400 is the number of the first iteration shown, 150 is the maximum number of iterations, and $\{0, 1\}$ is the starting point of the iteration. The range in the vertical axis is set to $\{-2.5, 2.5\}$.

```
In[93]:= BifurcationPlotND[HenonMap,
         {B, 0.1, 0.35},{0.,1.},
         Steps → 300,
         Iterates → 150,
         FirstIt → 400];
```

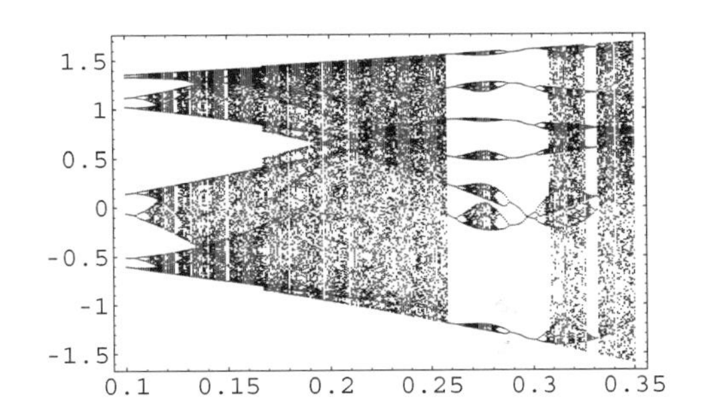

Zooming in, we get an interesting "bubble".

```
In[94]:= BifurcationPlotND[HenonMap,
         {B, 0.26, 0.29},{0.,1.},
         PlotRange → {1.15,1.3},
         Steps → 300,Iterates → 150,
         FirstIt → 400];
```

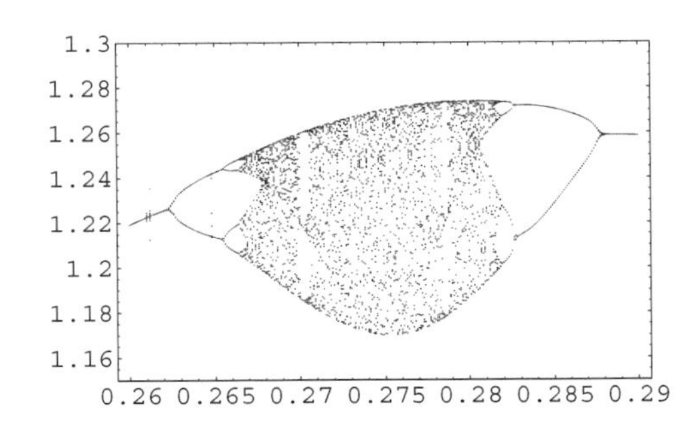

Here we have more numerical and visual evidence of the chaotic behavior of the Henon map for $A = 1.4, B = 0.3$.

```
In[95]:= A = 1.4; B = 0.3;

In[96]:= Lnum = LyapunovNumbers[HenonMap,
              {1.,1.},1200];

In[97]:= ListPlot[Lnum, PlotJoined→ True,
          PlotRange→ {1.2, 2.},
          Frame→ True]
```

Note that the Lyapunov numbers are clearly larger than 1.

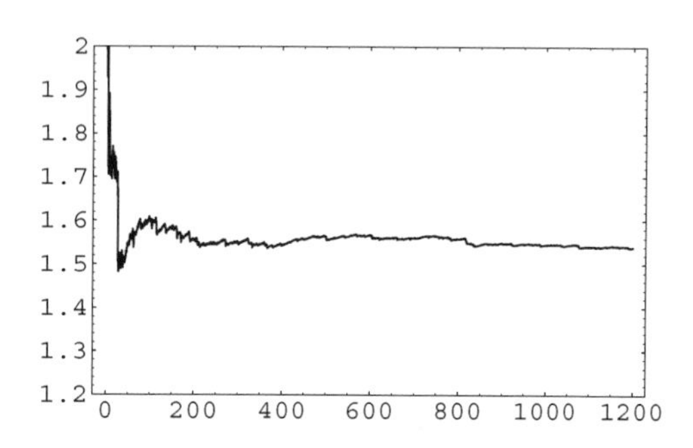

Let us calculate the box dimension for the values of the parameters that give the chaotic behavior.

```
In[98]:= BoxDimension[HenonMap[
            7000, 0.001, {0.1, 0.2}]
Out[98]= 1.28172
```

We now look at semicycles of an orbit with respect to one of the fixed points.

```
In[99]:= orb4 = Orbit[HenonMap, {1., 1.}, 20];

In[100]:= fp2
Out[100]= {0.883896, 0.883896}
```

This is semicycle information for `orb4` with respect to the equilibrium point `fp[[2]]`.

```
In[101]:= SemicycleTest[orb4, fp2[[1]]]
Positions and Maxima of Positive Semicycles
{3, {3, 1.21`}, {5, 1.74171`}, {9, 0.90814},
{11, 1.12126}, {13, 1.60279}, {17, 1.3710}, {19,
1.70382}}
Positions and Minima of Negative Semicycles
{{2, 0.7}, {4, 0.14590}, {6, -1.58979}, {10,
0.74241}, {12, 0.36549}, {14, -1.05928}, {16,
0.50651}, {18, -0.32786}, {20, -1.60138}}
Lengths of Positive Semicycles
{1, 1, 1, 1, 1, 1, 1, 1}
Lengths of Negative Semicycles
{-1, -1, -3, -1, -1, -3, -1, -2}
Positions and Lengths of both Positive and
Negative Semicycles
{1, -1, 1, -1, 1, -3, 1, -1, 1, -1, 1, -3, 1,
-1, 1, -2}
```

Similar analysis may be performed on the fixed point `fp1`. It appears that we have complicated dynamics with sensitive dependence on initial conditions. This confirms the known fact that for these values of the parameters A and B, Henon's map has very complicated, chaotic dynamics.

2.13 Invariants

An important tool for the investigation of stability and asymptotic behavior of some difference equations is the exploration of their invariants. When an invariant is known, in many cases it may be used to solve the difference equation in exact form, find the equilibrium points (in cases where the equilibrium points depend on the initial conditions), construct Lyapunov functions, etc. These subjects are treated in Sections 2.15 and 5.5, and in Chapter 4.

One of the characteristics of the nonhyperbolic equilibrium points for linear systems is the existence of invariants. Consider the linear difference equation

$$x_{n+1} = \frac{p x_n + q x_{n-1}}{p+q} \tag{2.57}$$

where p and q are positive parameters. It is easy to show that for any two initial values $x_{-1} = \alpha$, $x_0 = \beta$, $\alpha \neq \beta$ the corresponding solution of (2.57) converges to the equilibrium $\bar{x}$ (use the principle of nested intervals). However, the equilibrium point of (2.57) is any number, hence its value cannot be determined from this equation. To determine the value of the equilibrium point we can either use the formula for the explicit solution of (2.57), or notice that the expression

$$I(x, y) = (p+q)x + qy \tag{2.58}$$

has the property that

$$I(x_{n+1}, x_n) = I(x_n, x_{n-1}), \tag{2.59}$$

for every $n = 0, 1, \ldots$. Consequently,

$$I(x_n, x_{n-1}) = I(x_0, x_{-1}), \tag{2.60}$$

for every $n = 0, 1, \ldots$. Taking $n \to \infty$ in (2.60), we obtain

$$\bar{x} = \frac{(p+q)x_0 + qx_{-1}}{p+2q}.$$

The expression $I(x_n, x_{n-1})$ defined in (2.58) has the property of invariance with respect to the forward shift given by (2.59), and it is called an **invariant** or **first integral** of (2.57). The role of invariants in the theory of difference equations is similar to the role played by the first integrals in the theory of differential equations.

DEFINITION 2.19 *A nonconstant continuous function $I : R^2 \to R$ is an invariant for system (2.1) if*

$$I(x_{n+1}, y_{n+1}) = I(x_n, y_n)$$

for every $n = 0, 1, \ldots$.

The following basic result for linear systems and equations has been established recently in [BH].

THEOREM 2.19

System (2.3) possesses a continuous invariant $I : R^2 \to R$ *if and only if one of the following two conditions holds:*

(1) A has an eigenvalue $|\lambda| = 1$.

(2) A has eigenvalues λ_1 *and* λ_2 *with* $|\lambda_1| > 1$ *and* $0 < |\lambda_2| < 1$.

The two conditions in Theorem 2.19 are equivalent to the statement that neither all eigenvalues of A lie inside the unit circle nor all eigenvalues of A lie outside the unit circle. Another statement, also equivalent to the conditions in Theorem 2.19, is that the equilibrium point is either nonhyperbolic or is a saddle.

Note that one of the eigenvalues of (2.57) is 1, so Theorem 2.19 applies.

Example 2.9 (A nonhyperbolic equilibrium with monotonic convergence) Consider the linear system

$$\mathbf{z}_{n+1} = \mathbf{A}\mathbf{z}_n \tag{2.61}$$

where

$$\mathbf{A} = \begin{pmatrix} a & 0 \\ 0 & 1 \end{pmatrix},$$

with $a \in (0, 1)$. The positive orbit is given by

$$\mathbf{A}^n \mathbf{z}_0 = a^n x_0 \mathbf{v}^1 + y_0 \mathbf{v}^2.$$

It is clear that the origin is a stable equilibrium. Observe that in addition to the origin every point on the y-axis is an equilibrium. Furthermore, positive orbits converge to $(0, y_0)^T$ in a monotonic way in the direction of eigenvector $\mathbf{v}^1$.

If we assume that $a \in (-1, 0)$, then the convergence is oscillatory in the direction of $\mathbf{v}^1$. Notice that in view of Theorem 2.19 this system possesses an invariant. ∎

Example 2.10 (A nonhyperbolic equilibrium) Consider system (2.61) where

$$\mathbf{A} = \begin{pmatrix} 1 & 0 \\ 0 & -1 \end{pmatrix}.$$

The positive orbit is given by

$$\mathbf{A}^n\mathbf{z}_0 = x_0\mathbf{v}^1 + (-1)^n y_0\mathbf{v}^2.$$

It is clear that the origin is a stable equilibrium. Observe that in addition to the origin every point on the x-axis is an equilibrium. Furthermore, positive orbits converge to a period-two solution $\{(x_0, y_0)^T, (x_0, -y_0)^T\}$.

Notice that in view of Theorem 2.19 this system possesses an invariant. ∎

Example 2.11 (Complex conjugate eigenvalues) Consider the linear system (2.20) where $\mathbf{A}$ has the form

$$\mathbf{A} = \begin{pmatrix} \alpha & \beta \\ \beta & -\alpha \end{pmatrix}.$$

The eigenvalues of $\mathbf{A}$ are

$$\alpha \pm i\beta = \rho(\cos\omega \pm i\sin\omega),$$

where

$$\rho = \sqrt{\alpha^2 + \beta^2} \quad \text{and} \quad -\pi < \omega \le \pi.$$

Thus, $\mathbf{A}$ becomes

$$\mathbf{A} = \rho\begin{pmatrix} \cos\omega & \sin\omega \\ -\sin\omega & \cos\omega \end{pmatrix},$$

which is the well-known rotation matrix, see [B]. It is easy to check that

$$\mathbf{A}^n = \rho^n\begin{pmatrix} \cos(n\omega) & \sin(n\omega) \\ -\sin(n\omega) & \cos(n\omega) \end{pmatrix},$$

which is again the rotation matrix with angle of rotation $n\omega$. It is clear that $\mathbf{A}^n\mathbf{z}_0$ is obtained as the composition of rotation by angle $n\omega$ and then by multiplication by ρ^n. Consequently, if $\rho < 1$, the origin is asymptotically stable, and if $\rho > 1$, the origin is unstable.

The case $\rho = 1$ is of special interest. In view of Theorem 2.19, our system has an invariant. In fact, this invariant is

$$I(x_n, y_n) = x_n^2 + y_n^2.$$

Therefore, the orbit through the point $\mathbf{z}_0$ belongs to the circle with the radius $\|\mathbf{z}_0\|$. The asymptotic behavior of the orbit on such a circle depends on ω. To see this, observe that the positive orbit is given by

$$\mathbf{A}^n\mathbf{z}_0 = (\cos(n\omega)x_0 + \sin(n\omega)y_0)\mathbf{v}^1 + (-\sin(n\omega)x_0 + \cos(n\omega)y_0)\mathbf{v}^2.$$

Clearly, each orbit on the circle is periodic if $\omega/2\pi$ is rational, and dense if $\frac{\omega}{2\pi}$ is irrational. ∎

2.14 Lyapunov Functions, Stability, and Invariants

Consider system (2.1) where $D \subset R^2$ and $\mathbf{F} : D \to R^2$ is continuous.

DEFINITION 2.20 *The function $V : R^2 \to R$ is said to be a* **Lyapunov function** *on a subset D of R^2 if*

1. *V is continuous on D*
2. *$\Delta V(\mathbf{x}) = V(\mathbf{F}(\mathbf{x})) - V(\mathbf{x}) \leq 0$ when $\mathbf{x}$ and $\mathbf{F}(\mathbf{x}) \in D$.*

Let $B(\mathbf{a}, r)$ denote the open ball in R^2 of radius r and center $\mathbf{a}$

$$B(\mathbf{a}, r) = \{\mathbf{x} \in R^2 : \|\mathbf{x} - \mathbf{a}\| < r\}$$

DEFINITION 2.21 *We say that the real function V is* **positive definite** *at $\overline{\mathbf{x}}$ if*

1. *$V(\overline{\mathbf{x}}) = 0$*
2. *$V(\mathbf{x}) > 0$ for all $\mathbf{x} \in B(\overline{\mathbf{x}}, r)$, for some $r > 0$.*

Now we have the main result of this section, see [E1] and [LT]:

THEOREM 2.20 (Lyapunov Stability Theorem)
Assume that V is a Lyapunov function for (2.1) on a neighborhood D of the equilibrium point $\overline{\mathbf{x}}$, and that V is positive definite at $\overline{\mathbf{x}}$. Then,

i. $\overline{\mathbf{x}}$ is stable.

ii. If $\Delta V(x) < 0$ for all $\mathbf{x}, \mathbf{f}(\mathbf{x}) \in D$ and $\mathbf{x} \neq \overline{\mathbf{x}}$, then $\overline{\mathbf{x}}$ is asymptotically stable.

iii. If $D = R^2$ and

$$V(\mathbf{x}) \to \infty \quad as \quad \|\mathbf{x}\| \to \infty \tag{2.62}$$

then $\overline{\mathbf{x}}$ is globally asymptotically stable.

In the case of system (2.1) we present the following result which establishes a connection between invariants, Lyapunov functions, and the stability of equilibrium points, [Ku].

THEOREM 2.21
Consider the difference equation (2.1) where $D \subset R^2$ and $\mathbf{f} : D \to D$ is continuous. Suppose that $\overline{x}$ is an equilibrium point and that $I : R^2 \to R$ is a continuous

invariant of (2.1). If I attains an isolated local minimum (respectively, maximum) value at $\overline{\mathbf{x}}$, then the function

$$L(x) := I(x) - I(\overline{x}) \quad (\textit{respectively,} \quad L(x) := -I(x) + I(\overline{x}))$$

is a Lyapunov function. Consequently, the equilibrium $\overline{x}$ is stable.

We present next two applications of Theorem 2.21.

Example 2.12 Lyness' equation (see [KoĹ], [KLTT], and [Z]):

$$x_{n+1} = \frac{a + x_n}{x_{n-1}},$$

where $a > 0$ is a parameter and $x_1 > 0, x_0 > 0$ has an invariant:

$$I(x_n, x_{n-1}) = \left(1 + \frac{1}{x_n}\right)\left(1 + \frac{1}{x_{n-1}}\right)(a + x_n + x_{n-1}).$$

The equilibrium $p = \frac{1 \pm \sqrt{1+4a}}{2}$ satisfies $p^2 - p - a = 0$.

The necessary conditions for the extremum give

$$\frac{\partial I}{\partial x} = \left(1 + \frac{1}{y}\right)\left(1 - \frac{y + a}{x^2}\right) = 0,$$

$$\frac{\partial I}{\partial y} = \left(1 + \frac{1}{x}\right)\left(1 - \frac{x + a}{y^2}\right) = 0,$$

which leads to $x = y$ and to the equation $x^2 - x - a = 0$. Then the critical points are exactly the equilibrium points. The Hessian at the positive critical point $p = \frac{1+\sqrt{1+4a}}{2}$ is

$$H = \begin{pmatrix} A & B \\ B & C \end{pmatrix},$$

where

$$A = \frac{\partial^2 I}{\partial x^2}(p, p) = 2\frac{(p+1)(p+a)}{p^4},$$

$$B = \frac{\partial^2 I}{\partial x \partial y}(p, p) = \frac{a-2p^2}{p^4},$$

$$C = \frac{\partial^2 I}{\partial y^2}(p, p) = A.$$

Clearly $A > 0$, and

$$\det H = AC - B^2 = (3a + 2(a+1)p)\left(a + 2(a+1)p + 4p^2\right)p^{-8} > 0,$$

which implies that the invariant I attains a minimum at (p, p). Then,

$$\min\{I(x, y) : (x, y) \in D\} = I(p, p) = \frac{(p+1)^2(a+2p)}{p^2},$$

and by Theorem 2.21 we have that

$$V(x, y) = I(x, y) - \frac{(p+1)^2(a+2p)}{p^2}$$

is a Lyapunov function and that p is stable. See Section 2.15 for a more complete study of Lyness' equation. ∎

Example 2.13 The Gumovski-Mira equation [GM], [Mi], and [La]

$$x_{n+1} = \frac{2ax_n}{1+x_n^2} - x_{n-1}, \tag{2.63}$$

where $a > 1$ is a parameter, has the invariant

$$I(x_n, x_{n-1}) = x_n{}^2 x_{n-1}{}^2 + x_n{}^2 + x_{n-1}{}^2 - 2ax_n x_{n-1}. \tag{2.64}$$

Equation (2.63) has three equilibrium points $p = \sqrt{a-1} > 0$ and $p_- = -\sqrt{a-1} < 0$ if $a > 1$ and 0 for all values of the parameter a. Here we study the stability of p.

The necessary conditions for the extremum of the invariant (2.64) give

$$\frac{\partial I}{\partial x} = 2xy^2 + 2x - 2ay = 0,$$

and

$$\frac{\partial I}{\partial y} = 2x^2y + 2y - 2ax = 0,$$

which immediately lead to $x = y = p$. This shows that the critical point is equal to the equilibrium point. The Hessian at the point (p, p) is

$$H = \begin{pmatrix} A & B \\ B & A \end{pmatrix},$$

where

$$A = \frac{\partial^2 I}{\partial x^2}(p, p) = \frac{\partial^2 I}{\partial y^2}(p, p) = 2p^2 + 2,$$

and

$$B = \frac{\partial^2 I}{\partial x \partial y}(p, p) = 4p^2 - 2a.$$

Clearly $A > 0$, and

$$det\ H = A^2 - B^2 = (A-B)(A+B) = 4(a+1-p^2)(3p^2+1-a) = 16(a-1) > 0.$$

This shows, that the Hessian H is positive definite at (p, p), which implies that the invariant I attains a minimum at (p, p). Then,

$$\min \{I(x, y) : (x, y) \in D\} = I(p, p) = -(a-1)^2,$$

and by Theorem 2.21, the function

$$V(x, y) = I(x, y) + (a-1)^2,$$

is a Lyapunov function and p is stable. ∎

2.15 *Dynamica* Session on Lyness' Map

In this section we use *Dynamica* to study Lyness' Equation

$$x_{n+1} = \frac{A + x_n}{x_{n-1}} \tag{2.65}$$

We shall classify fixed points, find basic characteristics of semicycles such as maxima and length of positive semicycles and minima and length of negative semicycles, find invariants and corresponding Lyapunov functions, and plot bifurcation diagrams.

Equation (2.65) was discovered by Lyness in 1942 [L1], [L2] in connection with a problem in number theory. Later, the same equation appeared in frieze patterns (see Conway and Coxeter [CC]). See also [KL], pp. 133–134. Two discoveries of Lyness were the periodicity with period 5 of all solutions for the value of constant $A = 1$ and the invariant

$$I(x_n, x_{n-1}) = \left(1 + \frac{1}{x_n}\right)\left(1 + \frac{1}{x_{n-1}}\right)\left(A + x_n + x_{n-1}\right) \tag{2.66}$$

which remains constant along the solutions of this equation. In this section we provide the computer-aided proofs of both statements.

In recent years this equation has attracted a lot of attention from specialists in the field of discrete dynamical systems and difference equations. One reason for this interest is the rich dynamics that this equation possesses. All known results have been obtained in the case of the nonnegative value of parameter A, and, with the exception of [FJL] and [Ku], for positive initial conditions.

In [KoL] and [KLR] the boundedness of all solutions and the periodicity of all solutions of (2.65) were resolved, and basic characteristics of semicycles were found. The fact that no solution except equilibrium has a limit was discovered in [GJKL].

In this section we provide the computer-aided proof of stability of positive equilibrium by finding the Lyapunov function. There are still many open problems related to Lyness' equation. Some of them are formulated in [KL] and some of them are mentioned at the end of this section.

Let us consider the problem of finding and analyzing the equilibrium (fixed) points of Lyness' map, which has been defined in *Dynamica* in terms of the parameter A.

Lyness' map is already defined in *Dynamica*

```
In[1]:= <<Dynamica`;

In[2]:= LynessMap[x,y]
Out[2]= {y, (A + y)/x}
```

We have two fixed points.

```
In[3]:= Solve[LynessMap[x, y] == x, y, x, y]
```

$$Out[3]= \left\{\left\{y \to \frac{1}{2}\left(1-\sqrt{1+4A}\right),\ x \to \frac{1}{2}\left(1-\sqrt{1+4A}\right)\right\},\ \left\{y \to \frac{1}{2}\left(1+\sqrt{1+4A}\right),\ x \to \frac{1}{2}\left(1+\sqrt{1+4A}\right)\right\}\right\}$$

```
In[4]:= {fp1, fp2} = {x, y} /. %;
```

This is the Jacobian matrix of the map at `fp1`.

```
In[5]:= MatrixForm[jaclyn1]
```

$$Out[5]= \begin{pmatrix} 0 & 1 \\ -1 & -\frac{2}{-1+\sqrt{1+4A}} \end{pmatrix}$$

The eigenvalues of the Jacobian at `fp1`.

```
In[6]:= ev = Eigenvalues[jaclyn1]
```

$$Out[6]= \left\{\frac{-1-\sqrt{-1-4A+2\sqrt{1+4A}}}{-1+\sqrt{1+4A}},\ \frac{-1+\sqrt{-1-4A+2\sqrt{1+4A}}}{-1+\sqrt{1+4A}}\right\}$$

The characteristic polynomial shows that the product of the eigenvalues is one.

```
In[7]:= CharacteristicPolynomial[
          jaclyn1, x]
```

$$Out[7]= 1+\frac{2x}{-1+\sqrt{1+4A}}+x^2$$

To determine whether the eigenvalues are real or complex, we may investigate the sign of the discriminant of the characteristic polynomial.

$$In[8]:= \texttt{disc1} = \left(\frac{2}{-1+\sqrt{1+4A}}\right)^2 - 4;$$

A plot shows that the discriminant may be negative or positive, depending on A.

```
In[9]:= Plot[disc1, {A, 0, 2},
          PlotRange → {-4, 4}];
```

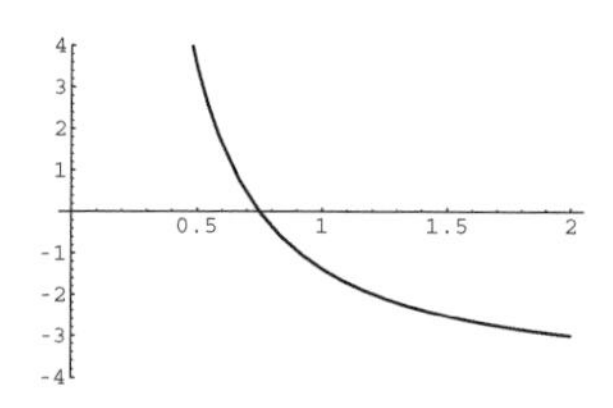

The discriminant is zero at these values of A. Since A is positive to begin with, this leaves only one value of A where the change of sign occurs.

```
In[10]:= Solve[disc == 0, A]
```

$$Out[10]= \left\{\left\{A \to -\frac{1}{4}\right\}, \left\{A \to \frac{3}{4}\right\}\right\}$$

These are the eigenvalues of the Jacobian at `fp1` for the critical value of A.

```
In[11]:= ev /. A → 3/4
Out[11]= {-1, -1}
```

The fixed point `fp1` for the critical value of A.

```
In[12]:= fp1 /. A → 3/4
```

$$Out[12]= \left\{-\frac{1}{2}, -\frac{1}{2}\right\}$$

We conclude the following about the Jacobian of the map at `fp1`.

i. If $A < \frac{3}{4}$, then there are two real eigenvalues, one inside and the other one outside the unit circle. Thus `fp1` is a saddle.

ii. If $A = \frac{3}{4}$, then both eigenvalues are equal to -1. Thus `fp1` is a hyperbolic fixed point of parabolic type, see Section 2.18.

iii. If $A > \frac{3}{4}$, then the eigenvalues form a complex-conjugate pair. Since their product is equal to 1, then both eigenvalues are on the unit circle. Consequently, `fp1` is a hyperbolic fixed point of elliptic type, see Section 2.18.

This is the characteristic polynomial of the Jacobian of the map at `fp2`.

```
In[13]:= CharacteristicPolynomial[
             jaclyn2, x]
```

$$Out[13]= 1 - \frac{2\,x}{1 + \sqrt{1 + 4\,A}} + x^2$$

The discriminant of the characteristic polynomial is clearly negative for all values of $A > 0$. Thus we conclude that the point `fp2` is nonhyperbolic of elliptic type.

$$In[14]:= \text{disc2} = \text{Together}\left[\left(\frac{2}{1 + \sqrt{1 + 4\,A}}\right)^2 - 4\right]$$

$$Out[14]= -\frac{4\left(1 + 4\,A + 2\sqrt{1 + 4\,A}\right)}{\left(1 + \sqrt{1 + 4\,A}\right)^2}$$

As an example, we choose a specific value for A. These are the fixed points.

```
In[15]:= A = 2.;

In[16]:= {fp1, fp2}
Out[16]= {{-1., -1.}, {2., 2.}}
```

Note that the eigenvalues have modulus 1.

```
In[17]:= Eigenvalues[jaclyn1]
Out[17]= {-0.5 + 0.866025 i, -0.5 - 0.866025 i}

In[18]:= Abs[%]
Out[18]= {1., 1.}

In[19]:= Eigenvalues[jaclyn2]
Out[19]= {0.25 + 0.968246 i, 0.25 - 0.968246 i}

In[20]:= Abs[%]
Out[20]= {1., 1.}
```

Note that a linearized stability analysis does not give any information about the stability of nonhyperbolic equilibria or asymptotic behavior of solutions. To study the stability of Lyness' equation one must use higher-order nonlinear terms, invariants, Lyapunov functions, etc. The study of stability by using higher-order nonlinear terms in Birkhoff normal form is accomplished with KAM–theory, see [HK] and [KLTT]. The study of stability by using invariants and Lyapunov functions was accomplished in [Ku] and [Z], and will be demonstrated later in this session. Next, we present some numerical explorations.

Here we generate an orbit of Lyness' map with 10 points that has $(0,0)$ as the initial point. Recall that $A = 2$ for now.

```
In[21]:= orb = Orbit[LynessMap, 1., 2., 100];

In[22]:= Take[orb, {1, 10}]
Out[22]= {{1., 2.}, {2., 4.}, {4., 3.},
          {3., 1.25}, {1.25, 1.08333},
          {1.08333, 2.46667},
          {2.46667, 4.12308},
          {4.12308, 2.48233},
          {2.48233, 1.08713},
          {1.08713, 1.24364}}
```

This is a plot of 100 points of the solution.

```
In[23]:= OrbitPlot[orb];
```

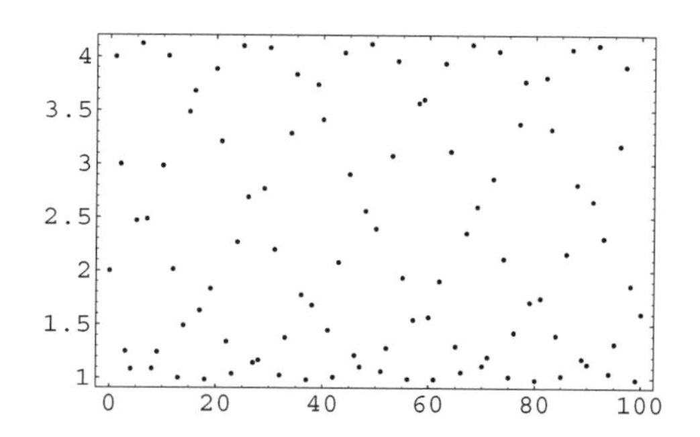

This is the corresponding time series plot.

```
In[24]:= TimeSeriesPlot[orb];
```

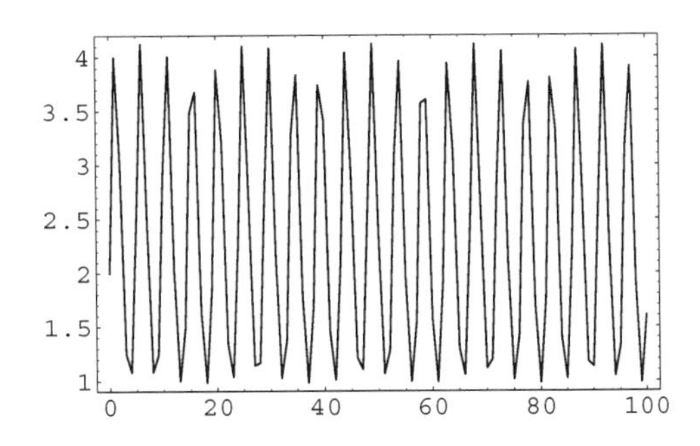

A phase portrait plot shows that the orbit is contained in what appears to be a smooth curve.

```
In[25]:= PhasePortrait[orb,
              PlotRange → {{0, 5}, {0, 5}}]
```

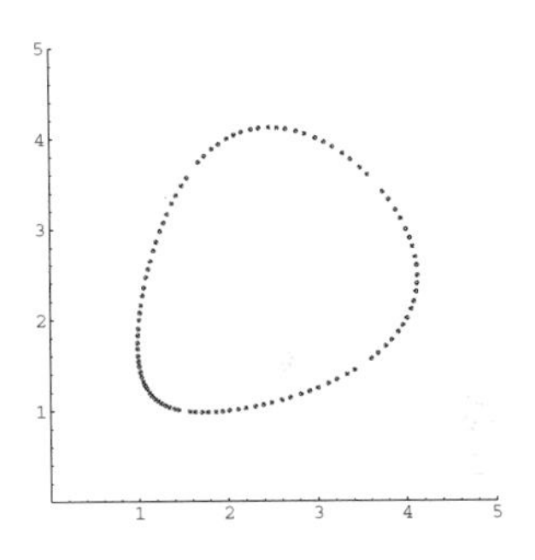

Now we find a rational invariant for Lyness' equation, which we shall use to find the corresponding Lyapunov function.

Begin by clearing the constant A.

```
In[26]:= Clear[A];
```

Lyness' equation has already been defined in *Dynamica*.

```
In[27]:= Lyness
```

$$\text{Out[27]= } x[1+n] == \frac{A + x[n]}{x[-1+n]}$$

An invariant for the equation has the form of a rational function of `x[n]` and `x[n-1]`.

```
In[28]:= RationalInvariant[Lyness]
```

$$\text{Out[28]= } c[1] + \frac{(1+A)\, c[7]}{x[-1+n]} + c[7]\, x[-1+n] + \frac{(1+A)\, c[7]}{x[n]} + \frac{A\, c[7]}{x[-1+n]\, x[n]} + \frac{c[7]\, x[-1+n]}{x[n]} + c[7]\, x[n] + \frac{c[7]\, x[n]}{x[-1+n]}$$

`RationalInvariant` succeeded in finding invariants. We may set `c[7]` to be one (this is just a normalization). Also, following Lyness in [L1], we set `c[1] = A+2`, to obtain an invariant that can be factored.

```
In[29]:= invLyness = %/.{c[7] → 1, c[1] → A+2}
```

$$\text{Out[29]= } 2 + A + \frac{1 + A}{x[-1+n]} + x[-1+n] + \frac{1 + A}{x[n]} + \frac{A}{x[-1+n]\,x[n]} + \frac{x[-1+n]}{x[n]} + x[n] + \frac{x[n]}{x[-1+n]}$$

For convenience, we write the invariant in terms of variables x and y.

```
In[30]:= invLyness = invLyness /.
            {x[n] → x, x[n - 1] → y}
```

$$\text{Out[30]= } 2 + A + \frac{1 + A}{x} + x + \frac{1 + A}{y} + \frac{A}{x\,y} + \frac{x}{y} + y + \frac{y}{x}$$

The invariant is indeed factorable.

```
In[31]:= Factor[invLyness]
```

$$\text{Out[31]= } \frac{(1 + x)\,(1 + y)\,(A + x + y)}{x\,y}$$

```
In[32]:= invLyness /. A → 2.
```

$$\text{Out[32]= } 4. + \frac{3.}{x} + x + \frac{3.}{y} + \frac{2.}{x\,y} + \frac{x}{y} + y + \frac{y}{x}$$

Here is a plot of the invariant for $A = 2$.

```
In[33]:= Plot3D[Factor[invLyness] /.
           A → 2., {x, 0.001, 5},
          {y, 0.001, 5},
          PlotRange → {0, 100},
          PlotPoints → 30];
```

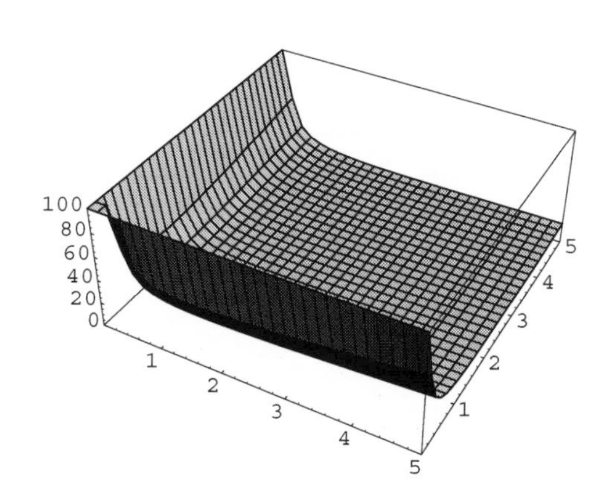

`FixedPointTest` is applied here to determine whether local minima and/or the maxima of the invariant are attained at the equilibrium point(s).

```
In[34]:= FixedPointTest[invLyness,
            LynessMap[{x,y}], {x,y}]
The point number  1  is
{- (1 - Sqrt[1 + 4 A]), - (1 - Sqrt[1 + 4 A])}
Principal Minors of the Hessian for this
point:
d1[  1  ] =  -(16 (A (-5 + Sqrt[1 + 4 A]) + 2
(-1 + Sqrt[1 + 4 A])))/(-1 + Sqrt[1 + 4 A])
d2[  1  ] =  -(256 (-4 A + 6 (-1 + Sqrt[1 + 4
A]) + A (-41 + 10 Sqrt[1 + 4 A]) + A (-34 + 22
Sqrt[1 + 4 A])))/(-1 + Sqrt[1 + 4 A])
1 1 The point number  2  is
{- (1 + Sqrt[1 + 4 A]),  - (1 + Sqrt[1 + 4 A])}
Principal Minors of the Hessian for this
point:
d1[  2  ] =  (16 (2 (1 + Sqrt[1 + 4 A]) + A (5
+ Sqrt[1 + 4 A])))/(1 + Sqrt[1 + 4 A])
d2[  2  ] =  (256 (4 A + 6 (1 + Sqrt[1 + 4
A]) + A (41 + 10 Sqrt[1 + 4 A]) + A (34 + 22
Sqrt[1 + 4 A])))/(1 + Sqrt[1 + 4 A])
```

From the output above we see that since `d1[2]` and `d2[2]` are both positive for positive A, then the invariant has a strict local minimum at the point `fp[2]`.

A plot of `d1[2]` for values of A between $-1/4$ and 0 helps to determine positivity.

```
In[35]:= Plot[d1[2], {A, -1/4, 0.1},
            PlotRange → {0, 10}];
```

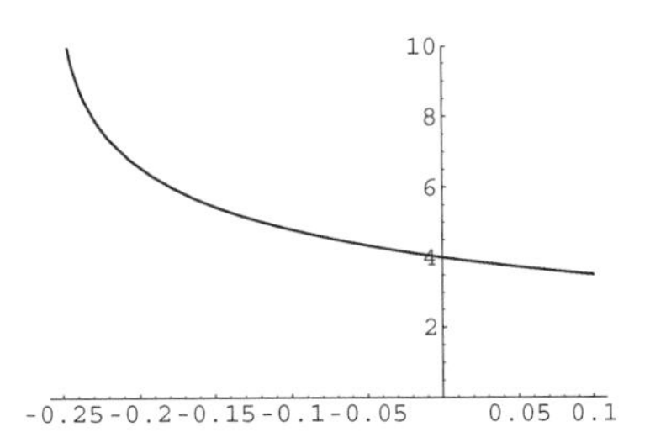

A plot of `d2[2]` for values of A between $-1/4$ and 0 shows that this quantity is positive.

```
In[36]:= Plot[d2[2], {A, -1/4, 0.1}];
```

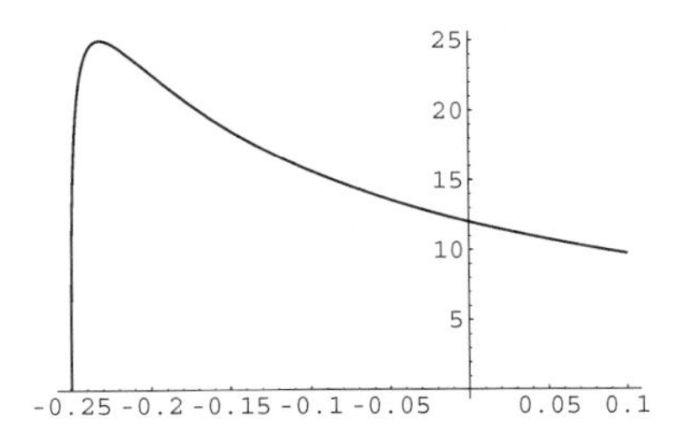

Therefore, `fp[2]` is a strict local minimum for all $A > -\frac{1}{4}$.

A Lyapunov function is computed as follows.

```
In[37]:= Clear[A];

In[38]:= lyapunovfunction =
            Simplify[invLyness-
   (invLyness/.{x → fp[2][[1]],
            y → fp[2][[2]]})]
```

Out[38]= $-3-\sqrt{1+4A}-\frac{4A}{\left(1+\sqrt{1+4A}\right)^2}-\frac{4(1+A)}{1+\sqrt{1+4A}}+\frac{1+A}{x}+x+\frac{1+A}{y}+\frac{A}{xy}+\frac{x}{y}+y+\frac{y}{x}$

This is a plot of the Lyapunov function for the case $A = 2$ and fixed point $(2, 2)$.

```
In[39]:= A = 2.0;

In[40]:= fp[2]
Out[40]= {2., 2.}

In[41]:= lf = lyapunovfunction
```

Out[41]= $-9.5+\frac{3.}{x}+x+\frac{3.}{y}+\frac{2.}{xy}+\frac{x}{y}+y+\frac{y}{x}$

```
In[42]:= Plot3D[lf, {x, .4, 3.}, {y, .4, 3.}];
```

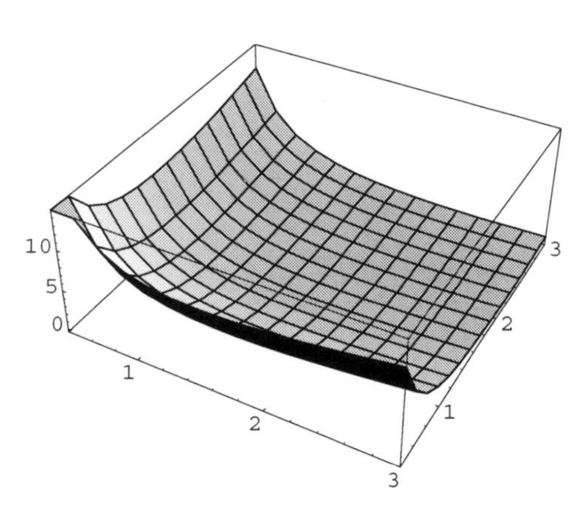

This is a contour plot for the case $A = 2$ and fixed point $(2, 2)$.

```
In[43]:= ContourPlot[lf, {x, .4, 3.},
         {y, .4, 3.}, Contours → 18,
         ContourShading → False]
```

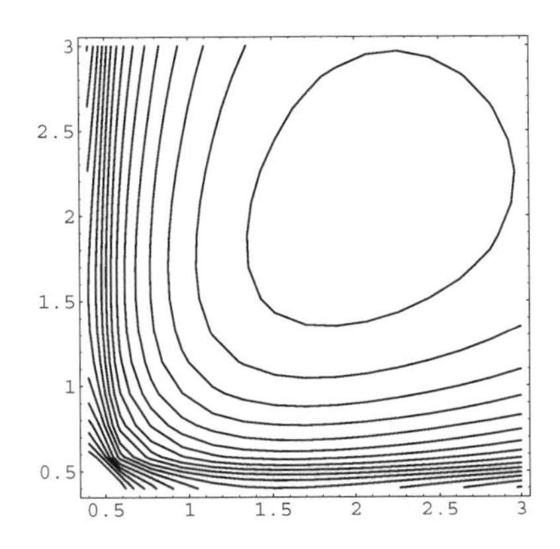

As an exercise, the reader may perform a stability analysis for `fp[1]`.

Now we study briefly the special case $A = 1$, and then follow with simulations for finding maxima and minima of positive and negative semicycles of solutions of (2.65). We remind the reader that the case $A = 1$ is special in that all solutions are periodic with period 5.

The case $A = 2$ is now considered. Here is the output of the `SemicycleTest` function applied to the orbit consisting of 20 points (the number of points chosen here is a small number to save space).

```
In[44]:= A = 2.;

In[45]:= orb1 = Orbit[LynessMap,
              2., 1.7, 20];

In[46]:= SemicycleTest[orb1, fp[2][[2]]]
Positions and Maxima of Positive Semicycles
{5, {5, 2.30524}, {9, 2.32960`}, {14, 2.36302},
{19, 2.36058}}
Positions and Minima of Negative Semicycles
{{2, 1.7`}, {7, 1.69223}, {12, 1.711572}, {16,
1.74947}}
Lengths of Positive Semicycles
{1, 2, 2, 3, 3}
Lengths of Negative Semicycles
{-2, -3, -2, -2, -1}
Positions and Lengths of both Positive and
Negative Semicycles
{1, -2, 2, -3, 2, -2, 3, -2, 3, -1}
```

This is time series plot of 40 points of the solution.

```
In[47]:= TimeSeriesPlot[LynessMap,
           {2.,1.7},40,,Axes → True,
           AxesOrigin → {0,2}];
```

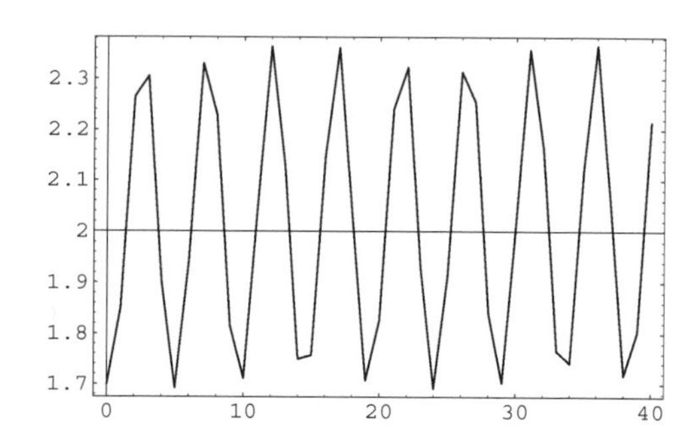

So all semicycles have length 2 or 3. If many simulations of this kind were performed, it would be the case that the semicycles of all solutions have length 2 or 3 and that the maxima of positive semicycles and minima of negative semicycles have a special location. This fact, proved in [KL], may be used to obtain some global properties of the solutions of this equation.

The complex behavior of Lyness' equation for different values of the parameter A can be seen from the bifurcation diagrams. We begin by resetting the value of A as p, which is the variable used as parameter by `BifurcationPlotND`.

This bifurcation plot for parameter values near 1 is interesting.

```
In[48]:= BifurcationPlotND[LynessMap,
           {A, 0.95, 1.05},{1.2,1.},
           Steps → 300,
           Iterates → 150,
           FirstIt → 400];
```

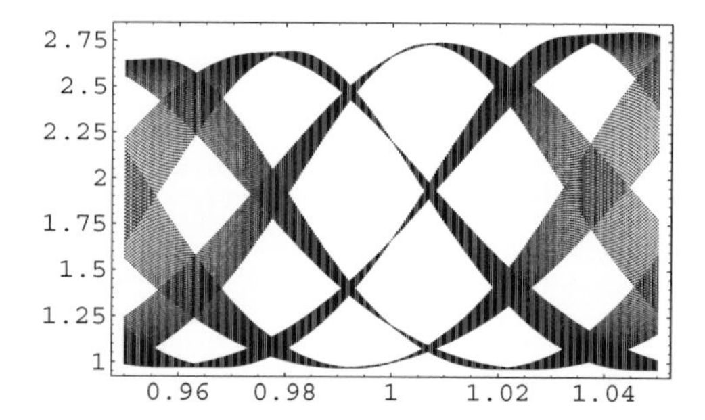

Zooming in near 1 gives visual confirmation of the period-five character of solutions when $A = 1$.

```
In[49]:= BifurcationPlotND[LynessMap,
           {A, 0.99, 1.01}, {1.2, 1.},
           Steps → 300,
           Iterates → 150,
           FirstIt → 400];
```

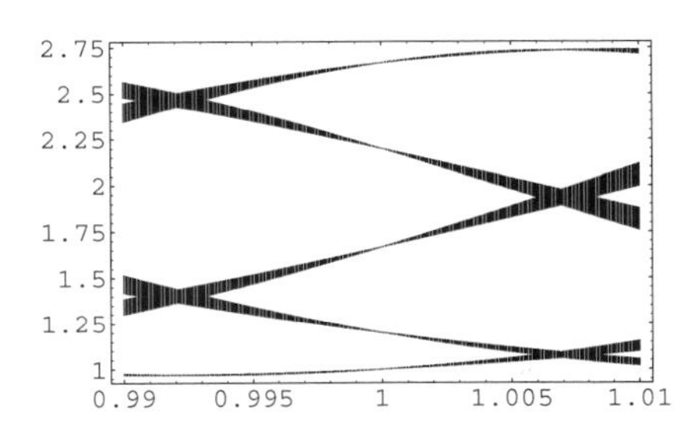

Using *Dynamica* `PoincarePlot2D` function we can show that the Lyness' map, with the value of parameter $A = 1$, have all solutions periodic with period 5.

Here we apply 10 times `PoincarePlot2D` on an initial circle to show the evolution of this circle under Lyness' map. We see only five different shapes because every solution is periodic with period five.

```
In[50]:= Lyness[{x_, y_}] := {y, (1 + y) /x};

In[51]:= PoincarePlot2D[Lyness, 10,
           Center → (1, 1)];
```

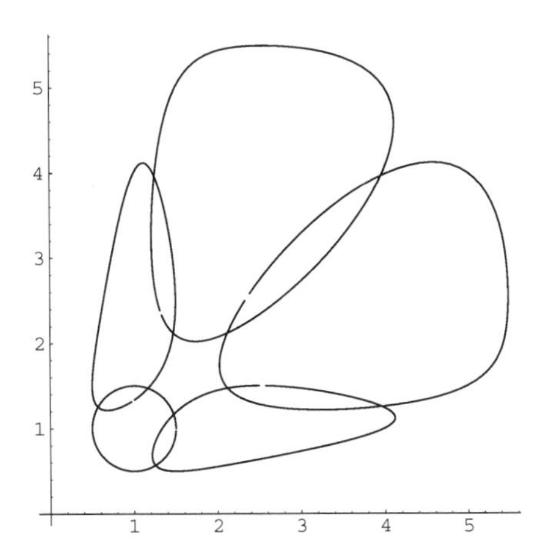

This plot suggests that there is complicated behavior for $A = 1.4$.

```
In[52]:= A = 1.4;

In[53]:= L1 = LyapunovNumbers[LynessMap,
            {1., 1.}, 400];

In[54]:= ListPlot[L1, PlotJoined → True];
```

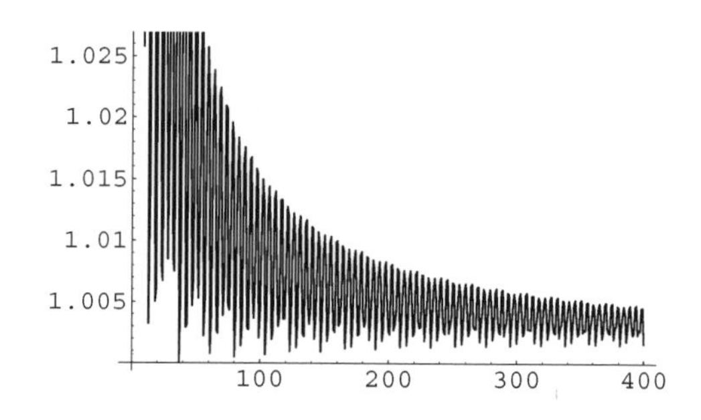

The case $A = 2$ also has complicated behavior.

```
In[55]:= A = 2.0;

In[56]:= L2 = LyapunovNumbers[LynessMap,
            {1., 1.}, 400];

In[57]:= ListPlot[L2, PlotJoined → True];
```

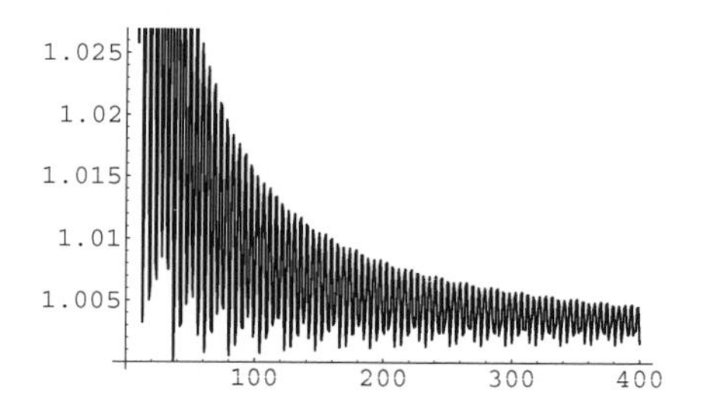

Here we calculate the box dimension of Lyness' map. The answer suggests possibly complicated behavior.

```
In[58]:= A = 4.;

In[59]:= BoxDimension[LynessMap[{x, y}],
            7000, .001, {1., 2.}]
Out[59]= 1.28172
```

Now we give an overview of the results and observations about Lyness' equation that were obtained so far. Some of these results have been obtained by using *Dynamica*. We assume that the parameter A and the initial values are positive.

1. Both equilibrium points of Lyness' equation are nonhyperbolic equilibrium points of the elliptic type.

2. Lyness' equation possesses an invariant (established in [L1]).

3. Both equilibrium points of Lyness' equation are stable (established in [KLTT] for positive equilibrium and in [Ku] for negative equilibrium as well. See also Section 2.14.

4. Every positive solution of Lyness' equation, with $A = 1$, is periodic with period 5 (established in [L1]).

5. Every positive nontrivial solution of Lyness' equation is strictly oscillatory about the positive equilibrium. Furthermore, every semicycle of a nontrivial solution contains either two or three terms (established in [KoL] and [KLR]).

6. For all $A > 0$, only the equilibrium solution have a limit (established in [GJKL]).

It is interesting to note that we were able to obtain results from 1 to 4 by using *Dynamica.*

The following is a list of possible research problems.

1. Find all possible periods of solutions of Lyness' equation for different values of parameter A (see [Z] for some results in this direction).

2. Investigate the stability nature of the periodic solutions.

3. Determine for which values of A the solution of Lyness' equation is chaotic.

4. Investigate the existence of solutions and the stability of the equilibrium points for negative values of parameters and initial conditions.

2.16 Dissipative Maps and Systems

In this section we discuss a special class of difference equations systems and the corresponding maps known as **dissipative systems** and **maps**. Roughly speaking, a dissipative system has the property that there exists a bounded subset B of R^2 that the positive orbit of every initial point eventually enters. The term dissipative is motivated by the class of problems in physics for which there is a loss of energy phenomenon, see [HK], [S1] and [SH]. The formal definition is the following.

DEFINITION 2.22 *A map* $\mathbf{F} : R^2 \to R^2$ *is said to be* **point dissipative** *if there is a bounded set* B *with the property that for every* $x \in R^2$*, there is a* *positive integer* $n_0 = n_0(x)$ *such that* $\mathbf{F}^n(\mathbf{x}) \in B$ *for all* $n \geq n_0$. *The set* B *is called an* **absorbing set**.

This definition extends in a straightforward way to the general case of maps $\mathbf{F} : R^k \to R^k$. The property that all positive orbits $\gamma^+(\mathbf{x}_{n_0})$ belong to the bounded set B is equivalent to the existence of **absorbing** or **attracting** intervals for second order difference equations.

DEFINITION 2.23 *An* **invariant interval** *for the difference equation*

$$x_{n+1} = f(x_n, x_{n-1}), \quad n = 0, 1, \ldots \tag{2.67}$$

is an interval I *with the property that if two consecutive terms of a solution fall in* I, *then all the subsequent terms of the solution also belong to* I.

An **absorbing** *or* **attracting** *interval for the difference equation (2.67) is an invariant interval* J *with the property that all solutions of (2.67) eventually enter* J.

In other words, I is an invariant interval for the difference equation (2.67) if $x_{N-1}, x_N \in I$ for some $N \geq 0$, then $x_n \in I$ for every $n > N$. One can similarly define **invariant set** for system (2.1). The difference equation (2.67) is called **dissipative** if it possesses an attracting interval. The dynamics of dissipative maps can be very simple, such as global attractivity of the equilibrium, or very complex such as chaos [HK] and [KL]. From a book [KL] and in a series of papers referred therein, general stability results for equation (2.67) and system (2.1) were used. Here we mention some of these results. The first two results were established in [KLS] and [KL].

THEOREM 2.22
Let $[a, b]$ *be an interval, and suppose that* $f : [a, b] \times [a, b] \to [a, b]$ *is a continuous function that has the following properties:*

a. $f(x, y)$ *is nondecreasing in* $x \in [a, b]$ *for each* $y \in [a, b]$, *and* $f(x, y)$ *is nonincreasing in* $y \in [a, b]$ *for each* $x \in [a, b]$;

b. *All solutions* $(m, M) \in [a, b] \times [a, b]$ *of the system*

$$f(x, y) = x \quad \text{and} \quad f(y, x) = y,$$

   satisfy $m = M$.

Then (2.67) has a unique equilibrium $\bar{x} \in [a, b]$ *and every solution of (2.67) that enters* $[a, b]$ *converges to* $\bar{x}$.

THEOREM 2.23
Let $[a, b]$ *be an interval, and suppose that* $f : [a, b] \times [a, b] \to [a, b]$ *is a continuous function that has the following properties.*

a. *$f(x, y)$ is nonincreasing in $x \in [a, b]$ for each $y \in [a, b]$, and $f(x, y)$ is nondecreasing in $y \in [a, b]$ for each $x \in [a, b]$.*

b. *The difference equation (2.67) has no solutions of prime period two in $[a, b]$.*

Then (2.67) has a unique equilibrium $\overline{x} \in [a, b]$, and every solution of (2.67) that enters $[a, b]$ converges to $\overline{x}$.

Results of this kind have wide applicability to a large class of equations including rational equations such as

$$x_{n+1} = \frac{\alpha + \beta x_n + \gamma x_{n-1}}{A + B x_n + C x_{n-1}},$$

see [KL], [KLP1], [KLP2], and [KLS].

Note that if equation (2.67) is dissipative and has $[a, b]$ as an absorbing interval, then Theorems 2.22 and 2.23 are global results, that is, all solutions of equation (2.67) converge to the equilibrium. A similar result, valid for systems, has been proved in [KN1].

Very often, the best strategy for obtaining global attractivity results for (2.1) is to work in the regions where the functions $f(x, y)$ and $g(x, y)$ are monotonic in their arguments. In this direction there are sixteen theorems depending on the monotonic character of the functions f and g. Here we present two of these theorems from [KN1].

THEOREM 2.24

Let $[a, b]$ be an interval and let $f, g : [a, b] \times [a, b] \to [a, b]$ be continuous functions that have the following properties.

a. *$f(x, y)$ and $g(x, y)$ are nondecreasing functions in each of their arguments;*

b. *The system of equations (2.1) has a unique positive equilibrium $(\overline{x}, \overline{y})$.*

Then every positive solution of system (2.1) converges to $(\overline{x}, \overline{y})$.

This theorem has been succesfully applied to the convergence problem of solutions of the system

$$x_{n+1} = \frac{a + x_n}{b + c x_n + y_n}$$
$$y_{n+1} = \frac{d + x_n}{e + x_n + f y_n}, \tag{2.68}$$

where a, b, c, d, e, f and the initial conditions x_0, y_0 are nonnegative constants such that the solution is defined for all $n \geq 0$, see [KN1].

A similar result is the following.

THEOREM 2.25
Let $[a,b]$ be an interval and let

$$f, g : [a,b] \times [a,b] \to [a,b]$$

be continuous functions that have the following properties.

a. *$f(x,y)$ is nondecreasing in $x \in [a,b]$ for each $y \in [a,b]$, and it is non-increasing in $y \in [a,b]$ for each $x \in [a,b]$; $g(x,y)$ is nonincreasing in $x \in [a,b]$ for each $y \in [a,b]$ and it is nondecreasing in $y \in [a,b]$ for each $x \in [a,b]$.*

b. *All solutions of the system*

$$\begin{cases} m = f(m, M'), & m' = f(m', M) \\ M = g(m', M), & M' = g(m, M') \end{cases} \tag{2.69}$$

satisfy $m = m'$ and $M = M'$.

Then (2.1) has a unique positive equlibrium $(\bar{x}, \bar{y}) \in [a,b] \times [a,b]$ and every positive solution of (2.1) converges to $(\bar{x}, \bar{y})$.

Note that if (2.1) is dissipative, then Theorems 2.24 and 2.25 are global results, that is, all solutions of (2.1) converge to the equilibrium.

2.17 *Dynamica* Session on a Rational Difference Equation

In this section we investigate the rational difference equation

$$x_{n+1} = \frac{Ax_n + x_{n-1}}{Bx_n + x_{n-1}} \tag{2.70}$$

for positive values of A and B, and for positive x_0, x_1. This equation was studied in [KLS] and [KL].

Equation (2.70) is dissipative. To see this, set

$$f(x,y) = \frac{Ax + y}{Bx + y},$$

and note that if $A < B$ the function $f(x,y)$ is decreasing in x and increasing in y for positive x and y. Clearly,

$$\frac{A}{B} = f(x,0) \leq f(x,y) \leq f(0,y) = 1, \quad \text{for all } x, y > 0.$$

Hence, for any (x, y), we have that $f(x, y)$ belongs to the interval $[\frac{A}{B}, 1]$. That is, $\left[\frac{A}{B}, 1\right]$ is an absorbing interval. Likewise, if $A > B$ the function $f(x, y)$ is increasing in x and decreasing in y, therefore

$$1 = f(0, y) \leq f(x, y) \leq f(x, 0) = \frac{A}{B}, \quad \text{for all } x, y > 0.$$

and we conclude that $\left[1, \frac{A}{B}\right]$ is an absorbing interval. Finally, if $A = B$, then $f(x_0, y_0) = 1$ for all (x_0, y_0). Therefore $\{1\}$ is an absorbing set.

We conclude that equation (2.70) is dissipative for all positive values of A and B.

Next we use *Dynamica* to study local stability of equilibria and prime period-two solutions of equation (2.70).

2.17.1 Stability analysis of the equilibrium

To find and analyze the equilibrium points of equation (2.70), we need first to define the corresponding map.

```
In[1]:= << Dynamica`

In[2]:= ration = x[n + 1] ==
              A x[n] + x[n - 1])
              ------------------ ;
              B x[n] + x[n - 1]

In[3]:= rationmap = DEToMap[ration];

In[4]:= rationmap[{x, y}]
Out[4]= {y, (x + A y)/(x + B y)}
```

There is only one fixed point.

```
In[5]:= Solve[rationmap[x, y] == x, y, x, y]
Out[5]= {{x -> (1 + A)/(1 + B), y -> (1 + A)/(1 + B)}}

In[6]:= fp = {x, y} /. %[[1]];
Out[6]= {(1 + A)/(1 + B), (1 + A)/(1 + B)}
```

The location with respect to the unit circle of the eigenvalues of the Jacobian at the fixed point determines the stability type of the equation. Here we see that the formula of the eigenvalues in terms of A and B is complicated. Thus, rather than using the formula for the eigenvalues, we shall study stability by applying Theorem 2.12.

```
In[7]:= rationjac = Simplify[
            Jac[rationmap, fp]];

In[8]:= evals = Eigenvalues[rationjac]
Out[8]=
```

$$\left\{\frac{A - B - \sqrt{(-A + B)^2 - 4\,(A - B)\,(1 + A + B + A B)}}{2\,(1 + A + B + A B)},\right.$$
$$\left.\frac{A - B + \sqrt{(-A + B)^2 - 4\,(A - B)\,(1 + A + B + A B)}}{2\,(1 + A + B + A B)}\right\}$$

We shall need the trace and the determinant of the Jacobian J of the map at the fixed point.

```
In[9]:= trjac = Factor[
          rationjac[[1, 1]]+
            rationjac[[2, 2]]]
```

$$Out[9]= \frac{A - B}{(1 + A)\ (1 + B)}$$

```
In[10]:= detjac = Factor[Det[rationjac]]
```

$$Out[10]= \frac{A - B}{(1 + A)\ (1 + B)}$$

By Theorem 2.12, the fixed point is locally asymptotically stable if and only if

$$|\operatorname{tr} J| < 1 + \det J < 2$$

An equivalent expression in terms of A and B is produced easily as shown.

```
In[11]:= << Algebra`InequalitySolve`

In[12]:= InequalitySolve[
           Abs[trjac] < 1 + detjac < 2
             && A > 0 && B > 0, {A, B}]
```

$$Out[12]= 0 < A < 1 \&\& 0 < B < \frac{-1 - 3A}{-1 + A}$$
$$|| A \geq 1 \&\& B > 0$$

The region in the $A - B$ plane where the equilibrium is asymptotically stable may be seen in the plot as the region to the right of the curve.

```
In[13]:= Plot[(-1 - 3 A)/(-1 + A), {A, 0, 1},
           PlotRange → {0, 20},
           AxesLabel → {"A", "B"}]
```

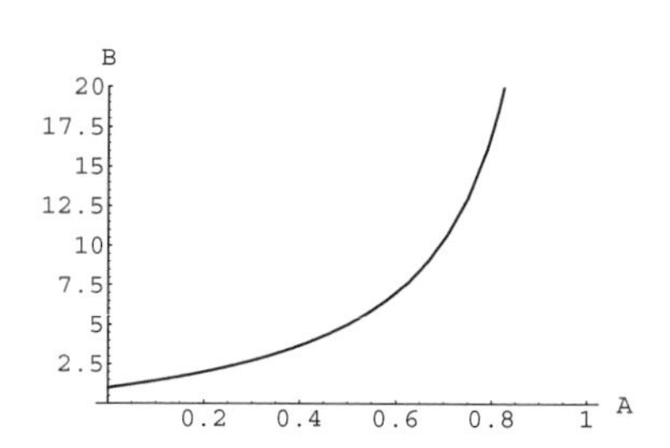

By Theorem 2.12, the fixed point is locally repelling if and only if

$$|\operatorname{tr} J| < 1 + \det J$$

and

$$|\det J| > 1$$

This calculation shows that there are no repelling fixed points.

```
In[14]:= InequalitySolve[
           Abs[trjac] < Abs[1 + detjac ]&&
             Abs[detjac] > 1&&
             A > 0 && B > 0, {A, B}]
Out[14]= False
```

By Theorem 2.12, the fixed point is a saddle equilibrum point if and only if

$$|\mathrm{tr}J| > 1 + \det J$$

and

$$(\mathrm{tr}\,J)^2 - 4(\det J)^2 > 0 \quad .$$

```
In[15]:= InequalitySolve[
           Abs[trjac] > Abs[1 + detjac]
             &&trjac^2 - 4 detjac > 0
             && A > 0 && B > 0, {A, B}]
Out[15]= 0 < A < 1&&B > (-1 - 3 A)/(-1 + A)
```

By Theorem 2.12, the fixed point is nonhyperbolic if and only if one of the following two conditions holds

$$(\mathrm{tr}J)^2 = (1 + \det J)^2$$

or

$$|\det J| = 1$$

```
In[16]:= Solve[trjac^2 == (1 + detjac)^2, A]
Out[16]= {{A -> (-1 + B)/(3 + B)}}
In[17]:= Solve[detjac^2 == 1, A]
Out[17]= { }
```

The reason $|\det J| = 1$ has no solutions in A, B, is that $|\det(J)|$ is always smaller than 1.

```
In[18]:= InequalitySolve[
           Abs[detjac] < 1, {A, B}]
Out[18]= True
```

It is of interest to determine the values of the eigenvalues of J at nonhyperbolic equilibrium points.

```
In[19]:= Eigenvalues[
           rationjac /. {A -> (-1 + B)/(3 + B)}]
Out[19]= {-1, 1/2}
```

The following result collects what has been obtained so far.

PROPOSITION 2.2

Let A, B be positive. Then, equation (2.70) has an equilibrium point, which is unique and given by

$$\bar{x} = \frac{1 + A}{1 + B}$$

Local stability of the equilibrium is given by the following cases.

i. $\bar{x}$ is locally asymptotically stable if and only if

$$0 < A < 1 \quad \textit{and} \quad 0 < B < \frac{-1 - 3A}{-1 + A}, \quad \textit{or}, \quad A \geq 1 \quad \textit{and} \quad B > 0$$

ii. $\bar{x}$ is locally repelling for no values of A and B.

iii. $\bar{x}$ is a saddle point if and only if

$$0 < A < 1 \quad \textit{and} \quad B > \frac{-1 - 3A}{-1 + A}$$

iv. $\overline{x}$ *is nonhyperbolic if and only if*

$$A = \frac{-1+B}{3+B}$$

In this case, the eigenvalues of the Jacobian of the map at $(\overline{x}, \overline{x})$ *are* $\lambda_1 = -1$ *and* $\lambda_2 = \frac{1}{2}$.

As an example, consider specific values of A and B. Condition (i) of Proposition 2.2 implies that the equilibrium is locally asymptotically stable. This can also be seen from the calculation of the eigenvalues of the Jacobian of the map.

```
In[20]:= A = 2; B = 1;

In[21]:= evals
```

$$\text{Out[21]=} \left\{\frac{1}{12}\left(1 - i\sqrt{23}\right), \frac{1}{12}\left(1 + i\sqrt{23}\right)\right\}$$

```
In[22]:= Abs[evals]
```

$$\text{Out[22]=} \left\{\frac{1}{\sqrt{6}}, \frac{1}{\sqrt{6}}\right\}$$

This is another case where (i) of Proposition (2.2) is satisfied.

```
In[23]:= A = 2; B = 3;

In[24]:= evals
```

$$\text{Out[24]=} \left\{-\frac{1}{3}, \frac{1}{4}\right\}$$

2.17.2 Prime Period-Two Solutions

In this section we determine conditions on A and B under which equation (2.70) has prime period-two positive solutions

$$\phi, \psi, \phi, \psi, \ldots \tag{2.71}$$

If (2.70) has a periodic solution (2.71), we may substitute ϕ and ψ in (2.70) to obtain these equations.

$$\text{In[25]:= eq1} = \psi == \frac{\psi + A\phi}{\psi + B\phi};$$

$$\text{In[26]:= eq2} = \phi == \frac{\phi + A\psi}{\phi + B\psi};$$

Solving for ϕ or ψ gives formulas with radicals. Instead, we eliminate one of the variables.

```
In[27]:= eq3 = Eliminate[{eq1, eq2}, φ]
```

$$\text{Out[27]=} A^3 - 2A^2B\psi + A\left(-1 + \psi + B\psi - B\psi^2 + B^2\psi^2\right) == \psi\left(1 - B - 2\psi + B\psi + B^2\psi + \psi^2 - B^2\psi^2\right)$$

We factor the difference of the two sides of eq3. This helps with the analysis.

```
In[28]:= factors =
           Factor[eq3[[2]] - eq3[[1]]]
```

$$\text{Out[28]=} -(1 + A - \psi - B\psi)\left(-A + A^2 - \psi + A\psi + B\psi - AB\psi + \psi^2 - B\psi^2\right)$$

Setting the first factor equal to zero gives, after some calculations, $\phi = \psi$, which is a case we are ruling out. Thus we do not need to further consider the first factor.

```
In[29]:= Solve[1 + A - ψ - B ψ == 0, ψ]
```

$$\text{Out[29]=}\ \left\{\left\{\psi \to \frac{-1-A}{-1-B}\right\}\right\}$$

The second factor is quadratic in ψ.

```
In[30]:= Collect[ - A + A^2 - ψ + A ψ + B ψ - A B ψ+
             ψ^2 - B ψ^2, ψ]
```

$$\text{Out[30]=}\ -A + A^2 + (-1 + A + B - A B)\,\psi + (1 - B)\,\psi^2$$

Note that by symmetry, a similar equation is satisfied by ϕ. We may put together what we have so far to the following statement.

PROPOSITION 2.3

Let A, B, ϕ, ψ be positive, with $\phi \neq \psi$. Then, the sequence

$$\phi, \psi, \phi, \psi, \ldots$$

is a solution to (2.70) if and only if $B \neq 1$ and the quadratic equation

$$z^2 + (A-1)z - \frac{(A-1)A}{B-1} = 0 \tag{2.72}$$

has exactly two solutions $z = \phi$ and $z = \psi$ which are positive and distinct.

Moreover, in either case, we have

$$\psi + \phi = 1 - A \quad \textit{and} \quad \phi\psi = -\frac{(A-1)A}{B-1} \tag{2.73}$$

These are the coefficients of the polynomial $p(z) = z^2 + p_1 z + p_0$ given by the left-hand side of equation (2.72).

```
In[31]:= p1 = - (1 - A);

         p0 = A (A - 1) / (1 - B);
```

The roots of $p(z)$ are positive and distinct if and only if $p_1 < 0$, $p_0 > 0$, and $p_1^2 - 4p_0 > 0$.

```
In[32]:= InequalitySolve[
            p1^2 - 4p0 > 0
             && A > 0 && B > 0
             && p1 < 0 && p0 > 0, {A, B}]
```

$$\text{Out[32]=}\ 0 < A < 1 \,\&\&\, B > \frac{-1-3A}{-1+A}$$

Thus we have the following corollary.

COROLLARY 2.2

If equation (2.70) has a positive prime period-two solution, then

$$0 < A < 1 \quad and \quad B > \frac{1 + 3A}{1 - A} \tag{2.74}$$

Conversely, if (2.74) holds, then equation (2.70) has a unique positive prime period-two solution $\{x_n\}$, where $x_0 = \phi$, and $x_1 = \psi$ are solutions to equation (2.72).

We now turn to the study of the linearized stability of period-two solutions to (2.70). We shall see that all prime period-two solutions are locally asymptotically stable.

PROPOSITION 2.4

Positive, prime period-two solutions to equation (2.70) are locally asymptotically stable.

The second iterate of the map is given by this expression.

```
In[33]:= rationmap2[{x_, y_}] = Simplify[
           rationmap[rationmap[{x, y}]]]
```

$$Out[33]= \left\{\frac{x + A y}{x + B y}, \frac{A x + A^2 y + y (x + B y)}{x y + B (x + y (A + y))}\right\}$$

Here we evaluate the Jacobian of the second iterate of the map at $\{\psi, \phi\}$.

```
In[34]:= J2 = Simplify[
           Jac[rationmap2, {ψ, ϕ}]];
```

Period-two solutions to (2.70) are fixed points of the second iterate of the associated map, `rationmap`. For local asymptotic stability of the period-two solutions, it is required that the eigenvalues of J2 be located inside the unit circle.

To find the locations of the eigenvalues, we compute the characteristic polynomial of J2.

```
In[35]:= charpol =
           CharacteristicPolynomial[J2, t]
```

Note that the characteristic polynomial of J2 has the form

$$p(t) = t^2 + c_1 t + c_0,$$

These are c_1 and c_0.

```
In[36]:= c0 = charpol /. t → 0 //Simplify
```

$$Out[36]= \frac{(A - B)^2 \phi (A \phi + \psi)}{(B \phi + \psi) (A B \phi + \phi \psi + B (\phi^2 + \psi))^2}$$

```
In[37]:= c1 = Coefficient[charpol, t];
```

We shall write all conditions in terms of ψ and ϕ. For this we need this replacement rule.

```
In[38]:= Solve[{φ + ψ == -b1,
            φ ψ == b0}, {A, B}]
```

$$Out[38]= \left\{\left\{B \to \frac{\phi - \phi^2 + \psi - \phi\psi - \psi^2}{\phi\psi}, A \to 1 - \phi - \psi\right\}\right\}$$

```
In[39]:= rep = %[[1]];
```

This is condition 2.73 in terms of ϕ and ψ.

```
In[40]:= InequalitySolve[
          Abs[c1] < 1 + c0 < 2&& A > 0 &&
          B > 0 /. rep, {φ, ψ} ]//Simplify
```

$$Out[40]= 0 < \phi \le \frac{1}{2} \&\& (\phi > \psi \&\& \psi > 0 \;||\; \phi < \psi \&\& \phi + \psi < 1) \;||\; \phi < 1 \&\& \phi > \frac{1}{2} \&\& \psi > 0 \&\& \phi + \psi < 1$$

The logical statement we obtained above is identical to the one we obtain when condition (2.74) of Proposition 2.2 is stated in terms of ϕ and ψ.

$$In[41]:= \text{Simplify[InequalitySolve[} 0 < A < 1 \&\& B > \frac{-1 - 3A}{-1 + A} \&\& \psi > 0 \&\& \phi > 0 \;/.\; \text{rep}, \{\phi, \psi\}]]$$

$$Out[41]= 0 < \phi \le \frac{1}{2} \&\& (\phi > \psi \&\& \psi > 0 \;||\; \phi < \psi \&\& \phi + \psi < 1) \;||\; \phi < 1 \&\& \phi > \frac{1}{2} \&\& \psi > 0 \&\& \phi + \psi < 1$$

We conclude that all prime period-two solutions are asymptotically stable.

A case where (iii) of (2.2) is satisfied. We see that the equilibrium point is a saddle point.

```
In[42]:= A = 0.5; B = 6;

In[43]:= evals
Out[43]= {-1.03158, 0.507773}
```

By Proposition 2.2, there is a locally asymptotically stable prime period-two solution, given by solutions of the quadratic equation shown here.

$$In[44]:= \text{Solve}[z^2 + (A - 1) z - \frac{(A - 1) A}{B - 1} == 0, z]$$

```
Out[44]= {{z → 0.138197}, {z → 0.361803}}

In[45]:= per2 = z/.%
Out[45]= {0.138197, 0.361803}
```

A few terms of the orbit of the prime period-two solution.

```
In[46]:= Orbit[rationmap,per2,4]
Out[46]= {{0.138197, 0.361803},
          {0.361803, 0.138197},
          {0.138197, 0.361803},
          {0.361803, 0.138197},
          {0.138197, 0.361803}}
```

2.17.3 Numerical Explorations

This is an orbit, and its phase portrait. The plot only hints at the structure of the attractor that seems to be a period-two solution.

```
In[47]:= orb1 = Orbit[rationmap, 2, 2, 400];

In[48]:= PhasePortrait[orb1,
           PlotRange → {{0, 0.7}, {0, 0.7}}]
```

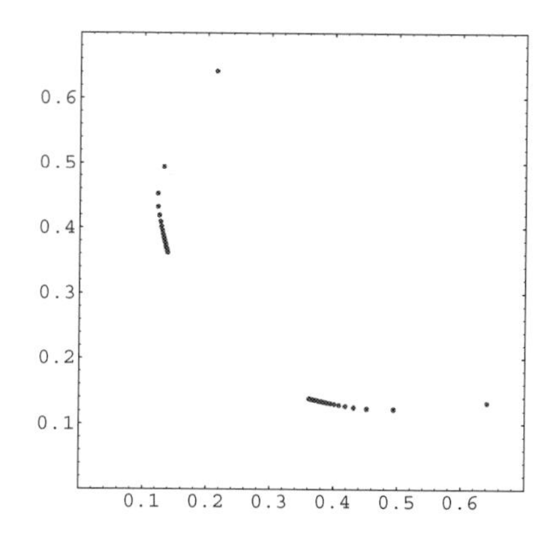

Next, we explore another aspect of the rational map, which may be helpful in understanding its dynamics. We now turn to plotting stable and unstable manifolds.

Periods of length 1 are just fixed points.

```
In[49]:= fp1 = PeriodicOrbit[rationmap, 1]
Out[49]= {{0.214286, 0.214286}}
```

The eigenvalues of the Jacobian at fp1 indicate that this equilibrium point is a saddle.

```
In[50]:= Multipliers[rationmap, fp1]
Out[50]= {-1.03158, 0.507773}
```

Period-two solutions and the corresponding multipliers. Note that if $\gamma^+(\phi, \psi)$ is a a period-two solution, then $\gamma^+(\psi, \psi)$ is a period-two solution too.

```
In[51]:= fp2 = PeriodicOrbit[rationmap, 2]
Out[51]= {{0.138197, 0.361803},
           {0.361803, 0.138197},
          {0.361803, 0.138197},
           {0.138197, 0.361803}}

In[52]:= Multipliers[rationmap, fp2[[2]]]
Out[52]= {1.33736, 0.0299096}
```

This is a plot of the unstable manifold at `fp1`.

```
In[53]:= wu = UnstableManifold[
              rationmap, fp1[[1]], 1, 10, 50];

In[54]:= PhasePortrait[{wu, fp1},
            Size → 0.01, 0.02,
            PlotRange → {{0.15, 0.3},
                  {0.15, 0.3}}]
```

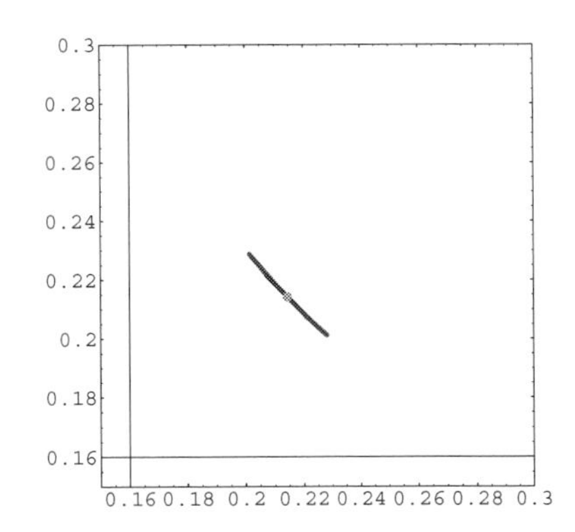

Next, we look at some phenomena involving both stable and unstable manifolds for different choices of parameters.

We shall find the stable manifold as the unstable manifold for the inverse map. First, we define the inverse map.

```
In[55]:= A = 0.5; B = 6;

In[56]:= Solve[rationmap[{u, v}] == {x, y},
            {u, v}]
```

$$Out[56]= \left\{\left\{u \to \frac{x\,(1. - 12.\,y)}{-2. + 2.\,y},\ v \to 1.\,x\right\}\right\}$$

```
In[57]:= rationinvmap[{x_, y_}] :=
               {u, v}/.%[[1]]
```

$$Out[57]= \left\{\frac{x\,(1. - 12.\,y)}{-2. + 2.\,y},\ x\right\}$$

To extend the manifold, the last input parameter of the `UnstableManifold` function is increased.

```
In[58]:= orb1 = Orbit[rationmap,
            {1, 2}, 5000];

In[59]:= wu = UnstableManifold[
            rationmap, fp1[[1]], 1, 6, 50];

In[60]:= ws = UnstableManifold[
            rationmapInv, fp1[[1]], 1, 6, 50];
```

Both the stable and unstable manifolds are shown here.

```
In[61]:= PhasePortrait[wu, ws, fp1, orb1,
            Size → {0.01, 0.01, 0.03, 0.01},
            PlotRange → {{0, 0.5}, {0, 0.5}}];
```

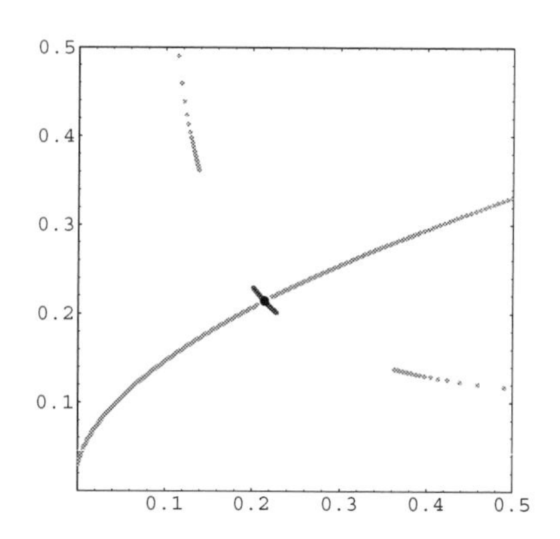

The complexity (or simplicity) of the behavior of equation (2.70) for different values of parameters A and B can be seen from the bifurcation diagrams. Theoretical results that support some numerical behavior seen in simulations are proven in [KLS].

Here, we fix the parameter B and vary A.

```
In[62]:= Clear[A];

In[63]:= B = 6;

In[64]:= BifurcationPlotND[rationmap,
           {A, 0.01, 1.0}, {0., 1.},
           Steps → 300,
           Iterates → 150,
           FirstIt → 400];
```

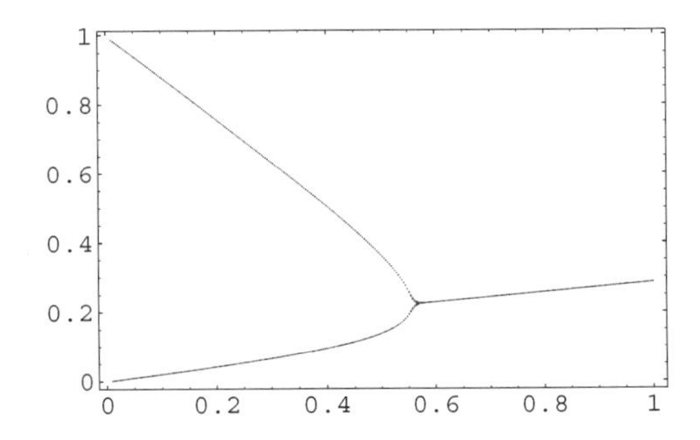

If more numerical experiments were performed, nothing qualitatively new would be produced. It seems that the asymptotically stable period-two solution is followed by asymptotically stable equilibrium, after a critical value of the parameter. This is a fact that was proven for certain values of parameters in [KLS].

Here we find maxima and minima of positive and negative semicycles of a solution.

```
In[65]:= A = 2.; B = 1.;

In[66]:= orb1 = Orbit[rationmap,
                 {2., 2.7}, 20];
```

Now, we look for semicycles relative to the equilibrium of the equation. We see that all semicycles are of length 2 or 3 before the solution becomes equal to the equilibrium.

```
In[67]:= fp
Out[67]= {1.5, 1.5}

In[68]:= SemicycleTest[orb1, fp[[1]]]
Positions and Maxima of Positive Semicycles
{2, {2, 2.7'}, {6, 1.5170},
{11, 1.50027}, {15, 1.50000},
{20, 1.50000}}
Positions and Minima of Negative Semicycles
{{4, 1.36834}, {9, 1.49831},
{13, 1.49996}, {18, 1.499999}}
Lengths of Positive Semicycles
{3, 2, 2, 2, 2}
Lengths of Negative Semicycles
{-2, -3, -2, -3}
Positions and Lengths of Positive and Negative
Semicycles
{3, -2, 2, -3, 2, -2, 2, -3, 2}
```

Here is the corresponding time series plot.

```
In[69]:= TimeSeriesPlot[orb1,
           AxesOrigin → {0, fp[[1]]},
           PlotRange → All]
```

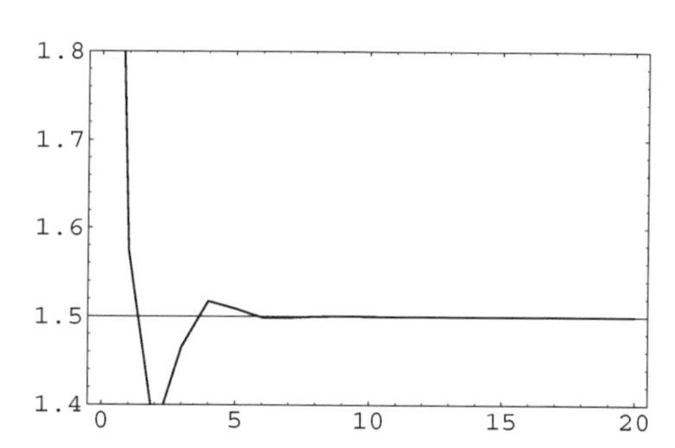

If many simulations of this kind are performed, one can find out that, if $A > B$, the semicycles of all solutions have length 2 or 3 and that the maxima of positive semicycles and minima of negative semicycles have a special location, see [KLS] and [KL]. This fact can be used to obtain some global properties of the solutions of this equation, such as global attractivity, see [KoL], [KL], and [KLS].

Based on what we have seen so far, it seems reasonable to pose the following problems:

1. In the case $A < B$, find the basin of attraction of the nonoscillatory solution.

2. Investigate the global stability nature of the period-two solution.

3. Assume that $A > B$. Then every solution of equation (2.70) has a finite limit. Prove or disprove.

4. Investigate the forbidden set and the stability of the equilibrium points for negative values of parameters and initial conditions.

2.18 Area-Preserving Maps and Systems

In this section, we consider a special class of maps and systems: area preserving. As the name suggests, an **area-preserving map** is a map that preserves the area of a planar region under the forward iterates of the map. The formal definition is as follows.

DEFINITION 2.24 *A map* $\mathbf{F} : R^2 \to R^2$ *is said to be area preserving if its Jacobian matrix* $J_\mathbf{F}$ *satisfies*

$$\det J_\mathbf{F}(\mathbf{x}) = 1 \quad \textit{for all points } \mathbf{x} \in R^2.$$

A detailed study of these maps may be found in [MK].

As our first example let us investigate the dynamics of linear area-preserving maps.

Example 2.14 In the case of the linear map $\mathbf{x} \to \mathbf{A}\mathbf{x}$ and the corresponding system, the condition for area preserving is equivalent to

$$\det \mathbf{A} = 1.$$

Since the determinant of a square matrix is the product of its eigenvalues λ_1 and λ_2, the last condition is equivalent to

$$\lambda_1 \lambda_2 = 1.$$

This condition leads to three qualitatively distinct cases:

(i) Saddle: λ_1 and λ_2 are both real of the same sign and $|\lambda_1| < 1 < |\lambda_2|$.

(ii) Parabolic: $\lambda_1 = \lambda_2 = 1$ or $\lambda_1 = \lambda_2 = -1$.

(iii) Elliptic: λ_1 are λ_2 are complex-conjugate and $|\lambda_1| = |\lambda_2| = 1$.

∎

The saddle case (i) is hyperbolic, and the linearization suffices to conclude instability. The linearization also gives information about the local stable and unstable manifolds. The remaining two cases are nonhyperbolic and the linearization is inconclusive. The elliptic case is one of the most frequent cases encountered in conservative mechanical systems exhibiting some stability and very complicated dynamics. The stability of such an equilibrium point is established by simplifying the nonlinear terms through appropriate coordinate transformations and bringing them into the so-called normal forms, where the stability (or instability) can be immediately established. One well-known normal form is given in the result given below, see [HK], pp. 484–494, but first we need to discuss notation.

Pairs (x, y) in the plane can be identified with complex numbers $z = x + iy$. Recall that $\bar{z} = x - iy$, and that

$$x = \frac{z + \bar{z}}{2}, \quad y = \frac{z - \bar{z}}{2i} \quad \text{and} \quad |z|^2 = z\bar{z} = x^2 + y^2.$$

By changing the two independent variables x and y to the variables z and $\bar{z}$, a function $F(x, y)$ can be realized as a function $\hat{F}(z, \bar{z})$ as follows:

$$F(x, y) = F\left(\frac{z + \bar{z}}{2}, \frac{z - \bar{z}}{2i}\right) = \hat{F}(z, \bar{z})$$

To avoid additional notation, we shall write $F(z, \bar{z})$ for $\hat{F}(z, \bar{z})$.

THEOREM 2.26 (Birkhoff Normal Form)

Let $\mathbf{F} : R^2 \to R^2$ *be an area-preserving* C^n *map (n-times continuously differentiable) with a fixed point at the origin whose complex-conjugate eigenvalues* λ *and* $\bar{\lambda}$ *are on the unit disk.*

Suppose there exists an integer q *such that*

$$4 \le q \le n + 1,$$

and suppose that the eigenvalues satisfy

$$\lambda^k \neq 1 \quad \text{for} \quad k = 1, 2, \ldots, q.$$

Let $r = [\frac{q}{2}]$ *be the integer part of* $\frac{q}{2}$.

Then there exists a smooth function $g(z, \bar{z})$ *that vanishes with its derivatives up to order* $r - 1$ *at* $z = 0$*, and there exists a real polynomial*

$$\alpha(w) = \alpha_1 w + \alpha_2 w^2 + \cdots + \alpha_r w^r$$

such that the map $\mathbf{F}$ *can be reduced to the normal form*

$$z \to \mathbf{F}(z, \bar{z}) = \lambda z e^{i\alpha(z\bar{z})} + g(z, \bar{z}).$$

Using Theorem 2.26 we can state the main stability result for an elliptic fixed point, known as the KAM Theorem (or Kolmogorov-Arnold-Moser Theorem), see [HK].

THEOREM 2.27 (KAM Theorem)
Let $\mathbf{F} : R^2 \to R^2$ be an area-preserving map with an elliptic fixed point at the origin satisfying the conditions of Theorem 2.26. If the polynomial $\alpha(|z|^2)$ is not identically zero, then the origin is a stable equilibrium point.

Let us restrict our attention to a special class of area-preserving maps defined in a ring which is well understood. Consider the system

$$\begin{aligned} r_{n+1} &= r_n \\ \theta_{n+1} &= \theta_n + h(r_n), \end{aligned} \tag{2.75}$$

in polar coordinates (r, θ) in a ring (annulus)

$$a \leq r \leq b,$$

with some appropriate conditions on h, that is given below. A map generated by system (2.75) is called a **twist map** because such a map leaves each circle of constant radius invariant and the orbits are simple rotations on these circles. We also impose an additional condition that h is not identically constant in the ring so that the angle of rotation is not constant.

Based on the results on the maps on circle and rotation numbers, it is easy to describe the dynamics of the twist map: on each circle for which the rotation number is a rational number the orbits are periodic, otherwise they are dense on the circle. The next theorem shows that under small area-preserving perturbations some of these invariant circles persist.

THEOREM 2.28 (Twist Theorem)
Consider in polar coordinates the system

$$\begin{aligned} r_{n+1} &= r_n + \epsilon g_1(r_n, \theta_n) \\ \theta_{n+1} &= \theta_n + \epsilon h(r_n) g_2(r_n, \theta_n), \end{aligned} \tag{2.76}$$

in a ring $a \leq r \leq b$ such that the functions $g_i, i = 1, 2$ are C^5, and $|\epsilon|$ is sufficiently small. Then, given any number ω between $h(a)$ and $h(b)$ incommensurable with 2π, and satisfying

$$\left| \frac{\omega}{2\pi} - \frac{p}{q} \right| \geq \frac{c}{|q|^{5/2}} \tag{2.77}$$

for all integers p and q, there exists a differentiable closed curve

$$\begin{aligned} r(s) &= T_1(\epsilon, s) \\ \theta(s) &= r + T_2(\epsilon, s) \end{aligned} \tag{2.78}$$

where T_1 and T_2 are 2π-periodic in s, which is an invariant of system (2.76). The positive orbits on the curve (2.78) are the rotations $s \to s + \omega$.

Notice that the unusual condition (2.77) (which is well-known in number theory, see [HaW]) requires that ω must be an irrational number which is "badly" approximated by the rational numbers. An implication of this theorem is that such an invariant circle of system (2.75) is slightly deformed to another closed curve under small area-preserving perturbations. Other invariant circles usually break up, although the Twist Theorem says nothing about them. Extensive visual and numerical evidence indicates that this break-up of circles may give rise to periodic orbits, and some of these periodic orbits can, in turn, be surrounded by invariant and disintegrating curves. This phenomenon may result in complex dynamics.

To illustrate some of these phenomena let us consider the example of May's host parasitoid model, which was treated in [LTT] and [MSHZ].

Example 2.15 (May's host parasitoid model) Consider the system

$$\begin{aligned} X_{n+1} &= \frac{\alpha X_n}{1+\beta Y_n} \\ Y_{n+1} &= \frac{\beta X_n Y_n}{1+\beta Y_n}, \quad n = 0, 1, \ldots, \end{aligned} \tag{2.79}$$

where α and β are positive numbers and the initial conditions X_0 and Y_0 are arbitrary positive numbers.

When $\alpha \in (1, \infty)$ and $\beta \in (0, \infty)$, this system is a special case of a host parasitoid model that was investigated by May in [M2], and by May and Hassell in [M3]. The system (2.79) has a zero equilibrium, and when $\alpha > 1$, it also has a positive equilibrium $(\frac{\alpha}{\beta}, \frac{\alpha}{\beta})$.

The change of variables $x_n = \beta X_n, y_n = \beta Y_n$ reduces (2.79) to

$$\begin{aligned} x_{n+1} &= \frac{\alpha x_n}{1+y_n} \\ y_{n+1} &= \frac{x_n y_n}{1+y_n}, \quad n = 0, 1, \ldots, \end{aligned} \tag{2.80}$$

so from now on we assume, without loss of generality, that $\beta = 1$, and investigate the behavior of solutions of (2.80). By eliminating x_n from the right-hand side of this system we obtain

$$x_{n+1} = \alpha \frac{y_{n+1}}{y_n}, \quad n = 0, 1, \ldots, \tag{2.81}$$

$$y_{n+1} = \frac{\alpha y_n^2}{(1+y_n)y_{n-1}}, \quad n = 1, 2, \ldots. \tag{2.82}$$

We also assume that $\alpha \geq 1$ in what follows. Equation (2.82) has a unique positive equilibrium $\bar{y}\alpha - 1$. However, this equation does not generate an area-preserving map, as we can see by computing the determinant of the corresponding Jacobian det $D(0,0) = 0$. The change of variables

$$y_n = (\alpha - 1)e^{u_n}$$

reduces (2.82) to

$$u_{n+1} = 2u_n - u_{n-1} + \ln \frac{\alpha}{(\alpha - 1)e^{u_n}}, \tag{2.83}$$

which is equivalent to the area-preserving map

$$\begin{aligned} u_{n+1} &= 2u_n - v_n + \frac{\alpha}{1 + (\alpha - 1)e^{u_n}} \\ v_{n+1} &= u_n, \quad n = 0, 1, \ldots. \end{aligned} \tag{2.84}$$

The positive equilibrium point $(\alpha, \alpha - 1)$ of system (2.80) corresponds to the equilibrium point $(0, 0)$ of system (2.84).

The linearized equation of (2.83) about zero is

$$z_{n+1} - \frac{\alpha + 1}{\alpha} z_n + z_{n-1} = 0, \quad n = 1, 2, \ldots. \tag{2.85}$$

The corresponding characteristic equation

$$\lambda^2 - \frac{\alpha + 1}{\alpha}\lambda + 1 = 0$$

has two complex-conjugate roots that lie on the unit disk:

$$\lambda_{\pm} = \frac{\alpha + 1}{2\alpha} + \frac{\sqrt{(\alpha - 1)(3\alpha + 1)}}{2\alpha}.$$

Then, zero is an elliptic equilibrium point of (2.83). In order to determine the stability character of the zero equilibrium we employ KAM theory, see Theorems 2.12 and 2.13 [Mr]. The computational procedure consists of the reduction of system (2.84) to its Birkhoff Normal Form. This is accomplished by successive application of the following three transformations.

Transformation 1:

The linear change of variables

$$\begin{aligned} u_n &= x_n + y_n \\ v_n &= \bar{\lambda} x_n + \lambda y_n \end{aligned} \tag{2.86}$$

transforms system (2.84) into

$$\begin{aligned} x_{n+1} &= \lambda x_n - \tfrac{\lambda}{\lambda^2 - 1}\left((\lambda - 1)^2(x_n + y_n) - \lambda \ln \tfrac{\alpha}{1+(\alpha-1)e^{x_n+y_n}}\right) \\ y_{n+1} &= \bar{\lambda} y_n - \tfrac{\lambda}{\lambda^2 - 1}\left((\bar{\lambda} - 1)^2(x_n + y_n) - \bar{\lambda} \ln \tfrac{\alpha}{1+(\alpha-1)e^{x_n+y_n}}\right), \end{aligned} \tag{2.87}$$

for $n = 0, 1, \ldots$ By expanding the function $\ln \frac{\alpha}{1+(\alpha-1)e^{x_n+y_n}}$ in power series of x_n+y_n and keeping only the terms up to order 3 in the expansion we obtain

$$\begin{aligned} x_{n+1} &= \lambda x_n + \sigma\left(\frac{(x_n+y_n)^2}{2} - \frac{\alpha-2}{\alpha}\frac{(x_n+y_n)^3}{3!}\right) + O((x_n+y_n)^4) \\ y_{n+1} &= \overline{\lambda} x_n - \overline{\sigma}\left(\frac{(x_n+y_n)^2}{2} - \frac{\alpha-2}{\alpha}\frac{(x_n+y_n)^3}{3!}\right) + O((x_n+y_n)^4), \end{aligned} \tag{2.88}$$

for $n = 0, 1, \ldots$, where

$$\sigma = \frac{\lambda}{\lambda - \overline{\lambda}} \frac{\alpha - 1}{\alpha}.$$

Transformation 2:

The change of variables

$$\begin{aligned} x_n &= \xi_n + \phi_2(\xi_n, \eta_n) + \phi_3(\xi_n, \eta_n) \\ y_n &= \eta_n + \psi_2(\xi_n, \eta_n) + \psi_3(\xi_n, \eta_n) \end{aligned} \tag{2.89}$$

where

$$\phi_k(\xi_n, \eta_n) = \sum_{m=0}^{k} a_{km}\xi^{k-m}\eta^m \quad \text{and} \quad \psi_k(\xi_n, \eta_n) = \sum_{m=0}^{k} \overline{a_{km}}\xi^m\eta^{k-m}, \quad k = 2, 3$$

and

$$\begin{aligned} a_{20} &= \frac{-1}{(\lambda-1)(\lambda-\overline{\lambda}}\frac{\alpha-1}{2\alpha^2} \\ a_{21} &= \frac{\lambda}{(\lambda-1)(\lambda-\overline{\lambda}}\frac{\alpha-1}{\alpha^2} \\ a_{22} &= \frac{-\lambda}{(\overline{\lambda}^2-\lambda)(\lambda-\overline{\lambda}}\frac{\alpha-1}{2\alpha^2} \\ a_{30} &= \frac{-\lambda}{(\lambda^2-1)^2}\frac{\alpha-1}{\alpha}\left(a_{20} + \overline{a_{22}} - \frac{\alpha-2}{5\alpha}\right) \\ a_{31} &= 0 \\ a_{32} &= \frac{\lambda^3}{(\lambda^2-1)^2}\frac{\alpha-1}{\alpha^2}\left(a_{22} + \overline{a_{20}} + 2Re(a_{21}) - \frac{\alpha-2}{2\alpha}\right) \\ a_{33} &= \frac{-\lambda}{(\lambda^2-1)(\overline{\lambda}^4-1)}\frac{\alpha-1}{\alpha^2}\left(a_{22} + \overline{a_{20}} - \frac{\alpha-2}{6\alpha}\right) \end{aligned}$$

reduces system (2.88) to the form

$$\begin{aligned} \xi_n &= \lambda\xi_n + \alpha_2\xi_n^2\eta_n + O_4(\xi_n, \eta_n) \\ \eta_n &= \overline{\lambda}\eta_n + \overline{\alpha_2}\xi_n\eta_n^2 + O_4(\xi_n, \eta_n), \end{aligned} \tag{2.90}$$

where

$$\alpha_2 = \frac{\alpha - 1}{2\alpha(2\alpha + 1)}\left(1 - \frac{\alpha + 1}{\sqrt{(\alpha - 1)(3\alpha + 1)}}\right),$$

and $O_4(\xi_n, \eta_n)$ is the expression that contains the terms of degree at least 4 in ξ_n and η_n.

The coefficients $a_{20}, a_{21}, a_{22}, a_{30}, \ldots$ above are determined by using the method of undetermined coefficients by substituting (2.89) and (2.90) in (2.88) and equating the coefficients of the terms with the same degree up to order 3.

Transformation 3:

The change of variables

$$\begin{aligned} \xi_n &= r_n + is_n \\ \eta_n &= r_n - is_n \end{aligned} \tag{2.91}$$

transforms system (2.90) into

$$\begin{aligned} r_{n+1} &= \zeta_1 r_n - \zeta_2 s_n + O_4(r_n, s_n) \\ s_{n+1} &= \zeta_2 r_n + \zeta_1 s_n + O_4(r_n, s_n), \end{aligned} \tag{2.92}$$

for $n = 0, 1, \ldots,$ where $O_4(r_n, s_n)$ is the expression that contains the terms of degree at least 4 in r_n and s_n and

$$\begin{aligned} \zeta_1 &= Re(\lambda) + Re(\alpha_2)(r_n^2 + s_n^2) \\ \zeta_2 &= Im(\lambda) + Im(\alpha_2)(r_n^2 + s_n^2). \end{aligned} \tag{2.93}$$

Furthermore, (2.92) can be written as

$$\begin{aligned} r_{n+1} &= \cos \omega r_n - \sin \omega s_n + O_4(r_n, s_n) \\ s_{n+1} &= \sin \omega r_n + \cos \omega s_n + O_4(r_n, s_n), \end{aligned} \tag{2.94}$$

where

$$\omega = \gamma_0 + \gamma_1(r_n^2 + s_n^2),$$

with

$$\cos \gamma_0 = Re(\lambda) \quad \text{and} \quad \gamma_1 = \frac{-1}{\sin \gamma_0} Re(\alpha_2). \tag{2.95}$$

The coefficients γ_0 and γ_1 are the twist coefficients and can be found by equating the formulas (2.92) and (2.94) and by expanding (2.92) in a power series of $r_n^2 + s_n^2$ keeping the terms up to order 3 in the expansion. System (2.94) is the Birkhoff Normal Form for system (2.84) and since the twist coefficients given by (2.95) are not zero, the polynomial

$$\omega(|z|^2) = \gamma_0 + \gamma_1 |z|^2$$

is not identically zero at the origin. Therefore, the KAM theorem 2.27 applies and we conclude that the origin is a stable equilibrium of (2.84). ∎

Some recent results on application of the KAM theorem to a general class of equations can be found in [PS1].

2.19 Biology Applications Projects

Problem 1.

A generalized Lotka-Volterra two-species system

$$\begin{aligned} x_{n+1} &= x_n e^{r_1(1-x_n)-s_1 y_n} \\ y_{n+1} &= y_n e^{r_2(1-y_n)-s_2 x_n}, \end{aligned} \tag{2.96}$$

was originally proposed by R. M. May [M1] as a model for competitive interaction in two dimensions. All the parameters r_1, r_2, s_1, and s_2 are nonnegative and the initial conditions x_0, y_0 are positive. The symmetric case $r_1 = r_2, s_1 = s_2$ was studied in detail in [JR]. This system has a unique positive equilibrium point as well as some periodic points of low periods. Studying the bifurcations of periodic points of low periods the period-doubling bifurcation phenomenon is indicated. A parameter region for stable Hopf bifurcation of a pair of symmetrically placed period-two points is determined.

A study of coexistence in a competitive interaction by Comins and Hassell [CH] reduces to the study of this system for $n = 2$.

Perform the following analysis for this system:

1. Find equilibrium points.
2. Check the linear stability of the equilibrium points.
3. Find periodic solutions of period 2 (if they exist).
4. Visualize the stable and unstable manifolds.
5. Plot the bifurcation diagram by fixing all parameters except one.

Problem 2.

A single locus, diallelic selection model with female and male viability differences is modeled by a two-dimensional system of difference equations

$$\begin{aligned} x_{n+1} &= \frac{a x_n y_n + x_n + y_n}{b + x_n + y_n} \\ y_{n+1} &= \frac{c x_n y_n + x_n + y_n}{d + x_n + y_n}, \end{aligned}$$

which proposed by J. F. Selgrade and M. Ziehe [SZ]. Here the parameters a, b, c, and d and the initial conditions x_0, y_0 are nonnegative. The variables are ratios of allele frequencies in each sex. Using the strong monotonicity of this system the authors showed that every initial genotypic structure converges to an equilibrium structure assuming that every equilibrium is hyperbolic.

Perform the following analysis for this system:

1. Find equilibrium points.
2. Check the linear stability of the equilibrium points.
3. Find periodic solutions of period-two (if they exist).
4. Visualize the stable and unstable manifolds.
5. Plot the bifurcation diagram by fixing all parameters except one.
6. Perform semicycle analysis for each component.

Problem 3.

A simple two-species system studied in [CK] and [CKS]

$$x_{n+1} = \frac{x_n}{a + y_n}$$

$$y_{n+1} = \frac{y_n}{b + x_n},$$

models competitive interaction. The parameters a and b and the initial conditions x_0, y_0 are nonnegative. This system exhibits a symmetry which can be seen by rewriting it as

$$x_{n+1} = f(a, x_n, y_n)$$
$$y_{n+1} = f(b, y_n, x_n,),$$

where $f(a, x_n, y_n) = \frac{x_n}{a+y_n}$. Performing linear stability analysis we conclude that there are nine cases depending whether $0 < a < 1, a = 1, a > 1$ and $0 < b < 1, b = 1, b > 1$. Because of the symmetry, three out of nine cases are analogues to other cases, hence there are six distinct cases to consider.

Perform the following analysis for this system:

1. Find equilibrium points.
2. Check the linear stability of the equilibrium points.
3. Find periodic solutions of period-two (if they exist).
4. Visualize the stable and unstable manifolds.
5. Plot the bifurcation diagram by fixing all parameters except one.
6. Plot phase plane trajectories and time series for different values of parameters. Try the cases $a = 1, b > 1$ and $a = 1, b < 1$ as well as their symmetric cases $a > 1, b = 1$ and $a < 1, b = 1$, respectively.
7. In the case $a = 1, b = 1$ find the invariant and the corresponding Lyapunov function. Use that to find the closed-form solution.

Problem 4.

Another two-species system

$$x_{n+1} = \frac{x_n}{(a + a_{11}x_n + a_{12}y_n)^{\alpha}}$$

$$y_{n+1} = \frac{y_n}{(b + b_{11}x_n + b_{12}y_n)^{\beta}}, \tag{2.97}$$

modeling competition was proposed by Hassell and Comins [HC]. Here all the parameters $a, b, a_{ij}, b_{ij}, \alpha$ and β and the initial conditions x_0, y_0 are nonnegative. The problem of coexistence of two competitive species has been addressed. The monotone character of this system with respect to an appropriately chosen cone has been used.

Perform the following analysis for this system:
1. Find equilibrium points.
2. Check the linear stability of the equilibrium points.
3. Visualize the stable and unstable manifolds.
4. Plot the bifurcation diagram by fixing all parameters except one.
5. Perform semicycle analysis for each component.

Problem 5.

Using both the exponential and rational nonlinearities of Hassell and Comins from [CH] and [HC], Bishir and Namkoong [BN] proposed the model for competition of the form

$$x_{n+1} = \frac{gx_n}{(a + x_n + \alpha y_n)^b}$$

$$y_{n+1} = y_n e^{(p-q(\beta x_n + y_n))},$$

where all the parameters $\alpha, \beta, a, b, g, p$, and q and the initial conditions (x_0, y_0) are positive. The problem of coexistence of two competitive species has been addressed. It has been proven that for some choice of the parameters two species coexist even though there is no interior equilibrium point. This is in contrast to the qualitative behavior of systems (2.96) and (2.97), where the discrete analogue of the "exclusion principle" for the Lotka-Volterra equations has been established: if there is no interior equilibrium point one of the species becomes extinct. See [FY1]. It has also been shown in [FY1] that if the species with the rational growth function dominates the species with the exponential growth function, then the latter becomes extinct.

Perform the following analysis for this system:
1. Find equilibrium points.
2. Check the linear stability of the equilibrium points.
3. Visualize the stable and unstable manifolds.
4. Plot the bifurcation diagram by fixing all parameters except one.
5. Perform semicycle analysis for each component.

Problem 6.

A two-dimensional system

$$\begin{aligned} x_{n+1} &= \alpha x_n + (1-\alpha)\frac{cx_n}{a + cx_n + y_n} \\ y_{n+1} &= \beta y_n + (1-\beta)\frac{dy_n}{b + x_n + dy_n}, \end{aligned} \tag{2.98}$$

modeling competition between two species of nematodes feeding on a plant crop was proposed by Jones and Perry [JP]. Here all the parameters α, β, a, b, c, and d and the initial conditions x_0, y_0 are nonnegative. The map corresponding to this system is a monotone of the first quadrant for all biologically relevant parameter values $(0 < \alpha, \beta < 1)$. In fact, the corresponding map is injective. See [S1].

Perform the following analysis for this system:

1. Find equilibrium points.
2. Check the linear stability of the equilibrium points.
3. Visualize the stable and unstable manifolds.
4. Plot the bifurcation diagram by fixing all parameters except one.
5. Perform semicycle analysis for each component.

Problem 7.

Another two-species system

$$\begin{aligned} x_{n+1} &= x_n f(c_{11}x_n + c_{12}y_n) \\ y_{n+1} &= y_n g(c_{21}x_n + c_{22}y_n), \end{aligned} \tag{2.99}$$

modeling competition was proposed by Franke and Yakubu [FY1], [FY1], and [FY3]. They gave sufficient conditions for one species to drive another species to extinction in models with exponential or rational growth functions or both. Roughly speaking, if a species with an exponential (rational) growth function weakly dominates another species with an exponential (rational) growth function, the former drives the latter to extinction. We say that species y_n weakly dominates species x_n if

$$\left\{(x,y) \in (0,\infty)^2 | \ f(c_{11}x + c_{12}y) \geq 1\right\} \subset \left\{(x,y) \in (0,\infty)^2 | \ g(c_{21}x + c_{22}y) > 1\right\}.$$

In this case one can see that species y_n is growing whenever species x_n is nonincreasing. See also [KoL], pp. 199–203. Here, all the parameters c_{ij} and the initial conditions x_0, y_0 are positive and the functions f and g satisfy the following conditions:

(C_1) $f, g \in C[[0,\infty), [0,\infty)]$ are decreasing,
(C_2) $f(\bar{x}) = 1, g(\bar{y}) = 1$ for some $\bar{x}, \bar{y} \in (0,\infty)$.

The monotone character of this system with respect to the first quadrant has been used.

For different choices of functions f and g that satisfy the conditions (C_1) and (C_2) perform an analysis similar to one for system (2.98).

Remark. Let x_n and y_n, in system (2.97), denote the densities of two populations at the n-th generation. Let u_n and v_n denote the *weighted total densities* for x_n and y_n, i.e.,

$$\begin{aligned} u_n &= c_{11}x_n + c_{12}y_n \\ v_n &= c_{21}x_n + c_{22}y_n \end{aligned}$$

where $c_{ij} \geq 0$ is called the *interaction coefficient.* The interaction coefficient reflects the effect of the j-th population on the i-th population. The matrix

$$C = \begin{pmatrix} c_{11} & c_{12} \\ c_{21} & c_{22} \end{pmatrix}$$

is called the *interaction matrix.* The per-capita growth rate and the fitness functions f and g are a smooth functions of their variables. Then, system (2.99) can be represented as

$$\begin{aligned} x_{n+1} &= x_n f(u_n) \\ y_{n+1} &= y_n g(v_n). \end{aligned}$$

Such systems are said to be of the Kolmogorov type. If f and g are monotonically decreasing for all values of u_n and v_n, then x_n and y_n are referred to as *pioneer* populations. Examples of pioneer fitnesses are exponential (see Moran [Mo] and Ricker [Ri]), rational (see Hassell and Comins [HC]) and linear functions (see Selgrade and Roberds [SR1], [SR2]). If f and g are unimodal functions, that is, if they monotonically increase up to a unique maximum value and then decrease for all values of weighted total densities u_n and v_n, then x_n and y_n are referred to as *climax* populations. Examples of climax fitnesses are exponential and quadratic functions, see Cushing [C1], [C2] and Selgrade and Namkoong [SN1], [SN2].

Very recently these results have been extended by Chan and Franke [CF], showing that for the situation where one or more species are becoming extinct a reduction in the model can be performed by eliminating species that are becoming extinct. On the reduced model, the sufficient conditions for one of the remaining species to drive another remaining species to extinction are given in [CF]. The authors showed that extinction on the submodel occurs on the full model as well.

Problem 8.

Kolmogorov-type systems with a density-depending stocking ($a > 0$) or harvesting ($a < 0$) term to the pioneer difference equation

$$\begin{aligned} x_{n+1} &= x_n f(c_{11}x_n + c_{12}y_n) + ax_n \\ y_{n+1} &= y_n g(c_{21}x_n + c_{22}y_n). \end{aligned} \tag{2.100}$$

were considered recently by J. F. Selgrade [Se]. After establishing that the pioneer self-crowding parameter c_{11} can destabilize the equilibrium the author used stocking or harvesting of the pioneer population to restabilize the equilibrium. Using the amount of stocking or harvesting that is directly proportional to the current pioneer density is mathematically the simplest strategy. Using this approach rigorous results have been established in [Se]. The effect of harvesting on dynamic behavior in discrete-time population models has recently received a great deal of attention, see [BF], [BS], [HW], [S], and [OC].

An extensive bibliography relating to Kolmogorov-type systems can be found in [Se].

Related discrete competitive system with planting

$$\begin{aligned} x_{n+1} &= x_n e^{p_1 - q_1(x_n + y_n)} + \alpha x_n \\ y_{n+1} &= y_n e^{p_2 - q_2(x_n + y_n)}, \end{aligned} \tag{2.101}$$

modeling the growth of two discretely reproducing populations in competition with only species 1 being planted has been considered by Yakubu, see [Y1] and [Y2]. The parameters p_1, p_2, q_1, q_2, and α are positive with the variable planting coefficient $\alpha \in (0, 1)$. There are a couple of interesting open problems and a conjecture posed in [Y2]. Numerical simulations indicate that this system has very complex behavior. The computer simulations indicate that for some values of parameters the period-two solution is a global attractor and there is a range of parameters with a chaotic behavior, but the proofs are yet to be found.

Perform the following analysis for system (2.101):
1. Find equilibrium points.
2. Check the linear stability of the equilibrium points.
3. Visualize the stable and unstable manifolds.
4. Plot the bifurcation diagram by fixing all parameters except one.
6. Compute Lyapunov exponents.
5. Perform semicycle analysis for each component.

Try a similar analysis for system (2.100) by choosing the pioneer and climax populations to be either rational or exponential functions.

Problem 9.

A generalized Lotka-Volterra two-species system

$$\begin{aligned} x_{n+1} &= x_n e^{r(1-x_n)+sy_n} \\ y_{n+1} &= y_n e^{r(1-y_n)+sx_n}, \end{aligned}$$

was proposed by W. Krawcewicz and T. D. Rogers [KR] as a model for cooperative interaction in two dimensions. The parameters r and s are nonnegative and the initial conditions x_0, y_0 are positive. This system has a unique positive equilibrium point as well as some periodic points of low periods. Studying bifurcations of periodic points of low periods, the period-doubling bifurcation phenomenon may be visually confirmed.

A more general system for cooperative interaction in two dimensions is

$$\begin{aligned} x_{n+1} &= x_n e^{r_1(1-x_n)+s_1y_n} \\ y_{n+1} &= y_n e^{r_2(1-y_n)+s_2x_n}, \end{aligned}$$

where all the parameters r_1, r_2, s_1, and s_2 are nonnegative and the initial conditions x_0, y_0 are positive.

Perform the following analysis for this system:

1. Find equilibrium points.
2. Check the linear stability of the equilibrium points.
3. Visualize the stable and unstable manifolds.
4. Plot bifurcation diagrams by fixing all parameters except one.

Problem 10.

A generalized Lotka-Volterra two-species system with variable growth rates r_n and s_n and variable intensities of intraspecific action (such as competition or predator–prey interaction) a_n, b_n, c_n, and d_n:

$$\begin{aligned} x_{n+1} &= x_n e^{r_n - a_n x_n - b_n y_n} \\ y_{n+1} &= y_n e^{s_n - c_n x_n - d_n y_n}, \end{aligned} \tag{2.102}$$

was considered by W. Wang and L. Zhengyi [WZ]. The initial conditions x_0, y_0 are positive.

A global asymptotic stability result has been proved in [WZ] for the general N-th order system of the form

$$x_i(n+1) = x_i(n) \exp(r_i(n) - \sum_{j=1}^{N} a_{ij}(n) x_j(n)), \; i = 1, \ldots, N,$$

where $x_i(n)$ is the density of i-th population at n-th generation, $r_i(n)$ is the growth rate of i-th population at n-th generation, and $a_{ij}(n)$ measures the interactions of i-th and j-th populations at the n-th generation.

Here are some challenging research projects where *Dynamica* simulations may be helpful:

1. Study system (2.102) assuming that all coefficients a_n, b_n, c_n, d_n, r_n and s_n are asymptotically constant, that is, the coefficients approach a limit as $n \to \infty$. For instance, assume that $a_n = a + \frac{1}{n+1}$ or $a_n = a + \frac{\sin n}{n+1}$, $n = 0, 1, \ldots$, and likewise for the other coefficients.

2. Study system (2.102) assuming that all coefficients a_n, b_n, c_n, d_n, r_n and s_n are periodic sequences to account for the environmental periodicity effects. In particular, study the case when the coefficients are periodic with period two.

3. Study system (2.102) assuming that at least one of the coefficients a_n, b_n, c_n, d_n, r_n and s_n is asymptotically constant and at least one is a periodic sequence with a prime period greater than 1. In particular, study the case when one of the coefficients is periodic with period two.

Problem 11.

A two-species system

$$x_{n+1} = x_n e^{r(1-x_n)+sy_n}$$

$$y_{n+1} = y_n e^{r(1-y_n)+sx_n},$$

was considered by Smith [S1] as a model for cooperative interaction between the population of juveniles x_n and adults y_n. Juveniles survive to become adults with density-dependent probability and adults leave juvenile offspring at a density-dependent rate.

The parameters r and s and the initial conditions x_0, y_0 are positive.

Perform the following analysis for this system:

1. Find equilibrium points.
2. Check the linear stability of the equilibrium points.
3. Find period-two solutions.
4. Determine the linear stability of period-two solutions.
5. Plot bifurcation diagrams by fixing all parameters except one.

Problem 12.

The two-species system

$$x_{n+1} = Ax_n \frac{y_n}{1+y_n}$$

$$y_{n+1} = By_n \frac{x_n}{1+x_n},$$

was considered in [KN2] as a model for cooperative interaction in two dimension. The parameters A and B are positive and the initial conditions x_0, y_0 are positive. This system always has $(0,0)$ as the equilibrium point and if $A > 1, B > 1$ it also has a unique positive equilibrium point. Based on local stability analysis one concludes that there are seven different cases depending on whether the values of parameters A and B are smaller, equal or greater than 1.

Perform the following analysis for this system:

1. Find equilibrium points.
2. Check the linear stability of the equilibrium points.
3. Visualize the stable and unstable manifolds.
4. Plot bifurcation diagrams by fixing one parameter.
5. Perform semicycle analysis for each component.

Problem 13.

A modified Nicholson-Bailey two-species system

$$x_{n+1} = x_n e^{(c-\frac{c}{K}x_n-ay_n)}$$

$$y_{n+1} = bx_n(1-e^{-ay_n}),$$

models predator–prey interaction, see [BFL1] and [BFL2]. All the parameters $a, b, c,$ and K and the initial conditions x_0, y_0 are positive. This system has a unique positive equilibrium point as well as some periodic points of low periods. Studying the bifurcations of periodic points of low periods the period-doubling bifurcation phenomenon is indicated. A parameter region for the stable Naimark-Sacker (discrete version of Hopf) bifurcation can be determined.

Perform the following analysis for this system:
1. Find equilibrium points.
2. Check the linear stability of the equilibrium points.
3. Visualize the stable and unstable manifolds.
4. Plot the bifurcation diagram by fixing all parameters except one.
6. Compute Lyapunov exponents.
5. Perform semicycle analysis for each component.

Problem 14.

Another modified Nicholson-Bailey two-species system

$$\begin{aligned} H_{n+1} &= FH_n e^{-aP_n} \\ P_{n+1} &= H_n(1 - e^{-aP_n}), \end{aligned} \tag{2.103}$$

models host-parasitoid interaction, see [NB], [HM], [HCM], and [M]. Here H_n and P_n are the number of hosts and parasitoids, respectively, in the n-th generation. F is the finite rate of increase of the host population (in the absence of a parasitoid, the host population grows geometrically with constant ratio F). The expression e^{-aP_n} (zero term of a discrete exponential distribution) is the fraction of hosts escaping parasitism. The fraction of hosts parasitized exactly k times is according to the discrete exponential distribution given by

$$\frac{(aP_n)^k}{k!} e^{-aP_n}.$$

Starting with the assumptions of the Nicholson-Bailey model and assuming that k parasitization events on a single host on average result in the emergence of α_k parasitoids of the next generation ($0 \le \alpha_k \le k$), one may obtain the following generalized model

$$\begin{aligned} H_{n+1} &= FH_n e^{-aP_n} \\ P_{n+1} &= H_n \phi(P_n) e^{-aP_n}, \end{aligned} \tag{2.104}$$

where

$$\phi(P_n) = \sum_{k=1}^{\infty} \alpha_k \frac{(aP_n)^k}{k!}.$$

This system was considered in [MSHZ] in the special case where $\phi(P_n) = aP_n$. In this case, using simple logarithmic substitution the system can be reduced to the form

$$x_{n+1} = x_n + \ln F(1 - e^{-y_n})$$

$$y_{n+1} = x_n + y_n + \ln F(1 - e^{-y_n}).$$

The equilibrium point is transformed into the origin and the transformed system is area preserving because

$$\frac{\partial(x_{n+1}, y_{n+1})}{\partial(x_n, y_n)} = 1.$$

The equilibrium point is elliptic for the value $\ln F \in (0, 4)$ and the KAM theory has been used to analyze this system, see [MSHZ].

Analyze this model using *Dynamica*.

Problem 15.

A system of difference equations

$$x_{n+1} = \frac{\alpha x_n}{1 + \beta y_n}$$

$$y_{n+1} = \frac{\beta x_n y_n}{1 + \beta y_n},$$

where

$$\alpha \in (1, \infty) \quad \text{and} \quad \beta \in (0, \infty)$$

with the positive initial conditions x_0, y_0 is a special case of a host-parasitoid model investigated by R. May [M2], R. May and M. P. Hassell [M3], and A. D. Taylor [T]. By employing KAM theory, it was shown in [LTT] that the positive equilibrium of this system is stable. It is remarkable that for the values of parameters $\alpha = \beta = 1$ one can find the invariant

$$I(x_n, y_n) = \frac{x_n}{y_n} + \frac{y_n}{x_n} + x_n + y_n$$

(use *Dynamica*) but we cannot apply Theorem 2.21 to find the related Lyapunov function.

Perform the following analysis for this system:

1. Find equilibrium points.
2. Check the linear stability of the equilibrium points.
3. Find periodic solutions of period two.
4. Determine the linear stability of period-two solutions.
5. Plot the bifurcation diagram by fixing all parameters except one.

6. Find an invariant of this equation.

Problem 16.
A plant-herbivore system

$$x_{n+1} = \frac{\alpha x_n}{e^{y_n} + \beta x_n}$$

$$y_{n+1} = \gamma(x_n + 1)y_n,$$

models the interaction of the apple twig borer (an insect pest of the grape vine) and grapes in the Texas High Plains, see [AHS], [ASTL], [KoL], and [KL]. The parameters α, β, and γ and the initial conditions x_0, y_0 are positive. More precisely the parameters are assumed to satisfy

$$\alpha \in (1, \infty), \quad \beta \in (0, \infty), \quad \text{and} \quad \gamma \in (0, 1).$$

This system has very complex behavior, which has been investigated mainly numerically and visually. There is a range of the values of parameters for which the equilibrium points on the x-axes are globally asymptotically stable, see [KoL]. The computer simulations indicate that for some parameter values there is chaotic behavior, but proofs are yet to be completed.

Perform the following analysis for this system:
1. Find equilibrium points.
2. Check the linear stability of the equilibrium points.
3. Visualize the stable and unstable manifolds.
4. Plot the bifurcation diagram by fixing all parameters except one.
6. Compute Lyapunov exponents.
5. Perform semicycle analysis for each component.

Problem 17.
A discrete system

$$x_{n+1} = py_n e^{-ay_n}$$

$$y_{n+1} = Rx_n e^{-bx_n},$$

modeling the competition in reproduction among the eggs and larvae was considered by Wilbur [W] and Kaitala et al. [KYH]. The model was initiated by Wilbur as an extension of the classical Riker model, see [Ri]. Wilbur's model has two state variables: the densities of eggs x_n and adults y_n. The assumption of the model is that during one time unit eggs develop into new adults, and adults give birth to new eggs. The parameters a, b, p, and R are positive.

This system is a special case of the system

$$\begin{aligned} x_{n+1} &= pf_a(y_n) \\ y_{n+1} &= Rf_b(x_n), \end{aligned}$$

where

$$f_c(u) = ue^{-cu}.$$

The variables in the system can be decoupled to give

$$\begin{aligned} x_{n+1} &= pf_a(Rf_b(x_{n-1})) \\ y_{n+1} &= Rf_b(pf_a(y_{n-1})), \end{aligned}$$

Two equations are identical, for the function f above, and essentially of order one in even-numbered and odd-numbered subsequences of the solution. The function on the right-hand side of this equation for some values of the parameters is bimodal function that gives rise to the possibility of the existence of alternative attractors with "large" basins of attractions. The alternative attractors may be stable equilibrium points, periodic solutions, and chaotic attractors. Numerical simulations indicate that this system has very complex behavior. For some parameter values there is the possibility of coexisting attractors. Each of the attractors may undergo some period-doubling bifurcations or period-doubling reversal to period-two solution, see [KYH] for the visual evidence.

Perform the following analysis for this system:

1. Find equilibrium points.
2. Check the linear stability of the equilibrium points.
3. Visualize the stable and unstable manifolds.
4. Plot the bifurcation diagram by fixing all parameters except one.
5. Perform semicycle analysis for each component.

Problem 18.

A system of difference equations

$$\begin{aligned} H_{n+1} &= \frac{(1-\epsilon)r_0 e^{-cr_p \frac{P_n}{H_n}} + r_0(1-d)\epsilon}{1+\left(\frac{H_n}{\beta_H}\right)^{\gamma_H}} \\ P_{n+1} &= \frac{r_P(1-\epsilon)P_n}{1+\left(\frac{P_n}{\beta_P H_n}\right)^{\gamma_P}}, \end{aligned} \tag{2.105}$$

modeling the parasitised host population was considered in several papers, see [KH], [KHG], and [KYH]. This model was motivated by the attempts to understand the complicated dynamics of small rodents living in seasonal environments. Here, P_n and H_n denote the population sizes of the parasite and the host, respectively, and ϵ denotes the frequency of immunised individuals in the host population. Parasites are assumed to attack all hosts at the same intensity. Perfect immunity, meaning that the parasites are able to reproduce only in nonimmunised hosts, is also assumed. Consequently, the frequency of nonimmunised individuals in the population is $1-\epsilon$. All parameters in this model are assumed to be positive and they have specific biological meaning. There is strong numerical and visual evidence that this system may have more than one attractor and that their natures may be different. For instance, one may have a locally asymptotically stable equilibrium point and periodic solutions of periods $3, 6$, and 12 or a locally asymptotically stable equilibrium point and chaotic attractor, etc. The corresponding basins of attraction seem to have the fractal nature.

Analyze system (2.105) using *Dynamica.*

Problem 19.

A system of difference equations

$$H_{n+1} = H_n e^{r(1-H_n) - \dfrac{aTP_n}{1 + aT_h H_n}}$$

$$P_{n+1} = H_n\left(1 - e^{-\frac{aTP_n}{1+aT_hH_n}}\right),$$

modeling the host–parasitoid interaction, was considered in several papers, see [H], [Roy], [Ro] and [KYH]. This model was motivated by attempts to understand the complicated dynamics of insect arthropod populations. Here P_n and H_n denote the population sizes in the n-th generation of the parasitoid and the host, respectively. All parameters in this model are assumed to be positive and they have specific biological meaning. There is strong numerical and visual evidence that this system may have more than one attractor and that their natures may be different. The corresponding basins of attraction seem to have a fractal nature. Even in the cases when all coexisting attractors are simple in nature (for example, a locally asymptotically stable equilibrium and a locally asymptotically stable periodic solution of a "small" period, such as $2, 3$ or 4) the fractal nature of their basins of attraction makes it difficult to determine on which of the attractors the solution will finally settle down. The situation is similar even for the simple basins of attraction for simple coexisting attractors if the initial conditions are taken in the boundaries between two or more basins of attraction.

Perform the following analysis for this system:

1. Find equilibrium points.
2. Check the linear stability of the equilibrium points.
3. Find periodic solutions of low periods $2, 3, 4$.
4. Visualize the stable and unstable manifolds for saddle points.
5. Plot the bifurcation diagram by fixing all parameters except one.
6. Compute Lyapunov exponents.

Problem 20.

A two-dimensional system of difference equations:

$$q_{n+1} = \frac{q_n f_A}{q_n f_A + (1 - q_n) f_a} \tag{2.106}$$

$$x_{n+1} = x_n(q_n f_A + (1 - q_n) f_a),$$

modeling the discrete time-evolution of allele frequency and population density was considered in [A], [CK], and [SR2]. In this model a diploid population with two alleles, A and a, at one locus is considered. The population density and the frequency of the A allele at n-th generation are denoted by x_n and q_n respectively. Hence, the population is divided into three subpopulations based on the genotypes AA, Aa, and aa. These genotypes have per-capita growth rate functions (fitnesses), f_{AA}, f_{Aa} and f_{aa}, which are functions of q_n and x_n. The allele fitnesses f_A and f_a are defined by

$$f_A = q_n f_{AA} + (1 - q_n) f_{Aa}, \quad f_a = q_n f_{Aa} + (1 - q_n) f_{aa}.$$

Assuming random mating and weak selection, one obtains system (2.106). The numerical evidence for period-doubling cascades has been presented in [A]. Carrying rigorous mathematical analysis period-doubling bifurcations has been established and illustrated in the case of exponential fitnesses, see [SR2].

Perform the following analysis for this system:
1. Find equilibrium points.
2. Check the linear stability of the equilibrium points.
3. Find periodic solutions of period-two (if they exist).
4. Visualize the stable and unstable manifolds.
5. Plot the bifurcation diagram by fixing all parameters except one.
6. Perform semicycle analysis for each component.

Problem 21.
A two-dimensional system of difference equations:

$$x_{n+1} = ax_n^b - d\alpha x_n^\nu y_n$$

$$y_{n+1} = y_n\left(\eta(p\alpha x_n^\nu - c) + 1\right),$$

models an open-access anchovy fishery, see [OC]. This model describes the change in resource abundance and the level of investment or effort engaged in at the fishery. Here, x_n is the biomass of anchovy in year n and y_n is the effort (hours fished) in year n, d is the biological discount factor, c is the cost per unit effort, p is the price of the anchovy, etc. Nonlinear optimization has been used to estimate some of these parameters. There is numerical and visual evidence for the existence of a stable equilibrium and stable periodic solutions, see [OC], but rigorous mathematical results are yet to be proved.

Perform the following analysis for this system:
1. Find equilibrium points.
2. Check the linear stability of the equilibrium points.
3. Find periodic solutions of period-two (if they exist).
4. Visualize the stable and unstable manifolds for saddle point equilibrium.
5. Plot the bifurcation diagram by fixing all parameters except one.
6. Perform semicycle analysis for each component.

Problem 22.

The discrete-time *SI* epidemic model, where S represents susceptibles and I represents infectives has the form

$$\begin{aligned} S_{n+1} &= S_n\left(1 - \alpha\tfrac{\Delta t}{N} I_n\right) \\ I_{n+1} &= I_n\left(1 + \alpha\tfrac{\Delta t}{N} S_n\right), \end{aligned} \tag{2.107}$$

with positive initial conditions $S_0 > 0$ and $I_0 > 0$ satisfying $S_0 + I_0 = N$. This model was considered in several papers, see [Al] and [AB] and references therein. The parameter $\alpha > 0$ is the contact rate, i.e., the average number of individuals with whom an infectious individual makes sufficient contact during a unit time interval. N is the total population size. The quantities S_n and I_n are the sizes of the susceptibles and infectives at time $n\Delta t$, respectively. Obviously, the sum of susceptibles and infectives remains constant throughout time and equals the total population, i.e.,

$$S_{n+1} + I_{n+1} = S_n + I_n = \ldots = S_0 + I_0. \tag{2.108}$$

If we assume that $\alpha\Delta t \le 1$, then $S_n > 0$ for all n, which implies that every solution of (2.107) converges monotonically to the equilibrium solution, i.e., $(S_n, I_n) \to (N, 0)$ as $n \to \infty$. The last statement means that, in the long run, the entire population becomes infected.

Perform the following analysis for this system:
1. Find equilibrium points.
2. Check the linear stability of the equilibrium points.
3. Prove the statement above on convergence to the equilibrium.
4. Use *Dynamica* to estimate the rate of convergence to the equilibrium.

Problem 23.

The discrete-time *SIS* epidemic model has been used to describe sexually transmitted diseases, see [Al] and [AB] and references therein. Individuals who are cured do not develop permanent immunity, but are immediately susceptible to the disease again. The system of difference equations has the form

$$\begin{aligned} S_{n+1} &= S_n\left(1 - \alpha\tfrac{\Delta t}{N} I_n\right) + \gamma\Delta t I_n \\ I_{n+1} &= I_n\left(1 - \gamma\delta t + \alpha\tfrac{\Delta t}{N} S_n\right), \end{aligned} \tag{2.109}$$

with positive initial conditions $S_0 > 0$ and $I_0 > 0$ satisfying $S_0 + I_0 = N$. The quantities S_n and I_n are the sizes of susceptibles and infectives at time $n\Delta t$, respectively. Here $\alpha > 0$ and $\gamma > 0$ are positive parameters and Δt is the time unit. The sum of population sizes S_n and I_n remains constant as (2.108) holds. Solutions of (2.109) are positive for all initial conditions if and only if

$$\gamma\Delta t \leq 1 \quad \text{and} \quad \alpha\Delta t < (1 + \sqrt{\gamma\Delta t})^2.$$

The dynamics of system (2.109) depends on the value of the basic reproductive rate $R = \alpha/\gamma$.

1. Show that if $R \leq 1$ every solution of (2.109) converges monotonically to the equilibrium, i.e., $(S_n, I_n) \to (N, 0)$ as $n \to \infty$.
2. Show that if $R > 1$, the substitutions

$$S_n = N - I_n, \quad x_n = \frac{\alpha\Delta I_n}{N(1 + (\alpha - \gamma)\Delta t)}$$

yield the logistic difference equation

$$x_{n+1} = px_n(1 - x_n), \tag{2.110}$$

where $p = 1 + (\alpha - \gamma)\Delta t$.

Perform the following analysis for this system:

1. Find equilibrium points.
2. Check the linear stability of the equilibrium points.
3. Prove the statement on the convergence to the equilibrium.
4. Use known results for (2.110) to discuss the possible behaviors of solutions of system (2.109).
5. Use *Dynamica* to plot bifurcation diagrams in the case where $R > 1$.

Discuss the generalization of system (2.109) of the form

$$\begin{aligned} S_{n+1} &= S_n(1 - \lambda_n) + (\beta + \gamma)I_n \\ I_{n+1} &= I_n(1 - (\beta + \gamma)) + \lambda_n S_n, \end{aligned} \tag{2.111}$$

where λ_n is the force of infection (number of contacts that result in infection per susceptible individual per unit time interval), β is the number of births or deaths per individual during the unit time interval, and γ is the removal number (number of individuals who recover in the unit time interval). Model (2.111) generalizes epidemic model (2.109) through the form of the force of infection. Assuming that $\lambda_n = \frac{\alpha}{N}I_n$ in (2.111) we obtain (2.109).

2.20 Applications in Economics

Cournot duopoly games

An **oligopoly** is a market structure where a few producers, each of considerable size, produce the same good or homogeneous goods, i.e., goods that are perfect substitutes. It is assumed that each company must take into account the actions of the competitors in choosing its own action, a property called *interdependence.* The first analysis of oligopoly was performed by A. Cournot [C] in the case of two companies and was naturally called *duopoly.* To explain the concept we can restrict our attention to this case. A duopoly can be modeled by assuming that at each discrete time period n two companies, denoted as C_1 and C_2, produce the quantities x_n and y_n respectively, and decide their productions for the next period, x_{n+1} and y_{n+1}, in order to maximize their expected profits. Because of interdependence, each profit depends on the price p_{n+1} at which the goods will be sold in period $n+1$, and such price depends on the total supply $Q_{n+1} = x_{n+1} + y_{n+1}$ according to a given demand function

$$p_{n+1} = D(Q_{n+1}).$$

For example, if the profit function of company C_1 is given by

$$P_1(x, y) = xD(x + y) - C(x),$$

or more precisely

$$P_1(x_n, y_n) = x_n D(x_n + y_n) - C(x_n),$$

where C represents the cost function, then its production for the period $n+1$ is decided by solving the optimization problem

$$x_{n+1} = \max P_1(x, y^e_{n+1}), \tag{2.112}$$

where y^e_{n+1} represents the expectation of company C_1 about the production decision of company C_2. By the assumptions from the original Cournot paper:

(i)

$$y^e_{n+1} = y_n,$$

that is, C_1 expects that the production of C_2 in the next period will be the same as in the current period, and

(ii) problem 2.112 has a unique solution

$$x_{n+1} = r_1(y^e_{n+1}) = r_1(y_n),$$

where r_1 is called the *reaction function.*

Since the reasoning of the companies is assumed to be identical we obtain that the time evolution of the considered duopoly system is described by the following system of difference equations:

$$\begin{aligned} x_{n+1} &= r_1(y_n) \\ y_{n+1} &= r_2(x_n), \end{aligned} \tag{2.113}$$

where

$$r_1 : X \to Y, \quad r_2 : Y \to X$$

are two reaction functions, obtained from two optimization problems:

$$\begin{aligned} \max P_1(x, y)_{x \in X} &= P_1(r_1(y), y), \ y \in Y; \\ \max P_2(x, y)_{y \in Y} &= P_2(x, r_2(x)), \ x \in X, \end{aligned} \tag{2.114}$$

where $X \subset [0, \infty)$ and $Y \subset [0, \infty)$ are the strategy sets from which the production choices x and y are taken. Define the mapping

$$T(x, y) = (r_1(y), r_2(x)). \tag{2.115}$$

Given an initial condition $(x_0, y_0) \in X \times Y$, a trajectory

$$\{(x_n, y_n)\}_{n=0}^{\infty} = \{T^n(x_0, y_0)\}_{n=0}^{\infty},$$

where T^n is the n-th iterate of the map (2.115), is known as the so-called *Cournot tatonnement*, see [BMG] and [L]. The practical interpretation of this process is the following: at time $n = 0$ company C_1 regards the output y_0 of company C_2 as fixed and, accordingly, chooses its strategy for the next period $n = 1$ by solving optimization problem (2.112), that is, by choosing its production level to be $x_1 = r_1(y_0)$. Likewise, by symmetry, company C_2 adjusts its level of production from the initial level y_0 to $y_1 = r_2(x_0)$, and so on. The fixed points of the map (2.115) are located at the intersections of two reaction functions and are called *Cournot-Nash equilibria* of the two-player game.

In most of the literature, including the original work of Cournot, the reaction functions r_1 and r_2 are assumed to be decreasing functions that intersect at an unique point of the positive quadrant $(\bar{x}, \bar{y})$, which satisfies:

$$\begin{aligned} \bar{x} &= r_1(\bar{y}) \\ \bar{y} &= r_2(\bar{x}). \end{aligned} \tag{2.116}$$

The equations in system (2.113) can be decoupled to give

$$\begin{aligned} x_{n+1} &= r_1(r_2(x_{n-1})) = F(x_{n-1}) \\ y_{n+1} &= r_2(r_1(y_{n-1})) = G(y_{n-1}), \end{aligned} \tag{2.117}$$

where $F = r_1 \circ r_2$ and $G = r_2 \circ r_1$ are decreasing functions. The equations in system (2.117) can be further decoupled to separate the dynamics of even-numbered and odd-numbered terms

$$\begin{aligned} u_{k+1} &= F(u_k) \\ v_{k+1} &= F(v_k) \\ w_{k+1} &= G(w_k) \\ z_{k+1} &= G(z_k), \end{aligned} \tag{2.118}$$

where we set:

$$u_k = x_{2k+1}, \quad v_k = x_{2k}, \quad w_k = y_{2k+1}, \quad z_k = y_{2k}, \quad k = 0, 1, 2, \ldots .$$

Then, the dynamics of the original system (2.113) is the same as two one-dimensional dynamics, which are, in general, well understood. By choosing the functions r_1 and r_2 appropriately we can obtain different dynamics for the system. For example, if r_1 and r_2 are decreasing functions, which is assumed in most of the literature, then the dynamics is quite simple. Assuming that the reaction functions r_1 and r_2 are unimodal would produce quite complex dynamics, with periodic and chaotic trajectories, sensitive dependence on initial conditions, etc. See the pioneering work [Ra], and more recent papers [P1] and [Ko]. In [P1] the reaction function is obtained by assuming a linear cost function and a rational demand function of the form $1/x$, which combines for an unimodal function and leads to complex dynamics. In [Ko], a linear demand function and a quadratic cost function are assumed leading to the standard logistic function:

$$r_1(x) = \mu x(1 - x).$$

In this case, Cournot map takes the form

$$C(x, y) = (r_1(y), r_2(x)), \tag{2.119}$$

where

$$r_2(x) = \nu x(1 - x).$$

In this case, the one-dimensional maps F and G take the form

$$F(x) = r_1 \circ r_2(x) = \mu\nu x(1 - x)(1 + \nu x(x - 1)),$$

$$G(y) = r_2 \circ r_1(x) = \mu\nu y(1 - y)(1 + \mu y(y - 1)).$$

In [Ko] the local stability of the fixed points in the symmetric case of identical producers, $\mu = \nu$, has been studied. In [BMG] some results about the periodic solutions and their basins of attractions have been established and exhaustive visual analysis has been performed in the nonsymmetric case.

Consider the system

$$\begin{aligned} x_{n+1} &= \mu y_n(1 - y_n) \\ y_{n+1} &= \nu x_n(1 - x_n), \end{aligned} \tag{2.120}$$

where $\mu > 0, \nu > 0$, and the initial conditions x_0 and y_0 are positive.

Perform the following analysis for this system:

1. Find equilibrium points.
2. Check the linear stability of the equilibrium points.
3. Find periodic solutions of period two and four.
4. Visualize the stable and unstable manifolds.

5. Plot the bifurcation diagram by fixing all parameters except one.
6. Compute Lyapunov exponents.
7. Perform semicycle analysis for each component.

For more applications of this theory including reaction functions with stronger coupling see [Ag], [AGP], [BGK], [BMG], [P], [P1], and [P2]. For example, some qualitative properties of the system

$$x_{n+1} = (1-\rho)x_n + \rho\mu y_n(1-y_n)$$

$$y_{n+1} = (1-\rho)y_n + \rho\mu x_n(1-x_n),$$

where ρ and μ are positive parameters, have been established in [Ag] . A similar phenomenon is modeled by the system of difference equations

$$x_{n+1} = A\sqrt{y_n} - 2y_n$$

$$y_{n+1} = B\sqrt{x_n + y_n} - x_n - y_n,$$

where A and B are positive parameters in [P2] . The last two systems of difference equations may be analyzed with *Dynamica*.

2.21 Exercises

Exercise 2.1
Use `RSolve` to find the general solution of the following difference equations:

1. $x_{n+1} - 5x_n + 6x_{n-1} = 0$
2. $x_{n+1} - 4x_n + 4x_{n-1} = 0$
3. $x_{n+1} - x_n + x_{n-1} = 0$
4. $x_{n+1} - 2x_n + x_{n-1} = n + n^2$

Exercise 2.2
Use `RSolve` to find the solution of the following initial value problems:

1. $x_{n+1} - x_n - 2x_{n-1} = 0, \quad x_{-1} = 1,\ x_0 = 2$
2. $x_{n+1} + 2x_n + x_{n-1} = 0, \quad x_{-1} = a,\ x_0 = b$
3. $x_{n+1} + 2x_n + 2x_{n-1} = 0 \quad x_{-1} = a,\ x_0 = b$

Exercise 2.3

Use `RSolve` to find the general solution of the following systems of difference equations:

1.
$$x_{n+1} = 8x_n - 3y_n$$
$$y_{n+1} = 6x_n - y_n$$

2.
$$x_{n+1} = x_n + y_n$$
$$y_{n+1} = -x_n + 3y_n$$

3.
$$x_{n+1} = -x_n + 2y_n$$
$$y_{n+1} = -5x_n + 5y_n$$

Exercise 2.4

Use `RSolve` to find the solution of the following initial value problems:

1.
$$x_{n+1} = x_n + y_n$$
$$y_{n+1} = 2x_n + y_n$$
$$x_0 = 1, y_0 = 1.$$

After you find the solution, check the values of the ratios y_n/x_n for $n = 1, 2, \ldots, 7$ and compare it to the value of $\sqrt{2}$.

2.
$$x_{n+1} = x_n + y_n$$
$$y_{n+1} = 5x_n + y_n$$
$$x_0 = 1, y_0 = 1.$$

After you find the solution, check the values of the ratios y_n/x_n for $n = 2, \ldots, 10$ and compare it to the value of $\sqrt{5}$.

3.
$$x_{n+1} = x_n + y_n$$
$$y_{n+1} = ax_n + y_n$$
$$x_0 = 1, y_0 = 1,$$

where a is a positive number. Find the connection between the sequence of the values of the ratios y_n/x_n for $n = 1, 2, \ldots$ and $\sqrt{a}$.

Exercise 2.5

Consider the difference equation

$$x_{n+1} = A + Bx_{n-1} - |x_n|, \tag{2.121}$$

where A and B are parameters. This equation is a piecewise linear version of the Henon equation and is known as the Lozi equation - see [K] and references therein.

1. Find the equilibrium solutions and the period-two solutions of equation (2.121) and investigate their local stability.
2. Use *Dynamica* to study the semicycles.
3. Use *Dynamica* to plot the bifurcation diagram and Lyapunov numbers.

Exercise 2.6

Consider the difference equation

$$x_{n+1} = x_n\left(A(1 - x_n) - B(Ax_{n-1}(1 - x_{n-1}) - x_n\right), \tag{2.122}$$

where A and B are the parameters. This equation is a model for the predator–prey interaction and has an interesting dynamics, see [KJ] and references therein.

Exercise 2.7

Consider the system of difference equations

$$\begin{aligned} x_{n+1} &= x_n \cos\theta - (y_n - x_n^2)\sin\theta \\ y_{n+1} &= x_n \sin\theta - (y_n - x_n^2)\cos\theta \end{aligned} \tag{2.123}$$

where A is a parameter. This equation is known as the Cremona equation, and the corresponding map is known as a Cremona map, see [K] and [KJ] and references therein.

1. Find the equilibrium solutions of equation (2.123) and investigate their local stability.
2. Show that the map generated by this equation is area preserving.
3. Use *Dynamica* to find a nontrivial invariant. Use this invariant to try to find a Lyapunov function.

4. Use *Dynamica* to study semicycles.
5. Use *Dynamica* to plot bifurcation diagram and the Lyapunov numbers.

Exercise 2.8

Consider the difference equation

$$x_{n+1} = -x_{n-1} + 2Ax_n + 4(1-A)\frac{x_n^2}{1+x_n^2}, \tag{2.124}$$

where A is a parameter. This equation is known as Mira's equation, and the corresponding map is known as Mira's map, see [GM] and [Mi].

1. Find the equilibrium solutions and the period-two solutions of equation (2.124) and investigate their local stability.
2. Use *Dynamica* to find a nontrivial invariant. Use this invariant to find the Lyapunov function.
3. Use *Dynamica* to study semicycles.
4. Use *Dynamica* to plot bifurcation diagrams and Lyapunov numbers.

Exercise 2.9

Consider the difference equation

$$x_{n+1} = A\frac{x_n^2}{(1+x_n)x_{n-1}}, \tag{2.125}$$

where A is a parameter. This equation is known as May's host parasitoid equation, and the corresponding map is known as May's map, see [LTT].

1. Find the equilibrium solutions and the period-two solutions of equation (2.125) and investigate their local stability.
2. Use *Dynamica* to find a nontrivial invariant.
3. Use *Dynamica* to study semicycles.
4. Use *Dynamica* to plot bifurcation diagrams and Lyapunov numbers.

Exercise 2.10

Consider the difference equation

$$x_{n+1} = \frac{px_n + qx_{n-1}}{1+x_n}, \quad x_{-1} > 0, x_0 > 0, n = 0, 1, \ldots, \tag{2.126}$$

where $p > 0$ and $q > 0$ are the parameters. This equation is one of the second-order rational difference equations considered in [KL] and [KLP1].

1. Find equilibrium solutions of equation (2.126).
2. Investigate local stability of the equilibrium solutions.
3. Use *Dynamica* to study semicycles.
4. Find invariant intervals.

Exercise 2.11

Consider the difference equation

$$x_{n+1} = \frac{p + x_{n-1}}{qx_n + x_{n-1}}, \qquad x_{-1} > 0, x_0 > 0, n = 0, 1, \ldots, \tag{2.127}$$

where $p > 0$ and $q > 0$ are the parameters. This equation is one of the second-order rational difference equations considered in [KL] and [KKLT].

1. Find the equilibrium solutions of equation (2.127) and investigate their local stability.
2. Find the period-two solutions.
3. Use *Dynamica* to study semicycles.
4. Find invariant intervals.

Exercise 2.12

Consider the difference equation

$$x_{n+1} = \frac{p + x_n}{1 + x_n + rx_{n-1}}, \qquad x_{-1} > 0, x_0 > 0, n = 0, 1, \ldots, \tag{2.128}$$

where $p > 0$ and $q > 0$ are the parameters. This equation was considered in [KL] and [KLMR].

1. Find the equilibrium solutions of equation (2.128) and investigate their local stability.
2. Prove that all solutions of this equation are bounded.
3. Use *Dynamica* to study semicycles. Try to prove some of the observed regularities about semicycles.
4. Find invariant intervals.
5. Try to prove the global attractivity of the equilibrium solutions in the case where $p \geq q + q^2 r$ or $q > p(1 + r)$.

Exercise 2.13

Consider the difference equation

$$x_{n+1} = -x_{n-1} + \frac{x_n - A}{2B^2 x_n^2}, \quad x_{-1} \neq 0, x_0 \neq 0, n = 0, 1, \ldots, \tag{2.129}$$

where A and $B \neq 0$ are the parameters. This equation is known as one of the discrete versions of the Korteweg–deVries equation, see [QRT1] and [QRT2].

1. Find the equilibrium solutions of equation (2.129) and investigate their local stability.
2. Find the period-two solutions of equation (2.129) and investigate their local stability.
3. Use *Dynamica* to find a nontrivial invariant. Use this invariant to find the Lyapunov function.
4. Use *Dynamica* to study semicycles.
5. Use *Dynamica* to plot bifurcation diagrams.

Exercise 2.14

Consider the difference equation

$$x_{n+1} = \frac{x_n^k + a}{x_n^\ell x_{n-1}}, \quad x_{-1} > 0, x_0 > 0, n = 0, 1, \ldots, \tag{2.130}$$

where k, ℓ and a are positive. This equation has been investigated in detail in [HT].

1. Find the equilibrium solutions of equation (2.130) and investigate their local stability.
2. Find the period-two solutions of equation (2.130) and investigate their local stability.
3. Use *Dynamica* to find a nontrivial invariant of parameter $a = (2^{k-\ell-2} - 1)/2^k$. Use this invariant to find the Lyapunov function.
4. Use *Dynamica* to study semicycles.
5. Use *Dynamica* to plot bifurcation diagrams.

Exercise 2.15

Consider the difference equation

$$x_{n+1} = -x_n - x_{n-1} + b + \frac{c}{x_n}, \quad x_{-1} > 0, x_0 > 0, n = 0, 1, \ldots, \tag{2.131}$$

where b and c are parameters. This equation is the simplest form of the discrete Painlevé equations. It has been investigated by many authors, see [FIK] and [BTR].

1. Find the equilibrium solutions of equation (2.131) and investigate their local stability.
2. Use *Dynamica* to find a nontrivial invariant. Use this invariant to find the Lyapunov function.
3. Use *Dynamica* to plot the orbits together with the nontrivial invariants.
4. Use *Dynamica* to plot bifurcation diagrams.

Exercise 2.16
Consider the system of difference equations

$$\begin{aligned} x_{n+1} &= (1-a)x_n - y_n^2 \\ y_{n+1} &= (1+b)y_n + x_n y_n \end{aligned} \tag{2.132}$$

where a and b are parameters. This system is obtained from a discretization of the differential equation which appears in hydrodynamics and is known as Burgers equation, and the corresponding map is known as Burgers map, see [MW] and references therein.

1. Find the equilibrium solutions of equation (2.132) and investigate their local stability.
2. Check if Burgers map has a period-two point.
3. Use *Dynamica* to visualize local stable and unstable manifolds.
4. Use *Dynamica* to plot bifurcation diagrams and the Lyapunov numbers.
5. Use *Dynamica* to find, if possible, a nontrivial invariant.

Exercise 2.17
Consider the system of difference equations

$$\begin{aligned} x_{n+1} &= (a - x_n - by_n)x_n \\ y_{n+1} &= (a - cx_n - y_n)y_n \end{aligned} \tag{2.133}$$

where $a, b,$ and c are parameters. This system has an interesting dynamics, see [EMTW].

1. Find the equilibrium solutions of equation (2.133) and investigate their local stability.
2. Check if (2.133) has a period-two point.
3. Use *Dynamica* to visualize local stable and unstable manifolds.
4. Use *Dynamica* to plot bifurcation diagrams and the Lyapunov numbers.

Exercise 2.18

Consider the system of difference equations

$$\begin{aligned} x_{n+1} &= a + x_n^2 - y_n^2 \\ y_{n+1} &= b + 2x_n y_n \end{aligned} \tag{2.134}$$

where a and b are parameters. The map generated by this system is known as the Mandelbrot map, see [B], [E2], [Mt], and [PJS]. It is customary to represent system (2.134) in the complex plane as

$$z_{n+1} = z_n^2 + c,$$

where $c = a + ib$. The dynamics of this map is highly complicated and is described in detail in [D1], [E2], and [PJS].

1. Find the equilibrium solutions of equation (2.134) and investigate their local stability.
2. Find the period-two point of (2.134) and investigate its local stability. Visualize the stability regions of equilibrium points and period-two points in the parametric plane (a, b).
3. Find the inverse map of the Mandelbrot map.
4. Use *Dynamica* to plot bifurcation diagrams.

Exercise 2.19

Consider the system of difference equations

$$\begin{aligned} x_{n+1} &= A + B(x_n \cos(x_n^2 + y_n^2) - y_n \sin(x_n^2 + y_n^2)) \\ y_{n+1} &= B(x_n \sin(x_n^2 + y_n^2) + y_n \cos(x_n^2 + y_n^2)), \end{aligned} \tag{2.135}$$

where A and B are parameters with $|B| < 1$. The map generated by this system is one version of the so-called Ikeda map, see [ASY] and [Ly], and references therein. This equation was proposed as a model of the type of cell that might be used in an optical computer.

It is customary to represent system (2.135) in the complex plane as

$$z_{n+1} = A + Bz_n e^{i|z_n|^2},$$

where $z_n = x_n + iy_n$. The dynamics of this map is highly complicated and is described in detail in [E2] and [PJS].

1. Find the equilibrium solutions of equation (2.135) and investigate their local stability.

2. Find the period-two point of (2.135) and investigate its local stability. Visualize the stability regions of equilibrium points and period-two points in the parametric plane (A, B).

3. Find the inverse map of Ikeda map.

Exercise 2.20

Consider the system of difference equations

$$x_{n+1} = \begin{cases} \frac{x_n + x_{n-1}}{2} & \text{if } x_n + x_{n-1} \text{ is an even integer,} \\ \frac{b|x_n + x_{n-1}|+1}{2} & \text{if } x_n + x_{n-1} \text{ is an odd integer,} \end{cases} \quad n = 1, 2, \ldots \qquad (2.136)$$

where $b \geq 1$ is an odd integer and x_0 and x_1 are integers. This equation was considered in [Cl] as a two-dimensional analogue of the Collatz $3x+1$ equation (1.8).

1. Find the limit of every solution for $b = 1$.

2. Find the limit of every solution for $b = 3, 5$.

3. Show that for $b \geq 7$ there exist unbounded solutions.

Exercise 2.21

Consider the system of difference equations

$$x_{n+1} = \begin{cases} \frac{x_n + x_{n-1}}{2} & \text{if } x_n + x_{n-1} \text{ is an even integer,} \\ x_n - x_{n-1} & \text{if } x_n + x_{n-1} \text{ is an odd integer,} \end{cases} \quad n = 1, 2, \ldots \qquad (2.137)$$

where x_0 and x_1 are integers. This equation was considered in [CL1] as another two-dimensional analogue of the Collatz $3x + 1$ equation (1.8). Denote by $gcod(x_0, x_1)$ the greatest common odd divisor of x_0 and x_1.

Show that if $x_0 \neq x_1$ and $gcod(x_0, x_1) = 1$, then every solution converges to 1 or -1 or is eventually equal to the period-six solution $\{3, 2, -1, -3, -2, 1\}$. First use *Dynamica* to simulate this system and to get an idea of how to prove this result.

Exercise 2.22

Consider the system of difference equations

$$\begin{aligned} x_{n+1} &= \left\lfloor \frac{x_n + y_n}{2} \right\rfloor \\ y_{n+1} &= y_n - x_n, \end{aligned} \qquad n = 0, 1, \ldots \tag{2.138}$$

where x_0 and y_0 are integers and $\lfloor\ \rfloor$ denotes the greatest-integer (floor) function. This equation was considered in [CL2] as the mathematical model of a card game.

1. Find an invariant of a linear system obtained from (2.138) by omitting the floor function.

2. Find the inverse map of the map generated by the system (2.138).

3. Find all period-two and period-three solutions of (2.138).

Chapter 3

Systems of Difference Equations, Stability, and Semicycles

3.1 Introduction

In this chapter we present some basic results about the solutions of the system of difference equations of the form

$$\mathbf{x}_{n+1} = \mathbf{f}(\mathbf{x}_n), \quad n = 0, 1, \ldots \tag{3.1}$$

where $\mathbf{f} : R^k \to R^k$ is a given function.

Basic results and definitioins concerning the linear theory of systems of difference equations are introduced in Sections 3.2. Stability of equilibrium and periodic solutions of (3.1) is discussed in Section 3.3, and in Section 3.4 we present two criteria for stability of systems. Also, stability of linear systems of equations is considered further in 3.5.

We present in Section 3.8 basic results about the stability character of solutions of the k-th order difference equation

$$x_{n+1} = F(x_n, x_{n-1}, \ldots, x_{n+1-k}), \tag{3.2}$$

where F is a given function. Limit sets and invariant manifolds are discussed in Section 3.6. Section 3.9 presents briefly results on semicycles of solutions of (3.1) that are natural generalizations of some concepts we have introduced in Chapter 2. The *Dynamica* sessions include the solution of some linear systems by using *Mathematica*'s `RSolve` package in Section 3.2, and a semicycle analysis of certain equations using *Dynamica* in Section 3.10.

3.2 Linear Theory

In this section we present notions and standard results from the theory of linear systems and equations. Proofs of most statements given here can be found in introductory textbooks on difference equations such as [E1], [KP], and [LT]. Proofs of more advanced results can be found in [R].

As in the two-dimensional case we first consider the homogeneous system of the form

$$\mathbf{Z}_{n+1} = A\mathbf{Z}_n \tag{3.3}$$

where A is a constant, real $k \times k$ matrix and $\mathbf{Z}_n = (x_n^1, x_n^2, \ldots, x_n^k)^T$. Here $\mathbf{v}^T$ denotes the transpose of the vector $\mathbf{v}$. All basic notions of solution, general solution and fundamental matrix are the same as in the two-dimensional case given in Section 2.2. As in the two-dimensional case the solution of (3.3) has the form

$$\mathbf{Z}_n = A^n\mathbf{Z}_0. \tag{3.4}$$

Computation of A^n may be performed by transforming the matrix A into the Jordan Normal Form and then computing the n-th power of the transformed matrix. In the general case of a $k \times k$ matrix, the Jordan Normal Form is more complicated as there are more possibilities for the location of the eigenvalues relative to the unit circle.

3.2.1 Jordan Normal Form

In this section we review the basic facts about linear algebra and the Jordan Normal Form or Jordan Canonical Form, see [E1], [L], and [R]. Recall that for a real $k \times k$ matrix A, an eigenvalue of A is a real or complex number λ such that $A\mathbf{z} = \lambda\mathbf{z}$ for some nonzero vector $\mathbf{z} \in C^k$, that is,

$$(A - \lambda I)\mathbf{z} = \mathbf{0}.$$

This equation has a nonzero solution $\mathbf{z}$ if and only if λ satisfies

$$\det(A - \lambda I) = 0 \tag{3.5}$$

or

$$\lambda^k + a_1\lambda^{k-1} + \ldots + a_k = 0. \tag{3.6}$$

Each of equations (3.5) and (3.6) is called **the characteristic equation** of matrix A and its roots are called the **eigenvalues** of A (some of them may repeat and some may be complex numbers).

Let A be a $k \times k$ real matrix. First, suppose that there exists a $k \times k$-dimensional basis consisting of the (complex) eigenvectors $\mathbf{v}^1, \ldots, \mathbf{v}^k$. Defining $V = \left(\mathbf{v}^1, \ldots, \mathbf{v}^k\right)$, we obtain $AV = VD$ where $D = diag(\lambda_1, \ldots, \lambda_k)$, that is, D is a matrix having zeros off diagonal and $\lambda_1, \ldots, \lambda_k$ on the main diagonal. Thus, $D = V^{-1}AV$.

If an eigenvalue $\lambda_j = \alpha_j + i\beta_j$ is a complex number, then the corresponding eigenvector $\mathbf{v}^j = \mathbf{u}^j + i\mathbf{w}^j$ must be complex, too. Since A is real, the complex conjugate $\bar{\lambda}_j = \alpha_j - i\beta_j$ is also an eigenvalue and with the corresponding eigenvector $\bar{\mathbf{v}}^j = \mathbf{u}^j - i\mathbf{w}^j$. Since $A\mathbf{v}^j = \lambda\mathbf{v}^j$, we have

$$A\left(\mathbf{u}^j + i\mathbf{w}^j\right) = \left(\alpha_j\mathbf{u}^j - \beta_j\mathbf{w}^j\right) + i\left(\beta_j\mathbf{u}^j + \alpha_j\mathbf{w}^j\right).$$

Equating the real and imaginary parts we obtain

$$A\mathbf{u}^j = \alpha_j \mathbf{u}^j - \beta_j \mathbf{w}^j,$$

$$A\mathbf{w}^j = \beta_j \mathbf{u}^j + \alpha_j \mathbf{w}^j.$$

Using the vectors $\mathbf{u}^j$ and $\mathbf{w}^j$, instead of $\mathbf{v}^j$ and $\bar{\mathbf{v}}^j$ as the part of a vector basis yields the block matrix of the form

$$D_j = \begin{pmatrix} \alpha_j & \beta_j \\ -\beta_j & \alpha_j \end{pmatrix}. \tag{3.7}$$

Consequently, if A has a basis B consisting of complex eigenvectors, then there is a real basis B_R in terms of which $A = diag(B_1, \ldots, B_p)$ where each of the blocks B_j is either a single eigenvalue λ_j or is of the form D_j given above.

Finally, we consider the case of repeated eigenvalues where the eigenvectors do not span the whole space. If A has a characteristic polynomial $P(\lambda)$, then by substituting A for λ we obtain $P(A)\mathbf{v} = 0$ for all vectors $\mathbf{v}$. This result is known as the Cayley-Hamilton Theorem. In particular, if $\lambda_1, \ldots, \lambda_q$ are the distinct eigenvalues with multiplicities $m_1, \ldots, m_q$, then

$$S_j = \{\mathbf{v} \in C^k : (A - \lambda_j I)^{m_j} \mathbf{v} = \mathbf{0}\}$$

is a vector space of dimension m_j. Vectors in S_j are called **generalized eigenvectors**. Consider an eigenvalue $\lambda = \lambda_j$ with multiplicity m. Then there are vectors $\left(\mathbf{v}^1, \ldots, \mathbf{v}^m\right)$ such that $(A-\lambda_j I)\mathbf{v}^1 = \mathbf{0}$ and $(A-\lambda_j I)\mathbf{v}^j = \mathbf{v}^{j-1}$ for $2 \leq j \leq m$. The corresponding $m \times m$ block matrix has the form

$$J_i = \begin{pmatrix} \lambda & 1 & 0 & \ldots & 0 & 0 \\ 0 & \lambda & 1 & \ldots & 0 & 0 \\ 0 & 0 & \lambda & \ldots & 0 & 0 \\ \vdots & & & \ddots & & \vdots \\ 0 & 0 & 0 & \ldots & \lambda & 1 \\ 0 & 0 & 0 & \ldots & 0 & \lambda \end{pmatrix} \tag{3.8}$$

and is called a **Jordan block**.

If we use the real and imaginary parts of the eigenvectors for the complex eigenvalues to form the basis of real vectors in the way presented here, we obtain the **Real Jordan Canonical Form** form for matrix A,

$$A = diag(J_1, \ldots, J_m)$$

where the Jordan blocks J_l are of one of the following four types:

(i) $J_l = \lambda_l$ for some real eigenvalue λ_l;

(ii) J_l is of type (3.8) for some real eigenvalue λ_l;

(iii) J_l is of type (3.7) for some complex eigenvalue $\lambda_l = \alpha_l + i\beta_l$;

(iv) J_l is of type

$$J_l = \begin{pmatrix} D_j & I & 0 & \dots & 0 & 0 \\ 0 & D_j & I & \dots & 0 & 0 \\ 0 & 0 & D_j & \dots & 0 & 0 \\ \vdots & & & \ddots & & \vdots \\ 0 & 0 & 0 & \dots & D_j & I \\ 0 & 0 & 0 & \dots & 0 & D_j \end{pmatrix}$$

where D_j is of type (3.7) for some complex eigenvalue $\lambda_l = \alpha_l + i\beta_l$ and I is a 2×2 identity matrix.

THEOREM 3.1 (Jordan Normal Form)
Every $k \times k$ real matrix is similar to its Jordan Normal Form.

3.2.2 Linear Nonhomogeneous System

In this section we consider the initial value problem

$$\mathbf{x}_{n+1} = A\mathbf{x}_n + \mathbf{b}_n, \quad \mathbf{x}_0 = \mathbf{d} \;\; n = 0, 1, \dots \tag{3.9}$$

where $k \in \{1, 2 \dots\}$, $\mathbf{d} \in R^k$, A is a real $k \times k$ matrix, and $\mathbf{b}_n \in R^k$ for $n = 0, 1, \dots$.
The next formula gives an analogue of (2.8) for the two-dimensional systems.
The IVP (3.9) has the unique solution given by

$$\mathbf{x}_n = A^n\mathbf{d} + \sum_{i=0}^{n} A^{n-i}\mathbf{b}_i, \quad n = 0, 1, \dots. \tag{3.10}$$

In particular, if $\mathbf{b}_n = \mathbf{b}$, a constant vector, then we obtain:

$$\mathbf{x}_n = A^n\mathbf{d} + \sum_{i=0}^{n} A^{n-i}\mathbf{b}, \quad n = 0, 1, \dots.$$

The basic results for systems (3.3) and (3.9) can be extended to the more general linear systems of the form

$$\mathbf{Z}_{n+1} = A_n\mathbf{Z}_n, \quad n = 0, 1, \dots \tag{3.11}$$

and

$$\mathbf{Z}_{n+1} = A_n\mathbf{Z}_n + B_n, \quad \mathbf{Z}_0 = \mathbf{d}, \tag{3.12}$$

where A_n is a real $k \times k$ matrix for every $n = 1, 2, \dots$.

3.2.3 Linear Difference Equations of k-th Order

The linear difference equation of order k has the form

$$x_{n+1} + a_{k-1}x_n + \cdots + a_0 x_{n-k+1} = b_n, \quad n = 0, 1, \ldots \tag{3.13}$$

where $k \geq 1$. The coefficients $a_k, a_{k-1}, \ldots, a_0$ of (3.13) are real numbers, and the initial condition of (3.13) is $(x_0, x_{-1}, \ldots, x_{-k+1}) \in R^k$. We also assume that $a_0 \neq 0$.

It can be shown that a simple change of variables can transform (3.13) into an associated first order linear system of the form (3.9) where A is a real $k \times k$ matrix, $\{\mathbf{b}_n\}_{n=0}^{\infty}$ is a sequence in R^k, and $\mathbf{x}_0 \in R^k$. The most standard change of variables is:

$$\begin{aligned} x_n^1 &= x_{n-k+1} \\ x_n^2 &= x_{n-k+2} \\ &\vdots \\ x_n^k &= x_n, \quad , n = 0, 1, \ldots, \end{aligned}$$

which transforms (3.13) into the system of the form (3.9):

$$\begin{pmatrix} x_{n+1}^1 \\ x_{n+1}^2 \\ \vdots \\ x_{n+1}^{k-1} \\ x_{n+1}^k \end{pmatrix} = \begin{pmatrix} 0 & 1 & 0 & \cdots 0 & 0 \\ 0 & 0 & 1 & \cdots 0 & 0 \\ \vdots & \vdots & \vdots & \vdots & \vdots \\ 0 & 0 & 0 & \cdots 0 & 1 \\ -a_{k-1} & -a_{k-2} & -a_{k-3} & \cdots -a_1 & -a_0 \end{pmatrix} \begin{pmatrix} x_n^1 \\ x_n^2 \\ \vdots \\ x_n^{k-1} \\ x_n^k \end{pmatrix} + \begin{pmatrix} 0 \\ 0 \\ \vdots \\ 0 \\ b_n \end{pmatrix} \quad n = 0, 1, \ldots$$

The same change of variables can be applied to the general nonlinear difference equation of the form (3.2) to reduce it to the system of the form (3.1).

Mathematica has the built-in package `RSolve` that computes the solution to a given difference equation or a system of difference equations using the method of generating functions. The equations and systems that can be solved are mainly linear. With `RSolve` one can find the general solution of the system of equations, as well as particular solutions with prescribed initial conditions.

This loads the `RSolve` package.

```
In[70]:= << DiscreteMath`RSolve`
```

RSolve can solve a linear system of difference equations. Notice that it is possible to get the solution in pure function form by specifying $\{a, b\}$ rather than $\{a_n, b_n\}$.

```
In[71]:= soln = RSolve[
              {a[n + 1] - b[n] + 2a[n] == 1,
              a[n + 1] - 3b[n + 1] + b[n] == n,
              a[0] == b[0] == 0}, {a, b}, n]
```

Out[71]=
$$\left\{\left\{a \to \left(\frac{1}{50}\left(40 + \left(\frac{1}{3}\left(-2-\sqrt{10}\right)\right)^{\#1}\left(-20+\sqrt{10}\right) - 20\left(\frac{1}{3}\left(-2+\sqrt{10}\right)\right)^{\#1} - \sqrt{10}\left(\frac{1}{3}\left(-2+\sqrt{10}\right)\right)^{\#1} - 10\,\#1\right)\&\right), b \to \left(\frac{1}{25}3^{-\#1}\left(10\,3^{1+\#1} - 15\left(-2+\sqrt{10}\right)^{\#1} - 4\sqrt{10}\left(-2+\sqrt{10}\right)^{\#1} + \left(-2-\sqrt{10}\right)^{\#1} \times \left(-15+4\sqrt{10}\right) - 5\,3^{1+\#1}\,\#1\right)\&\right)\right\}\right\}$$

Here are the first 7 terms in the solution for a and b.

```
In[72]:= Table[(a[n]/.soln)[[1]],
              {n, 0, 7}]//Expand
```

Out[72]=
$$\left\{0, 1, -\frac{2}{3}, \frac{17}{9}, -\frac{80}{27}, \frac{395}{81}, -\frac{2222}{243}, \frac{10529}{729}\right\}$$

```
In[73]:= Table[(b[n]/.soln)[[1]],
              {n, 0, 7}]//Expand
```

Out[73]=
$$\left\{0, \frac{1}{3}, -\frac{4}{9}, -\frac{5}{27}, -\frac{166}{81}, -\frac{95}{243}, -\frac{3532}{729}, \frac{2623}{2187}\right\}$$

This is a homogeneous three-dimensional system of equations.

```
In[74]:= soln = RSolve[
         {a[n + 1] + 13a[n] + 12a[n] + 18c[n] == 0,
            b[n + 1] - 7a[n] - 7b[n] - 9c[n] == 0,
            c[n + 1] - 5a[n] - 4b[n] - 8c[n] == 0},
            {a[0] == 0, b[0] == 1, c[0] == 2},
            {a, b, c}, n]
```

```
Out[74]= {{a → (- 36 11^#1/53 + 1/30581 (9 2^(1-#1) (
    (577 - 63 √577) (- 21 + √577)^#1 +
    (- 21 - √577)^#1 (577 + 63 √577))) &),
  b → (9 11^(1+#1)/53 + 1/30581 (2^(-#1) (
    (- 21 - √577)^#1 (- 13271 + 247 √577) -
    (- 21 + √577)^#1 (13271 + 247 √577))) &),
  c → (72 11^#1/53 - 1/30581 (2^(-#1) (
    (- 21 - √577)^#1 (- 9809 + 625 √577) -
    (- 21 + √577)^#1 (9809 + 625 √577))) &)}}
```

These are the terms a_n of the solution for $2 \leq n \leq 5$.

```
In[75]:= Table[(a[n]/.soln)[[1]],
    {n, 2, 5}]//Expand
Out[75]= {540, -14940, 305964, -7220700}
```

3.3 Stability of Linear Systems

3.3.1 Norms of Vectors and Matrices

We start this section by introducing the notion of norms and matrices that will be needed in our definition of the stability of linear systems.

DEFINITION 3.1 *A nonnegative function on R^k is called a* **norm** *and is denoted by $\|\ \|$, if the following properties hold:*

(i) $\|\mathbf{x}\| = 0$ *only if* $\mathbf{x} = \mathbf{0}$;

(ii) $\|a\mathbf{x}\| = |a|\|\mathbf{x}\|$ *for all* $\mathbf{x} \in R^k$ *and* $a \in R$;

(iii) $\|\mathbf{x} + \mathbf{y}\| \leq \|\mathbf{x}\| + \|\mathbf{y}\|$ *for all* $\mathbf{x}, \mathbf{y} \in R^k$.

The three most commonly used norms on R^k are:

1.) the L_1 norm $\|\mathbf{x}\|_1 = \sum_{i=1}^{k} |x_i|$;

2.) the L_2 norm $\|\mathbf{x}\|_2 = \left(\sum_{i=1}^{k} x_i^2\right)^{1/2}$;

3.) the L_∞ norm $\|\mathbf{x}\|_\infty = \max_{1\le i\le k} |x_i|$.

It is a basic fact from linear algebra that all norms on R^k are equivalent. That is, that if $\| \ \|$ and $\| \ \|'$ are any two norms on R^k, then there exist constants $\rho, \sigma > 0$ such that

$$\rho\|x\| \le \|x\|' \le \sigma\|x\|, \quad \text{for every } x \in R^k.$$

An important consequence of the equivalence of the norms is that if $\{\mathbf{x}_n\}$ is a sequence in R^k, then

$$\|\mathbf{x}_n\| \to 0 \quad \text{if and only if} \quad \|\mathbf{x}_n\|' \to 0 \quad \text{as } n \to \infty.$$

For each vector norm $\| \ \|$ on R^k one may define an associated matrix norm of a $k \times k$ matrix A by

$$\|A\| = \max_{\|x\|\ne 0} \frac{\|A\mathbf{x}\|}{\|\mathbf{x}\|} = \max_{\|x\|\le 1} \|A\mathbf{x}\| = \max_{\|x\|=1} \|A\mathbf{x}\|. \tag{3.14}$$

Using this definition one may compute effectively $\|A\|$ relative to any vector norm.

In order to give an explicit formula for the matrix norm that corresponds to the L_2 vector norm we remind the reader about the notion of **spectral radius** of A, denoted as $\rho(A)$, which is defined to be the greatest modulus of the eigenvalues, that is,

$$\rho(A) = \max\{|\lambda| : \lambda \text{ is an eigenvalue of } A\}.$$

It follows from (3.14) that for any matrix norm of A

$$\rho(A) \le \|A\|. \tag{3.15}$$

Let us recall that the **transpose** of a matrix $A = (a_{ij})_{k\times k}$ is defined as $A^T = (a_{ji})_{k\times k}$.

Now we want to present the formulas of the norms of a matrix A associated with the three basic norms of the vectors.

norm	vector $\|\| x \|\|$	*matrix* $\|\| A \|\|$
L_1	$\sum_{i=1}^{k} \lvert x_i \rvert$	$\max_{1\le j\le k} \sum_{i=1}^{k} \lvert a_{ij} \rvert$
L_2	$\max_{1\le i\le k} \lvert x_i \rvert$	$\max_{1\le i\le k} \sum_{j=1}^{k} \lvert a_{ij} \rvert$
L_∞	$\left(\sum_{i=1}^{k} x_i^2\right)^{1/2}$	$\left(\rho(A^T A)\right)^{1/2}$

3.3.2 Basic Notions of Stability

Consider the system of difference equations

$$\mathbf{x}_{n+1} = A_n\mathbf{x}_n + \mathbf{b}_n, \quad n = 0, 1, \ldots \tag{3.16}$$

where $k \in \{1, 2, \ldots\}$, $\mathbf{x}_0 \in R^k$, and where for each $n \geq 0$, A_n is a real, nonsingular $k \times k$ matrix and $\mathbf{b}_n \in R^k$.

DEFINITION 3.2 *Let $\{\mathbf{x}_n\}_{n=n_0}^{\infty}$ be a solution of (3.16). We say that $\{\mathbf{x}_n\}_{n=n_0}^{\infty}$ is a* **stable** *solution of (3.16) if given $\varepsilon > 0$ and $n_1 \geq n_0$, there exists $\delta = \delta(\varepsilon, n_1) > 0$ such that if $\{\tilde{\mathbf{x}}_n\}_{n=n_1}^{\infty}$ is a solution of (3.16) with $\|\tilde{\mathbf{x}}_{n_1} - \mathbf{x}_{n_1}\| < \delta$, then $\|\tilde{\mathbf{x}}_n - \mathbf{x}_n\| < \varepsilon$ for all $n \geq n_1$.*

Note that if $\{\tilde{\mathbf{x}}_n\}_{n=n_0}^{\infty}$ and $\{\mathbf{x}_n\}_{n=n_0}^{\infty}$ are two solutions of (3.16), then for all $n \geq n_0$,

$$\tilde{\mathbf{x}}_{n+1} - \mathbf{x}_{n+1} = \mathbf{A}_n\left(\tilde{\mathbf{x}}_n - \mathbf{x}_n\right).$$

So if we set $\mathbf{y}_n = \tilde{\mathbf{x}}_n - \mathbf{x}_n$, then

$$\mathbf{y}_{n+1} = \mathbf{A}_n\mathbf{y}_n \quad , \quad n = n_0, n_0 + 1, \ldots.$$

Thus it is enough to study the stability of the zero solution of the last system. Assuming that $n_0 = 0$ we are going to study the stability of the zero solution of the system of difference equations:

$$\mathbf{x}_{n+1} = \mathbf{A}_n\mathbf{x}_n, \quad n = 0, 1, \ldots. \tag{3.17}$$

REMARK 3.1 We say that (3.17) is autonomous if $\mathbf{A}_n$ does not depend upon n; that is, there exists a constant real $k \times k$ matrix $\mathbf{A}$ such that

$$\mathbf{x}_{n+1} = \mathbf{A}\mathbf{x}_n, \quad n = 0, 1, \ldots. \tag{3.18}$$

∎

The following definitions of stability are standard, see [E1], [LT], and [L].

DEFINITION 3.3 (Stability for Linear System)

(a) The zero solution of (3.17) is called **stable** *if given $\varepsilon > 0$ and $n_0 \geq 0$, there exists $\delta = \delta(\varepsilon, n_0) > 0$ such that*

$$\|\mathbf{x}_{n_0}\| < \delta \quad \text{implies} \quad \|\mathbf{x}_n\| < \varepsilon \quad \text{for all } n \geq n_0.$$

(b) The zero solution of (3.17) is called **locally asymptotically stable** *if it is stable, and if for all $n_0 \geq 0$, there exists $K = K(n_0) > 0$ such that*

$$\|\mathbf{x}_{n_0}\| < K \quad \text{implies} \quad \lim_{n\to\infty} \|\mathbf{x}_n\| = 0.$$

(c) *The zero solution of (3.17) is called* **unstable** *if it is not stable.*

(d) *The zero solution of (3.17) is called a* **global attractor** *if for every solution* $\{\mathbf{x}_n\}_{n=n_0}^{\infty}$ *of Eq(3.17), we have*

$$\lim_{n\to\infty} \|\mathbf{x}_n\| = 0.$$

(e) *The zero solution of (3.17) is called* **globally asymptotically stable** *if it is stable and is a global attractor.*

The following result provides an important tool from linear algebra in our study of stability, see [E1].

THEOREM 3.2

Let A be a constant $k \times k$ matrix. Then the following statements are true.

1. $\lim_{n\to\infty} \|A^n\| = 0$ *if and only if the spectral radius of A is less than 1, that is,* $\rho(A) < 1$.

2. *There exists a constant $M > 0$ such that $\|A^n\| \leq M$ for all $n \geq 0$ if and only if every eigenvalue λ of A has $|\lambda| \leq 1$, and every eigenvalue λ of A with $|\lambda| = 1$ has multiplicity 1.*

3. $\lim_{n\to\infty} \|A^n\| = \infty$ *if and only if there exists an eigenvalue λ of A with $|\lambda| > 1$, or there exists an eigenvalue λ of A with $|\lambda| = 1$ and index greater than 1.*

We can now formulate the main result of this section.

THEOREM 3.3

Consider the difference equation (3.18). Then the following statements are true.

1. *The zero solution of (3.18) is stable but not locally asymptotically stable if and only if every eigenvalue λ of A has $|\lambda| \leq 1$, every eigenvalue λ of A with $|\lambda| = 1$ has index 1, and there exists at least one eigenvalue λ of A with $|\lambda| = 1$.*

2. *The zero solution of (3.18) is globally asymptotically stable if and only if $\rho(A) < 1$.*

3. *The zero solution of (3.18) is unstable if and only if there exists an eigenvalue λ of A with $|\lambda| > 1$, or there exists an eigenvalue λ of A with $|\lambda| = 1$ and index greater than 1.*

3.4 The Routh-Hurwitz and Schur-Cohn Criteria

In this section we give the Routh-Hurwitz and Schur-Cohn criteria, which are effective tools for determining asymptotic stability of linear systems.

THEOREM 3.4 (Routh-Hurwitz Criterion)
For real numbers $a_0, a_1, \ldots, a_k$, and $a_k > 0$, let

$$P(z) := a_k\lambda^k + a_{k-1}\lambda^{k-1} + \cdots + a_1\lambda + a_0. \tag{3.19}$$

Consider the polynomial equation

$$P(z) = 0. \tag{3.20}$$

For each $n = 1, 2, \ldots, k$, let Δ_n be the principal minor of order n of the $k \times k$ matrix

$$A = \begin{bmatrix} a_{k-1} & a_k & 0 & \cdots & 0 \\ a_{k-3} & a_{k-2} & a_{k-1} & \cdots & 0 \\ a_{k-5} & a_{k-4} & a_{k-3} & \cdots & 0 \\ \vdots & \vdots & \vdots & \ddots & \vdots \\ 0 & 0 & 0 & \cdots & a_0 \end{bmatrix}.$$

The following statements are true.

1. *A necessary and sufficient condition for all of the roots of (3.20) to have a negative real part is*

$$\Delta_n > 0 \quad \text{for} \quad n = 1, 2, \ldots, k.$$

2. *A necessary and sufficient condition for the existence of a root of (3.20) with a positive real part is*

$$\Delta_n < 0, \quad \textit{for some } n \in \{1, 2, \ldots, k\}.$$

Example 3.1 Show that a necessary and sufficient condition that all roots of the quadratic polynomial with real coefficients

$$\lambda^2 + p\lambda + q = 0$$

have a negative real part is

$$p > 0 \quad \text{and} \quad q > 0.$$

Note that

$$A = \begin{bmatrix} p & 0 \\ 1 & q \end{bmatrix}$$

and so a necessary and sufficient condition that all roots of the equation have a negative real part is

$$p > 0 \quad \text{and} \quad pq > 0$$

from which the result follows. ▮

The Routh-Hurwitz criterion is not immediately applicable to discrete dynamical systems and difference equations but it can be used in conjunction with the bilinear transformation

$$z = \frac{\lambda + 1}{\lambda - 1} \quad \text{or} \quad \lambda = \frac{z + 1}{z - 1}.$$

This transformation maps the region outside the unit circle in the λ-plane to the positive real half of a z-plane, and the region inside the unit circle in the λ-plane to the positive real half of a z-plane.

Thus we have the theorem.

THEOREM 3.5

Let $P(z)$ be the polynomial defined by (3.19), and let $Q(z)$ be the polynomial given by

$$Q(z) = (z-1)^k P\left(\frac{z+1}{z-1}\right)$$

Consider the equation

$$Q(z) = 0. \tag{3.21}$$

The following statements are true.

1. *All of the roots of (3.21) lie in the open unit disk $|\lambda| < 1$ if and only if all of the roots of (3.20) lie in the open left-half plane $Re(z) < 0$.*
2. *There exists a root λ of (3.21) with $|\lambda| > 1$ if and only if there exists a root of (3.20) that lies in the open right-half plane $Re(z) > 0$.*

An alternative, more direct, test for stability of discrete dynamical systems and difference equations is provided by the **Schur-Cohn criterion**, see [Ja], pp. 54–57. This test, like the Routh-Hurwitz criterion, is expressed in terms of determinants formed from coefficients of the characteristic equation.

THEOREM 3.6 (Schur-Cohn Criterion)

Consider the polynomial

$$Q(\lambda) = a_k\lambda^k + a_{k-1}\lambda^{k-1} + \cdots + a_1\lambda + a_0$$

where $a_\ell \in R$ for $1 \le \ell \le k$, and $a_k > 0$.

For $j = 1, 2, \ldots, k$, consider the order $2j$ determinant

$$D_j = \begin{vmatrix} a_0 & 0 & \cdots & 0 & | & a_k & a_{k-1} & \cdots & a_{k-j+1} \\ a_1 & a_0 & \cdots & 0 & | & 0 & a_k & \cdots & a_{k-j+2} \\ \vdots & \vdots & \ddots & \vdots & | & \vdots & \vdots & \ddots & \vdots \\ a_{j-1} & a_{j-2} & \cdots & a_0 & | & 0 & 0 & \cdots & a_k \\ -- & -- & -- & -- & & -- & -- & -- & -- \\ a_k & 0 & \cdots & 0 & | & a_0 & a_1 & \cdots & a_{j-1} \\ a_{k-1} & a_k & \cdots & 0 & | & 0 & a_0 & \cdots & a_{j-2} \\ \vdots & \vdots & \ddots & \vdots & | & \vdots & \vdots & \ddots & \vdots \\ a_{k-j+1} & a_{k-j+2} & \cdots & a_k & | & 0 & 0 & \cdots & a_0 \end{vmatrix} \tag{3.22}$$

A necessary and sufficient condition for $Q(\lambda)$ to have all its zeros inside the unit disk is

$$\begin{matrix} D_j < 0 & \text{for } j \text{ odd} \\ D_j > 0 & \text{for } j \text{ even} \end{matrix}, \quad j = 1, \ldots, k. \tag{3.23}$$

Next, we use the Schur-Cohn criterium to derive explicit necessary and sufficient conditions for the zero solution of second and third order linear homogeneous difference equations to be asymptotically stable. We are especially interested in these two cases because they are used in Chapters 2 and 5.

Example 3.2 Consider the second order linear difference equation with real coefficients

$$x_{n+2} + px_{n+1} + qx_n = 0, \quad n = 0, 1, \ldots, \quad x_0, x_1 \in R. \tag{3.24}$$

We claim that a necessary and sufficient condition for the zero solution of equation (3.24) to be asymptotically stable is that

$$|p| < 1 + q < 2. \tag{3.25}$$

Indeed, the determinants (3.22) are

$$D_1 = \begin{vmatrix} q & 1 \\ 1 & q \end{vmatrix} = q^2 - 1,$$

$$D_2 = \begin{vmatrix} q & 0 & 1 & p \\ p & q & 0 & 1 \\ 1 & 0 & q & p \\ p & 1 & 0 & q \end{vmatrix} = (q-1)^2((q+1)^2 - p)^2.$$

Note that $D_1 < 0$ and $D_2 > 0$ may be written as

$$q^2 < 1, \quad \text{and} \quad p^2 < (q+1)^2,$$

which is equivalent to condition (3.25).

Similarly, one can show that a necessary and sufficient condition that all roots of the cubic equation with real coefficients

$$\lambda^3 + a_1\lambda^2 + a_2\lambda + a_3 = 0$$

have a negative real part is

$$a_1 > 0, \ a_2 > 0, \ a_3 > 0, \quad \text{and} \quad a_1 a_2 > a_3.$$

By using similar technique as in Example 3.2 one can show that a necessary and sufficient condition that all roots of the cubic equation with real coefficients

$$\lambda^3 + a_2\lambda^2 + a_1\lambda + a_0 = 0$$

lie in the open disk $|\lambda| < 1$ is

$$|a_2 + a_0| < 1 + a_1 \ , \ |a_2 - 3a_0| < 3 - a_1, \quad \text{and} \quad a_0^2 + a_1 - a_0 a_2 < 1. \tag{3.26}$$

We leave the details of the proof to the reader. ∎

3.5 Nonlinear Systems and Stability

In this section we define the stability for the general nonlinear system of difference equations of the form

$$\mathbf{x}_{n+1} = \mathbf{f}(n, \mathbf{x}_n), \quad n = 0, 1, \ldots \tag{3.27}$$

where $\mathbf{x}_n \in R^k$ and $\mathbf{f} : Z \times R^k \to R^k$ is continuous in $\mathbf{x}$.

Equation (3.27) is said to be **autonomous** if the variable n does not appear explicitly in the right side of this equation, that is, if (3.27) has the form

$$\mathbf{x}_{n+1} = \mathbf{f}(\mathbf{x}_n), \quad n = 0, 1, \ldots. \tag{3.28}$$

A point $\overline{\mathbf{x}}$ is called an **equilibrium point** of equation (3.27) if $\overline{\mathbf{x}} = \mathbf{f}(n, \overline{\mathbf{x}})$ for all $n \geq N$. For all purposes of the stability theory we can assume that the equilibrium point is the zero solution. To see this, set $\mathbf{y}_n = \mathbf{x}_n - \overline{\mathbf{x}}$. Then (3.27) takes the form

$$\mathbf{y}_{n+1} = \mathbf{f}(n, \mathbf{y}_n + \overline{\mathbf{x}}) - \overline{\mathbf{x}} = \mathbf{g}(n, \mathbf{y}_n),$$

which has the zero solution as the equilibrium that corresponds to the equilibrium solution $\overline{\mathbf{x}}$ of (3.27). However, we will consider (3.27) since the above mentioned transformation may be technically complicated.

DEFINITION 3.4 (Stability for Nonlinear System)

(*a*) *An equilibrium solution of Eq(3.27) is called* **stable** (**uniformly stable**) *if given* $\varepsilon > 0$ *and* $n_0 \geq 0$, *there exists* $\delta = \delta(\varepsilon, n_0) > 0$ (*if* δ *is independent of* n_0) *such that*

$$\|\mathbf{x}_{n_0} - \overline{\mathbf{x}}\| < \delta \quad \text{implies} \quad \|\mathbf{x}_n - \overline{\mathbf{x}}\| < \varepsilon \quad \text{for all } n \geq n_0.$$

(*b*) *An equilibrium solution of (3.27) is called* **exponentially stable** *if there exist* $\delta > 0, M > 0, \eta > 0$ *such that*

$$\|\mathbf{x}_{n_0} - \overline{\mathbf{x}}\| \leq M\|\mathbf{x}_0 - \overline{\mathbf{x}}\|\eta^{n-n_0}, \quad \textit{whenever } \|\mathbf{x}_0 - \overline{\mathbf{x}}\| < \delta.$$

(*c*) *An equilibrium solution of (3.27) is called* **unstable** *if it is not stable.*

(*d*) *An equilibrium solution of (3.27) is called* **attractive** *if there exists* $K = K(n_0)$ *such that* $\|\mathbf{x}_0 - \overline{\mathbf{x}}\| < K$ *implies* $\lim_{n\to\infty} \mathbf{x}_n = \overline{\mathbf{x}}$, *and* **uniformly attractive** *if the choice of* K *is independent of* n_0.

(*e*) *An equilibrium solution of (3.27) is called* **asymptotically stable** *if it is stable and attractive, and* **uniformly asymptotically stable** *if it is uniformly stable and attractive.*

(*f*) *An equilibrium solution of (3.27) is called a* **global attractor** *if*

$$\lim_{n\to\infty} \|\mathbf{x}_n\| = \overline{\mathbf{x}}$$

for every solution $\{\mathbf{x}_n\}_{n=n_0}^{\infty}$ *of (3.27).*

(*g*) *The equilibrium solution of (3.27) is called* **globally asymptotically stable** *if it is stable and is a global attractor.*

REMARK 3.2 It can be shown for autonomous systems that stability (resp. asymptotic stability) is equivalent to uniform stability (resp. uniform asymptotic stability) of the system, see [E1]. ▮

3.5.1 Linearized Stability

The linearization method is probably the first method used in stability theory. It was originated by Lyapunov, Perron, and Poincaré in the study of the stability of differential equations, see [E1], [LT], and [L]. In this approach we consider (3.27) together with its linearization

$$\mathbf{z}_{n+1} = D_{\mathbf{f}}(n, \overline{\mathbf{x}})\, \mathbf{z}_n = A(n)\, \mathbf{z}_n, \tag{3.29}$$

where $A(n) = D_{\mathbf{f}}(n, \overline{\mathbf{x}})$ is the Jacobian matrix of $\mathbf{f}$ evaluated at an equilibrium solution $\overline{\mathbf{x}}$ of (3.27). Thus we can rewrite (3.27) as:

$$\mathbf{x}_{n+1} = A(n)\mathbf{x}_n + \mathbf{g}(n, \mathbf{x}_n), \tag{3.30}$$

where $\mathbf{g}(n, \mathbf{y}) = \mathbf{f}(n, \mathbf{y}) - A(n)\mathbf{y}$.

Note that the autonomous versions of equations (3.30) and (3.29) are, respectively,

$$\mathbf{y}_{n+1} = A\mathbf{y}_n + \mathbf{g}(\mathbf{y}_n), \tag{3.31}$$

and

$$\mathbf{z}_{n+1} = D_{\mathbf{f}}(\overline{\mathbf{x}})\mathbf{z}_n = A\mathbf{z}_n. \tag{3.32}$$

We shall need the following definition.

DEFINITION 3.5 *We say that $G(n, \mathbf{y}) = o(\mathbf{y})$ as $\|\mathbf{y}\| \to 0$ if, given $\varepsilon > 0$, there exists $\delta > 0$ such that $\|G(n, \mathbf{y})\| \le \varepsilon\|\mathbf{y}\|$ whenever $\|\mathbf{y}\| < \delta$ and $n \in Z^+$.*

Now we have our major stability results, known as the Linearized Stability Theorems [E1] and [LT].

THEOREM 3.7 (Linearized Stability Theorem)
Assume that $\mathbf{g}(n, \mathbf{y}) = o(\mathbf{y})$ as $\|\mathbf{y}\| \to 0$. If the zero equilibrium of equation (3.29) is uniformly asymptotically stable, then the zero equilibrium of equation (3.30) is exponentially stable.

In the special case of system (3.32) we have stronger results:

THEOREM 3.8

1. *If $\rho(A) < 1$, then the zero equilibrium of (3.32) is exponentially stable.*
2. *If $\|D_{bff}(\mathbf{0})\| < 1$, then the zero equilibrium of (3.32) is exponentially stable.*

A partial converse of this result is the following:

THEOREM 3.9

1. *If $\rho(A) = 1$, then the zero equilibrium of (3.32) may be stable or unstable.*
2. *If $\rho(A) > 1$, and $\mathbf{g}(\mathbf{y}) = o(\mathbf{y})$ as $\|\mathbf{y}\| \to 0$, then the zero equilibrium of (3.32) is unstable.*

3.5.2 Lyapunov Functions and Stability

Consider the autonomous system of difference equations

$$\mathbf{x}_{n+1} = \mathbf{f}(\mathbf{x}_n), \tag{3.33}$$

where $\mathbf{x}_n \in R^k$ and $\mathbf{f} : G \to R^k$, $G \subset R^k$ is continuous. Let $\overline{\mathbf{x}}$ be an equilibrium point of (3.33), that is, $\mathbf{f}(\overline{\mathbf{x}}) = \overline{\mathbf{x}}$.

DEFINITION 3.6 *The function $V : R^k \to R$ is said to be a* **Lyapunov function** *on a subset D of R^k if*

1. *V is continuous on D.*

2. *$\Delta V = V(\mathbf{f}(\mathbf{x})) - V(\mathbf{x}) \leq 0$ when $\mathbf{x}$ and $\mathbf{f}(\mathbf{x}) \in D$.*

Let $B(\mathbf{a}, r)$ denote the open ball in R^k of radius r and center $\mathbf{a}$, that is, $B(\mathbf{a}, r) = \{\mathbf{x} \in R^k : ||\mathbf{x} - \mathbf{a}|| < r\}$.

DEFINITION 3.7 *The real function V is* **positive definite** *at $\overline{\mathbf{x}}$ if*

1. *$V(\overline{\mathbf{x}}) = 0$.*

2. *$V(\mathbf{x}) > 0$ for all $\mathbf{x} \in B((\overline{\mathbf{x}}, r)$, for some $r > 0$.*

Now we present the main result of this section, see [E1] and [LT].

THEOREM 3.10 (Lyapunov Stability Theorem)
Assume that V is a Lyapunov function for (3.33) on a neighborhood D of the equilibrium point $\overline{\mathbf{x}}$ and that V is positive definite at $\overline{\mathbf{x}}$, then $\overline{\mathbf{x}}$ is stable. If in addition $\Delta V(x) < 0$ for all $\mathbf{x}, \mathbf{f}(\mathbf{x}) \in D$ and $\mathbf{x} \neq \overline{\mathbf{x}}$, then $\overline{\mathbf{x}}$ is asymptotically stable. Furthermore, if $D = R^k$ and

$$V(\mathbf{x}) \to \infty \quad as \quad ||\mathbf{x}|| \to \infty, \tag{3.34}$$

then $\overline{\mathbf{x}}$ is globally asymptotically stable.

Very often it is important to establish whether all solutions of a given system are bounded. As the following result shows, the Lyapunov function can help in this respect.

THEOREM 3.11
Assume that V is a Lyapunov function for (3.33) on the set $U = \{\mathbf{x} : ||\mathbf{x}|| > d$ for some $d > 0\}$ and that the condition (3.34) holds, then all solutions of (3.33) are bounded.

Similarly, using Lyapunov function one can obtain the conditions for instability, see [E1].

THEOREM 3.12
Assume that V is a Lyapunov function for (3.33) on the set $B(\mathbf{0}, r) = \{\mathbf{x} : \|\mathbf{x}\| < r\}$ and that ΔV is positive definite on $B(\mathbf{0}, r)$ such that there exists a sequence $a_k \to 0$ with $V(a_k) > 0$. Then the zero equilibrium of (3.33) is unstable.

It should be mentioned that there are many stability results similar to Theorems 3.10 to 3.12 for both autonomous and nonautonomous systems. Some of these results can be found in [E1] and [LT]. The major problem that arises when one tries to apply the Lyapunov method is that there is no systematic way of finding Lyapunov functions, and the known methods for finding these functions are heuristic. In the special case when the considered system possesses an invariant (or first integral), there is a systematic way of finding Lyapunov functions. In this case, one may use *Dynamica* to find the Lyapunov function and to prove stability or asymptotic stability. See Chapter 4 for numerous applications of this method and many solved examples.

3.6 Limit Sets and Invariant Manifolds

In the case of the autonomous system (3.33) we can say a little bit more about the local behavior of solutions in the neighborhood of the equilibrium solution. Let us consider first the linear autonomous system, that is,

$$\mathbf{x}_{n+1} = A\mathbf{x}_n, \tag{3.35}$$

where A is a given $k \times k$ matrix. This equation can be also considered as a linear map $\mathbf{x} \to A\mathbf{x}$ of R^k into itself. An orbit of this map is a sequence $\{\mathbf{x}_i\}_{-\infty}^{\infty}$, if A is not singular matrix, or $\{\mathbf{x}_i\}_{i=N}^{\infty}$, if A is a singular matrix, defined by (3.35). The equilibrium solution of (3.35) is called the **fixed point** of the map.

We define **stable**, **unstable**, and **center subspaces** as follows.

$E^u = span\{\mathbf{v}^u$:	$\mathbf{v}^u$ is a generalized eigenvector for an eigenvalue λ_u of A with $\lvert\lambda_u\rvert > 1,\}$
$E^s = span\{\mathbf{v}^s$:	$\mathbf{v}^s$ is a generalized eigenvector for an eigenvalue λ_s of A with $\lvert\lambda_s\rvert < 1,\}$
$E^c = span\{\mathbf{v}^c$:	$\mathbf{v}^c$ is a generalized eigenvector for an eigenvalue λ_c of A with $\lvert\lambda_c\rvert = 1.\}$

The orbits in E^s and E^u are characterized by contraction and expansion, respectively. Clearly, if A has no eigenvalues on the unit disk, then the eigenvalues alone serve to determine stability. In this case, $\mathbf{x} = \mathbf{0}$ is called a **hyperbolic fixed point**.

In general, for the nonlinear system

$$\mathbf{x}_{n+1} = \mathbf{f}(\mathbf{x}_n), \tag{3.36}$$

where $\mathbf{f} : R^k \to R^k$, one can define the **positive orbit** starting at some point $\mathbf{d}$ as the sequence $\gamma^+(\mathbf{d}) = \{\mathbf{f}^i(\mathbf{d})\}_{i=0}^{\infty}$. Similarly, if the map $\mathbf{f}$ is invertible we define the **negative orbit** starting at some point $\mathbf{d}$ as the sequence $\gamma^-(\mathbf{d}) = \{\mathbf{f}^{-1^i}(\mathbf{d})\}_{i=0}^{\infty}$. The **complete orbit** $\gamma(\mathbf{d})$ of a point $\mathbf{d}$ is the union of positive and negative orbits, i.e. $\gamma(\mathbf{d}) = \gamma^+(\mathbf{d}) \cup \gamma^-(\mathbf{d})$.

Throughout this book whenever we use the notion *orbit* it will always mean the *positive orbit*. Thus the *negative orbit* and the *complete orbit* will always be stated with an associated description.

A basic objective of dynamical systems theory is to predict the long-term behavior of a system based on the knowledge of its present state. An approach to this problem consists of determining the possible long-term behaviors of the system and determining which initial conditions lead to these long-term behaviors. In more technical terms we are interested in finding *invariant sets* and their *basins of attraction*. The nature of the invariant sets and their basins of attraction will greatly affect our ability to predict the long-term behavior of a system. Before defining these terms rigorously let us give an illustrative example.

Example 3.3 Consider the IVP

$$x_{n+1} = x_n^2, \quad x_0 = d \in R. \tag{3.37}$$

This equation generates the map $x^2 : R \to R^+$. Equation (3.37) has the explicit solution $x_n = d^{2^n}$, $n = 0, 1, \ldots$, which may be used to completely characterize the behavior of all solutions. We have

$$x_n \begin{cases} \to 0 & \text{if } -1 < d < 1 \\ = 1 & \text{if } d = \pm 1 \\ \to \infty & \text{if } |d| > 1. \end{cases}$$

From a dynamical systems point of view this equation (map) has two equilibrium (fixed) points, 0 and 1. The set of all initial conditions that converges to 0 is $(-1, 1)$. It is natural to call this set the basin of attraction of 0. This set is denoted as $\mathcal{B}(\{0\})$ and we have $\mathcal{B}(\{0\}) = (-1, 1)$. Similarly we have that $\mathcal{B}(\{1\}) = \{-1, 1\}$ and $\mathcal{B}(\{\infty\}) = (-\infty, -1) \cup (1, \infty)$. Thus, in this case, we have determined all limit sets and their basins of attraction. In general, this is not going to happen often. ∎

Example 3.4 Consider the difference equation

$$x_{n+1} = x_n^3, \quad x_0 = d \in R. \tag{3.38}$$

This equation generates the map $x^3 : R \to R$. The explicit solution of this equation $x_n = d^{3^n}, \quad n = 0, 1, \ldots$ completely characterizes the behavior of all solutions of this equation:

$$x_n \begin{cases} \to 0 & \text{if } -1 < d < 1 \\ = 1 & \text{if } d = 1 \\ = -1 & \text{if } d = -1 \\ \to \infty & \text{if } d > 1 \\ \to -\infty & \text{if } d < -1. \end{cases}$$

From a dynamical systems point of view this equation (map) has three equilibrium (fixed) points $0, -1$, and 1. The set of all initial conditions that converges to 0 is $(-1, 1)$. Thus, $\mathcal{B}(\{0\}) = (-1, 1)$. Similarly, we have that $\mathcal{B}(\{1\}) = \{1\}$ and $\mathcal{B}(\{-1\}) = \{-1\}$. Finally, $\mathcal{B}(\{\infty\}) = (1, \infty)$ and $\mathcal{B}(\{-\infty\}) = (-\infty, -1)$. Thus, in this case, we have five possible limit sets and we have determined entirely their basins of attraction. ▮

The previous two examples had simple limit sets and corresponding basins of attractions. As we have seen in Chapter 1 the limiting sets may consist of periodic orbits, dense orbits and the corresponding basins of attractions may be as complicated as the Cantor set. Two such examples are the logistic map and the tent map for certain values of parameters as well as any map conjugate to them. Now, we give the rigorous definitions of limit set and basin of attraction.

DEFINITION 3.8 *The map* $\mathbf{f} : R^k \to R^k$ *is called the* C^r-**diffeomorphism** *if it is* C^r*-differentiable (has continuous derivatives up to order r inclusive) and has* C^r*-differentiable inverse. A C-diffeomorphism is called a homeomorphism. A fixed point* $\overline{\mathbf{x}}$ *of a map* $\mathbf{f}$ *(resp. equilibrium solution of a system* (3.36)*) is called a* **hyperbolic fixed point** *if the Jacobian evaluated at the fixed point,* $D_{\mathbf{f}}(\overline{\mathbf{x}})$ *has no eigenvalues on the unit disk, that is, if* $E^c = \{\mathbf{0}\}$*. A hyperbolic fixed point* $\overline{\mathbf{x}}$ *is called a* **sink** *provided all the eigenvalues of* $D_{\mathbf{f}}(\overline{\mathbf{x}})$ *are inside the unit disk, i.e.,* $E^u = E^c = \{\mathbf{0}\}$*. Similarly as in the one-dimensional case, a hyperbolic fixed point* $\overline{\mathbf{x}}$ *is called a* **source** *provided all the eigenvalues of* $D_{\mathbf{f}}(\overline{\mathbf{x}})$ *are outside the unit disk, i.e.,* $E^s = E^c = \{\mathbf{0}\}$*. Finally, unlike in one-dimensional cases, a hyperbolic fixed point* $\overline{\mathbf{x}}$ *is called a* **saddle** *if* $E^u \neq \{\mathbf{0}\}$ *and* $E^s \neq \{\mathbf{0}\}$*.*

The case of saddle is not found in one-dimensional dynamics. There are powerful results that reveal the local behavior of the orbits near the hyperbolic fixed points. The first of them shows that in some neighborhood of a hyperbolic fixed point, the nonlinear map is conjugate to its linearization.

Let us mention that the definitions of the corresponding terms for the periodic point $\mathbf{p}$ of period m is similar, with $\mathbf{f}$ and $D_{\mathbf{f}}(\mathbf{p})$ replaced by $\mathbf{f}^m$ and $D_{\mathbf{f}^m}(\mathbf{p})$, respectively.

Let us introduce basic notions about invariant manifolds.

DEFINITION 3.9 *A set $B \subset R^k$ is said to be* **positively invariant** *under the mapping $\mathbf{f} : R^k \to R^k$ if $\mathbf{f}(B) \subset B$. Likewise B is said to be* **negatively invariant** *under the mapping $\mathbf{f} : R^k \to R^k$, which possesses an inverse mapping if $\mathbf{f}(B) \supset B$. A set $B \subset R^k$ is said to be* **invariant** *under the mapping $\mathbf{f} : R^k \to R^k$, which possesses an inverse mapping, if it is both positively invariant and negatively invariant, that is, if $\mathbf{f}(B) = B$.*

DEFINITION 3.10 *A point $\mathbf{x} \in R^k$ is said to be an ω-***limit point** *of $\mathbf{d} \in R^k$ under the mapping $\mathbf{f} : R^k \to R^k$ if there exists a sequence $\{n_i\}_{i=1}^{\infty}$, $n_i \to \infty$, with $\mathbf{f}^{n_i}(\mathbf{d}) \to \mathbf{x}$ as $i \to \infty$. The union of all such points for given point $\mathbf{d}$ is called the ω-***limit set** *of $\mathbf{d}$ and denoted as $\omega(\mathbf{d})$. Thus*

$$\omega(\mathbf{d}) = \{\mathbf{x} \in R^k : \mathbf{f}^{n_i}(\mathbf{d}) \to \mathbf{x} \quad \textit{for some} \quad n_i \to \infty\}.$$

*Likewise the ω-***limit set** *of a bounded set B is defined by*

$$\omega(B) = \{\mathbf{x} \in R^k : \mathbf{f}^{n_i}(\mathbf{d}) \to \mathbf{x} \quad \textit{for some} \quad n_i \to \infty \quad \textit{and}\ \mathbf{y}_i \in B\}.$$

In the case when the map is a diffeomorphism we can consider the negative orbit and define an analogue of the ω-**limit point** and the ω-**limit set**.

DEFINITION 3.11 *A point $\mathbf{x} \in R^k$ is said to be an α-***limit point** *of $\mathbf{d} \in R^k$ under the diffeomorphism $\mathbf{f} : R^k \to R^k$ if there exists a sequence $\{n_i\}_{i=1}^{\infty}$, $n_i \to \infty$, with $\mathbf{f}^{-n_i}(\mathbf{d}) \to \mathbf{x}$ as $i \to \infty$. The union of all such points for given point $\mathbf{d}$ is called the α-***limit set** *of $\mathbf{d}$ and denoted as $\alpha(\mathbf{d})$. Thus*

$$\alpha(\mathbf{d}) = \{\mathbf{x} \in R^k : \mathbf{f}^{-n_i}(\mathbf{d}) \to \mathbf{x} \quad \textit{for some} \quad n_i \to \infty\}.$$

*Similarly the α-***limit set** *of a bounded set B is defined by*

$$\alpha(B) = \{\mathbf{x} \in R^k : \mathbf{f}^{-n_i}(\mathbf{d}) \to \mathbf{x} \quad \textit{for some} \quad n_i \to \infty \quad \textit{and}\ \mathbf{y}_i \in B\}.$$

The closure of the set A is denoted as $\overline{A}$. The next result gives an important characterization of limit sets.

THEOREM 3.13
*The ω-***limit point** *of the mapping $\mathbf{f} : R^k \to R^k$ for the point $\mathbf{d}$ is given by*

$$\omega(\mathbf{d}) = \bigcap_{m \geq 0} \overline{\bigcup_{n \geq m} \mathbf{f}^n(\mathbf{d})}.$$

Similarly the ω-**limit set** *of the mapping* $\mathbf{f}$ *for the bounded set* B *is given by*

$$\omega(B) = \bigcap_{m\geq 0}\overline{\bigcup_{n\geq m}\mathbf{f}^n(B)}.$$

The next result states that ω-**limit set** of the continuous mapping is invariant and can be derived from the characterization given in Theorem 3.13.

THEOREM 3.14
The ω-**limit set** *of the continuous mapping* $\mathbf{f} : R^k \to R^k$ *for the bounded set* $B \in R^k$ *is a closed positively invariant set. If* $\bigcup_{n\geq 0}\mathbf{f}^n(B)$ *is bounded, then* $\omega(B)$ *is invariant.*

Now we define the notion of the basin of attraction. Let us recall that the bounded set in R^k is one that can be placed in the ball of finite radius and that the **compact set** in R^k is closed and bounded.

DEFINITION 3.12 *The* **basin of attraction** $\mathcal{B}(U)$ *of a compact invariant set* $U \in R^k$ *is the set of all points* $\mathbf{x} \in R^k$ *whose* ω*-limit points are in* U*, i. e.,*

$$\mathcal{B}(U) = \{\mathbf{d} : \omega(\mathbf{d}) \subset U\}.$$

Example 3.5 Now we revisit Examples 3.3 and 3.4. In the case of Example 3.3, $\omega((-1,1)) = \{0\}$, $\omega(\{-1,1\}) = \{1\}$, $\omega((1,\infty)) = \emptyset$, and $\omega((-\infty,-1)) = \emptyset$. In general, we have

$$\bigcup_{\mathbf{d}\in B}\omega(\{\mathbf{d}\}) \subset \omega(B),$$

and this dynamical system provides us an example where

$$\bigcup_{\mathbf{d}\in B}\omega(\{\mathbf{d}\}) \neq \omega(B).$$

Let $B = [-1,1]$. Clearly $\mathbf{f}^n(B) = B$. Hence, by Theorem 3.13 we have that $\omega(B) = B$. Now consider $\omega(d)$ for $d \in B$. If $d \in (-1,1)$, then $\omega(d) = 0$ and $\omega(1) = \omega(-1) = 1$. Hence

$$\bigcup_{d\in B}\omega(d) = \{0,1\} \neq [-1,1] = \omega(B).$$

In this case $\omega(B)$ consists of the union of the ω-limit sets of individual orbits and the complete orbits which connect different ω-limit sets. In particular, there are complete orbits connecting the fixed point 1 to both another fixed point 0 and ∞. Note that for an orbit connecting 1 to 0, 0 is the ω-limit set and 1 is the α-limit set of any point on the orbit.

In the case of Example 3.4 $\omega((-1,1)) = \{0\}$ and $\omega(\{-1,1\}) = \{-1,1\}$. In this case we have unique inverse function and so we can also find the α-limit sets as $\alpha(1) = (0,\infty)$ and $\alpha(-1) = (-\infty,0)$. In this case also the orbits exist that connect the fixed points 1, 0, and -1. ∎

DEFINITION 3.13 *A complete orbit through the point* $\mathbf{d} \in R^k$ *is called* **heteroclinic** *if* $\omega(\mathbf{d})$ *and* $\alpha(\mathbf{d})$ *are nonempty. It is called* **homoclinic** *if, in addition,* $\omega(\mathbf{d}) = \alpha(\mathbf{d})$. *Heteroclinic and homoclinic orbits are called* **connecting orbits**.

For the purposes of this book, the invariant manifolds may be considered as invariant sets which can be represented as a graph $G : X \to Y$ for two subspaces X and Y of R^k which sum gives R^k and $X \cap Y = \{0\}$. In this case we say that R^k is a direct sum of X and Y and denote it as $R^k = X \oplus Y$.

THEOREM 3.15 (Hartman-Grobman Theorem)
Let $\mathbf{f} : R^k \to R^k$ *be a* C^r*-diffeomorphism with a hyperbolic fixed point* $\overline{\mathbf{x}}$ *. Then there exist neighborhoods* U *of* $\overline{\mathbf{x}}$ *and* V *of* $\mathbf{0}$ *and a homeomorphism* $h : V \to U$ *such that* $\mathbf{f}$ *is* **conjugate** *to the Jacobian* $D_{\mathbf{f}}(\overline{\mathbf{x}})$, *that is,*

$$\mathbf{f}(\mathbf{h}(\mathbf{x})) = \mathbf{h}(\mathbf{D}_{\mathbf{f}}(\overline{\mathbf{x}}))$$

for all $\mathbf{x} \in V$.

In two dimensions, Hartman [R] proved the stronger result, that near a hyperbolic fixed point any C^2-diffeomorphism is C^1 conjugate (h in Theorem 3.15 can be chosen to be C^1) to its linearized map.

Let $\mathbf{f} : R^k \to R^k$ be a C^1-diffeomorphism. Assume that $\overline{\mathbf{x}} \in R^k$ is a hyperbolic fixed point of $\mathbf{f}$. Let U denote a neighborhood of $\overline{\mathbf{x}}$. The sets

$$W^s_{loc}(\overline{\mathbf{x}}) = \{\mathbf{x} \in U \; : \; \mathbf{f}^n(\mathbf{x}) \in U, \quad \text{for } n \geq 0 \quad \text{and} \quad \lim_{n\to\infty} \mathbf{f}^n(\mathbf{x}) = \overline{\mathbf{x}}, \}$$

and

$$W^u_{loc}(\overline{\mathbf{x}}) = \{\mathbf{x} \in U \; : \; \mathbf{f}^{-n}(\mathbf{x}) \in U, \quad \text{for } n \geq 0 \quad \text{and} \quad \lim_{n\to\infty} \mathbf{f}^{-n}(\mathbf{x}) = \overline{\mathbf{x}}, \}$$

are called **local stable** and **unstable manifold**, respectively.

The following theorem, called the Stable Manifold Theorem in R^n, explains the significance of $\bar{\mathbf{x}}$ being a saddle equilibrium, see [GH], pp. 16–19 and [R], p. 182.

THEOREM 3.16 (The Stable Manifold Theorem)
Let $\mathbf{f} : R^k \to R^k$ *be a* C^r*-diffeomorphism with a hyperbolic fixed point* $\overline{\mathbf{x}}$. *Then there exist a local stable manifold and a local unstable manifold* $W^s_{loc}(\overline{\mathbf{x}})$, $W^u_{loc}(\overline{\mathbf{x}})$ *tangent to the eigenspaces* E^s *and* E^u *of the corresponding linearized map* $D_{\mathbf{f}}(\overline{\mathbf{x}})$, *respectively, and of the corresponding dimension. In addition,* $W^s_{loc}(\overline{\mathbf{x}})$, $W^u_{loc}(\overline{\mathbf{x}})$ *are as smooth as the map* $\mathbf{f}$.

Global stable and unstable manifolds $W^s(\overline{\mathbf{x}})$, $W^u(\overline{\mathbf{x}})$ are defined, in the same way as in the two-dimensional case, as the unions of backward and forward iterates of the local manifolds, that is,

$$W^s(\overline{\mathbf{x}}) = \cup_{k\geq 0}\mathbf{f}^{-k}(W^s_{loc}(\overline{\mathbf{x}})),$$

$$W^u(\overline{\mathbf{x}}) = \cup_{k\geq 0}\mathbf{f}^{k}(W^u_{loc}(\overline{\mathbf{x}})).$$

While finding the local stable or unstable manifold numerically is relatively simple, finding the corresponding global manifolds is, in most cases, a formidable task. See [CKS], [KN2], and [SZ] for some examples of finding the global stable manifold. *Dynamica* has functions that can numerically find both stable and unstable manifolds numerically for the two-dimensional systems. See Sections 2.12 and 2.17 for *Dynamica* sessions.

Notice that Theorem 3.16 does not say anything about the nonhyperbolic fixed point. In this case, similar to for one-dimensional equations, the situation may become very complicated. One can show that under certain conditions in this case there exists an invariant manifold tangent to the eigenspace E^c. It is not surprising that these types of results are known as the center manifold results, see [GH], pp.127–138, [R], p. 200, and [SH], p. 62.

THEOREM 3.17 (The Center Manifold Theorem)

Let $\mathbf{f} : R^k \to R^k$ be a C^r-diffeomorphism with a nonhyperbolic fixed point $\overline{\mathbf{x}}$. Then there exists an invariant manifold $W^c_{loc}(\overline{\mathbf{x}})$ tangent to the eigenspaces E^c of the corresponding linearized map $D_{\mathbf{f}}(\overline{\mathbf{x}})$, called the **center manifold**, *which is locally a C^{r-1}-map.*

Example 3.6

Consider the system

$$\begin{aligned} x_{n+1} &= ax_n + x_n(y_n - x_n^2) \\ y_{n+1} &= (a^2 - c)y_n + (a^2 - 1)x_n^2, \end{aligned} \tag{3.39}$$

where a and c are the parameters. One of the equilibrium points of this system is $(0, 0)$ and the linearized system at this point is:

$$\begin{aligned} x_{n+1} &= ax_n \\ y_{n+1} &= (a^2 - c)y_n. \end{aligned} \tag{3.40}$$

Thus depending on the values of a and c we can have different cases for the eigenspaces E^s, E^u, and E^c. For example, if $|a| < 1$ and $a^2 - 1 < c < a^2 + 1$ then $(0, 0)$ is locally asymptotically stable and $dim\, E^s = 2$. If $|a| > 1$ and $a^2 - 1 < c < a^2 + 1$ or $|a| < 1$ and $|c - a^2| > 1$, then $(0, 0)$ is a saddle point and $dim\, E^s = dim\, E^u = 1$. If $|a| > 1$ and $|c - a^2| > 1$, then $(0, 0)$ is a source and $dim\, E^u = 2$. If $|a| = 1$ and/or $|c - a^2| = 1$, then $(0, 0)$ is a nonhyperbolic $dim\, E^c \geq 1$. In this case we can have any combination of the eigenspace E^c with any of the eigenspaces E^s, E^u, and E^c.

In the case when $|a| < 1$ and $|c - a^2| > 1$ it is clear from (3.40) that $E^s = span(1, 0)$ which is the x-axis and $E^u = span(0, 1)$ which is the y-axis.

In all cases we can establish the following relation for all solutions of system (3.39):

$$y_{n+1} - x_{n+1} = \left(y_n - x_n^2\right)\left(a^2 - c - 2ax_n^2 - x_n^2(y_n - x_n^2)\right), \quad n = 0, 1, \ldots \tag{3.41}$$

which immediately implies that

$$y_n - x_n^2 = \left(y_0 - x_0^2\right) H(x_n, y_n), \quad n \geq 1.$$

This shows that the parabola $y = x^2$ is an invariant set for (3.39) in the sense that if $y_0 = x_0^2$ then $y_n = x_n^2$ for every $n \geq 1$. For all initial points that satisfy $y_0 = x_0^2$ the first equation of (3.39) gives a linear equation $x_{n+1} = ax_n$ which has an explicit solution $x_n = a^n x_0$ and so $x_n \to 0$ as $n \to \infty$. Thus in this case we see that $y = x^2$ is a part of the global stable manifold W^s and is tangent to the corresponding eigenspace E^s, x axis, at the fixed point $(0, 0)$. Similarly one can show that the y-axis is a part of the unstable manifold W^u, which actually coincides with the corresponding linear eigenspace E^u.

In the case when $|a| < 1$ and $|c - a^2| = 1$ it is clear from (3.40) that $E^s = span(1, 0)$-x-axis and $E^c = span(0, 1)$-y-axis. Using (3.41) we can show that the part of the stable manifold $W^s((0, 0))$ is the parabola $y = x^2$ and that the part of the center manifold $W^c((0, 0))$ is the y-axis.

Similarly, the appropriate choice of conditions on parameters a and c will give all possible combinations of the three types of manifolds for the nonlinear system (3.39). ▮

REMARK 3.3 It is important to note that the behaviors of global stable and unstable manifolds of the fixed points of maps differs from the similar invariant manifolds of the equilibrium points of differential equations. In particular, it is possible for two invariant manifolds of a fixed point of a map to cross each other at some other point without coinciding. Such points have a special role in explaining the dynamics of the map and are called *homoclinic points*. ▮

DEFINITION 3.14 *Let* $\overline{\mathbf{x}}$ *be a saddle point of a diffeomorphism* $\mathbf{f}$. *A point* $\mathbf{h} \neq \overline{\mathbf{x}}$ *is called a* **homoclinic point** *for* $\mathbf{f}$ *if* $\mathbf{h} \in W^s(\overline{\mathbf{x}}) \cap W^u(\overline{\mathbf{x}}) \setminus \{\overline{\mathbf{x}}\}$, *that is, if*

$$\mathbf{f}^n(\mathbf{p}) \to \overline{\mathbf{x}} \quad \text{as} \quad n \to \infty \quad \text{and as} \quad n \to -\infty.$$

The homoclinic point $\mathbf{h}$ *is said to be* **transversal** *if the tangent vectors to* $W^s(\overline{\mathbf{x}})$ *and* $W^u(\overline{\mathbf{x}})$ *at* $\overline{\mathbf{x}}$ *do not coincide.*

The presence of a transversal homoclinic point is a clear sign of dynamical complexity (see [ASY]). This subject has been explored in Section 2.12.

3.7 Dissipative Maps

In this section we consider *dissipative* maps. Roughly speaking, a dissipative map has the property that there exists a bounded subset of R^k which is eventually entered by the positive orbit of every initial point. Physically, the term is motivated by the class of problems for which there is a loss of energy. The precise definition is the following:

DEFINITION 3.15 *A map* $\mathbf{f} : R^k \to R^k$ *is called* **dissipative** *if there is a bounded positively invariant set* $A \subset R^k$ *such that every bounded set* $B \subset R^k$ *eventually enters* A*, that is, there exists* $N(B, A) \geq 0$ *such that* $\mathbf{f}^n(B) \subset A$*. The set* A *is known as the* **absorbing set** *or* **attracting set**.

The absorbing set is not uniquely determined as any bounded set that contains the absorbing set is the absorbing set itself.

Example 3.7 The map generated by the difference equation

$$x_{n+1} = \frac{1}{1 + x_n^{2q}},$$

where $q = 1, 2, \ldots$ is dissipative and the interval $[0, 1]$ is an absorbing set as well as any other closed interval that contains $[0, 1]$.

The map generated by the difference equation

$$x_{n+1} = \frac{1}{1 + x_n^{2q_0} + x_{n-1}^{2q_1} + \ldots + x_{n-M}^{2q_M}},$$

where $q_i \in \{1, 2, \ldots\}, i = 0, 1, \ldots$ is dissipative and the interval $[0, 1]^{M+1}$ is an absorbing set. ∎

In the case of dissipative map with absorbing set A, the set of all possible asymptotic behaviors is captured in the set $\omega(A)$. Here are two important results that reveal the structure of the global attractors of a dissipative map.

THEOREM 3.18

The global attractor of the disipative map $\mathbf{f} : R^k \to R^k$ *with absorbing set* A *has the form*

$$\mathcal{A} = \omega(A) = \bigcap_{n\geq 0} \mathbf{f}^n(A).$$

THEOREM 3.19

The global attractor of the dissipative map $\mathbf{f} : R^k \to R^k$ *with absorbing set* A *consists of the union of all the bounded complete orbits of* $\mathbf{f}$*. In other words,* *the global attractor of the dissipative map is the maximal bounded invariant set under this map.*

An important consequence of this result is the following result:

COROLLARY 3.1
The global attractor $\mathcal{A}$ of the dissipative map $\mathbf{f} : R^k \to R^k$ with absorbing set A satisfies:

$$\cup_{\mathbf{x}\in\mathcal{F}} \overline{W^u(\mathbf{x})} \subset \mathcal{A},$$

where $\mathcal{F}$ denotes the set of fixed points of the map $\mathbf{f}$.

Example 3.8 The map generated by the difference equation

$$x_{n+1} = x_n^3,$$

has a heteroclinic orbit connecting 1 to 0. This heteroclinic orbit is the stable manifold of 0 and unstable manifold of 1.

In general, connecting orbits are formed from the intersections of any pair of the stable, unstable and center manifolds of limit sets. ∎

In spite of the fact that dissipative maps have the special property that all ω-limit sets are contained within a bounded absorbing set A, within this set the behavior may be very complicated. Recall the one-dimensional logistic and the tent maps with the parameter values that give chaos. Even in the cases when the global behavior is relatively simple, it may require ingenious methods to prove such behavior, see [KL] for many cases of two-dimensional dissipative maps with simple behavior generated by the difference equation

$$x_{n+1} = \frac{\alpha + \beta x_n + \gamma x_{n-1}}{A + Bx_n + Cx_{n-1}}, \quad n = 0, 1, \ldots . \tag{3.42}$$

3.8 Stability of Difference Equations

In this section, we present basic definitions for the general $k+1$-th order difference equation

$$x_{n+1} = f(x_n, x_{n-1}, \ldots, x_{n-k}), \quad n = 0, 1, \ldots . \tag{3.43}$$

As we have seen in Section 3.3 the results for equation (3.43) follow immediately from the corresponding results for the systems presented in Section 3.5. However, we will state the basic definitions of stability and formulate the linearized equation for the convenience of the reader and because of the fact that most of the applications of stability theory are in difference equations.

Let I be some interval of real numbers and let $f \in C^1[I^{k+1}, I]$. Let $\bar{x} \in I$ be an equilibrium point of the difference equation (3.43), that is,

$$\bar{x} = f(\bar{x}, \ldots, \bar{x}).$$

DEFINITION 3.16

(i) *An equilibrium $\bar{x}$ of equation (3.43) is called* **locally stable** *if for every $\varepsilon > 0$, there exists $\beta > 0$ such that if $x_0, \ldots, x_{-k} \in I$ with $|x_0 - \bar{x}| + \ldots + |x_{-k} - \bar{x}| < \beta$, then*

$$|x_n - \bar{x}| < \varepsilon \text{ for all } n \geq -k.$$

(ii) *An equilibrium $\bar{x}$ of equation (3.43) is called* **locally asymptotically stable** *if it is locally stable, and if there exists $\gamma > 0$ such that $x_0, \ldots, x_{-k} \in I$ with $|x_0 - \bar{x}| + \ldots + |x_{-k} - \bar{x}| < \gamma$, then*

$$\lim_{n \to \infty} x_n = \bar{x}.$$

(iii) *An equilibrium $\bar{x}$ of equation (3.43) is called a* **global attractor** *if for every $x_0, \ldots, x_{-k} \in I$ we have*

$$\lim_{n \to \infty} x_n = \bar{x}.$$

(iv) *An equilibrium $\bar{x}$ of equation (3.43) is called* **globally asymptotically stable** *if it is locally stable and a global attractor.*

(v) *An equilibrium $\bar{x}$ of equation (3.43) is called* **unstable** *if it is not stable.*

(vi) *An equilibrium $\bar{x}$ of equation (3.43) is called* **hyperbolic** *provided all the eigenvalues λ of the corresponding linearized equation at $\bar{x}$*

$$z_{n+1} - \frac{\partial f}{\partial u_1}(\bar{x}, \ldots, \bar{x})z_n - \frac{\partial f}{\partial u_2}(\bar{x}, \ldots, \bar{x})z_{n-1} - \ldots - \frac{\partial f}{\partial u_{k+1}}(\bar{x}, \ldots, \bar{x})z_{n-k} = 0 \quad (3.44)$$

are not on a unit circle.

(vii) *An hyperbolic equilibrium $\bar{x}$ of equation (3.44) is called a* **saddle** *provided there are eigenvalues λ of the corresponding linearized equation at $\bar{x}$ (3.44) which are inside and outside of unit circle.*

The main results for Eq(3.43) are the analogues of the Hartman-Grobman Theorem and the Stable Manifold Theorem that relate the stability type of this equation with the stability type of its linearized equation (3.44) at the hyperbolic equilibrium point.

Equation (3.43) is dissipative if there exists an **absorbing interval** $I = [a, b]$ with the property that if $k + 1$ consecutive terms of the solution fall in I then all the subsequent terms of the solution also belong to I.

3.9 Semicycle Analysis

In this section we present basic facts on semicycle analyis.

Let us consider the general nonlinear equation of the form

$$x_{n+1} = f(x_n, x_{n-1}, \ldots, x_{n-k}), \quad n = 0, 1, \ldots. \tag{3.45}$$

A **positive semicycle** of a solution $\{x_n\}$ of equation (3.45) consists of a "string" of terms $\{x_p, x_{p+1}, \ldots, x_m\}$, all greater than or equal to the equilibrium $\bar{x}$, with $l \geq -1$ and $m \leq \infty$ and such that

$$\text{either } p = -1, \quad \text{or } p > -1 \text{ and } x_{p-1} < \bar{x}$$

and

$$\text{either } m = \infty, \quad \text{or } m < \infty \text{ and } x_{m+1} < \bar{x}.$$

A **negative semicycle** of a solution $\{x_n\}$ of equation (3.45) consists of a "string" of terms $\{x_p, x_{p+1}, \ldots, x_m\}$, all less than the equilibrium $\bar{x}$, with $p \geq -1$ and $m \leq \infty$ and such that

$$\text{either } p = -1, \quad \text{or } p > -1 \quad \text{and } x_{l-1} \geq \bar{x}$$

and

$$\text{either } m = \infty, \quad \text{or} \quad m < \infty \text{ and } x_{m+1} \geq \bar{x}.$$

3.10 *Dynamica* Session on Semicycles

In this section we use *Dynamica* to compute and visualize semicycles.

3.10.1 Lyness' Equation

We now show that, for $A = 0$, every solution of Lyness' equation

$$x_{n+1} = \frac{A + x_n}{x_{n-1}}$$

is periodic with the same period.

It is known that for $A = 0$ all solutions are periodic with period six. We may verify this fact by computing 10 terms of the orbit starting with $\{a, b\}$.

```
In[1]:= << Dynamica`;

In[2]:= A = 0; Clear[x, y];

In[3]:= Simplify[Orbit[LynessMap,
            {a, b}, 10]]
```

$$Out[3]= \left\{\{a, b\}, \left\{b, \frac{b}{a}\right\}, \left\{\frac{b}{a}, \frac{1}{a}\right\}, \left\{\frac{1}{a}, \frac{1}{b}\right\}, \left\{\frac{1}{b}, \frac{a}{b}\right\}, \left\{\frac{a}{b}, a\right\}, \{a, b\}, \left\{b, \frac{b}{a}\right\}, \left\{\frac{b}{a}, \frac{1}{a}\right\}, \left\{\frac{1}{a}, \frac{1}{b}\right\}, \left\{\frac{1}{b}, \frac{a}{b}\right\}\right\}$$

Visualization of this fact can be done by using `PoincarePlot2D`, which is a function that takes the circle of radius 1/2 centered at (1, 1) and applies the map generated by the difference equation or the system of difference equations a prescribed number of times. The default values for the center and the radius can be changed.

Here we apply 20 times `PoincarePlot2D` on an initial circle centered at (2, 3) to show the evolution of this circle under Lyness' map produces only six different shapes, providing visual evidence that every solution is periodic with period six. One can use different initial shapes to produce similar plots.

```
In[4]:= fper6[{x_, y_}] := {y, y/x};

In[5]:= PoincarePlot2D[fper6, 20,
            Center → (2, 3)];
```

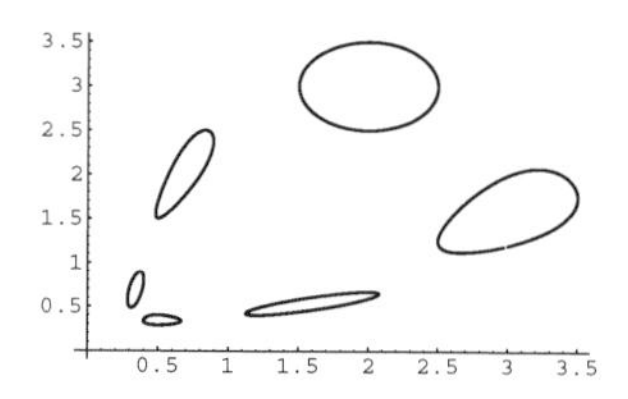

More about Lyness' equation can be found in Section 2.15.

3.10.2 Todd's Equation

We show symbolically that every solution of Todd's equation

$$x_{n+1} = \frac{A + x_n + x_{n-1}}{x_{n-2}}$$

for the specific value of A is periodic with the same period. A detailed study of semicycles of this equation is presented in Section 5.5.

It is known that for $A = 1$ all solutions of Todd's equation are periodic with period eight. Here we verify this fact.

```
In[6]:= A = 1; Clear[x, y];

In[7]:= Simplify[Orbit[ToddMap,
            {p, q, r}, 10]]
```

Out[7]=

$$\left\{\{p, q, r\}, \left\{q, r, \frac{1+q+r}{p}\right\}, \left\{r, \frac{1+q+r}{p}, \frac{1+p+q+r+pr}{pq}\right\}, \left\{\frac{1+q+r}{p}, \frac{1+p+q+r+pr}{pq}, \frac{(1+p+q)(1+q+r)}{pqr}\right\}, \left\{\frac{1+p+q+r+pr}{pq}, \frac{(1+p+q)(1+q+r)}{pqr}, \frac{1+p+q+r+pr}{qr}\right\}, \left\{\frac{(1+p+q)(1+q+r)}{pqr}, \frac{1+p+q+r+pr}{qr}, \frac{1+p+q}{r}\right\}, \left\{\frac{1+p+q+r+pr}{qr}, \frac{1+p+q}{r}, p\right\}, \left\{\frac{1+p+q}{r}, p, q\right\}, \{p, q, r\}, \left\{q, r, \frac{1+q+r}{p}\right\}, \left\{r, \frac{1+q+r}{p}, \frac{1+p+q+r+pr}{pq}\right\}\right\}$$

The evolution of a sphere under twenty iterations of Todd's map produces only eight different shapes, thus suggesting that every solution is periodic with period eight. One can use different initial shapes to produce similar plots.

```
In[8]:= gper8[{x_, y_, z_}] :=
            {y, z, (1 + y + z)/x};

In[9]:= PoincarePlot3D[gper8, 20,
          Axes → False];
```

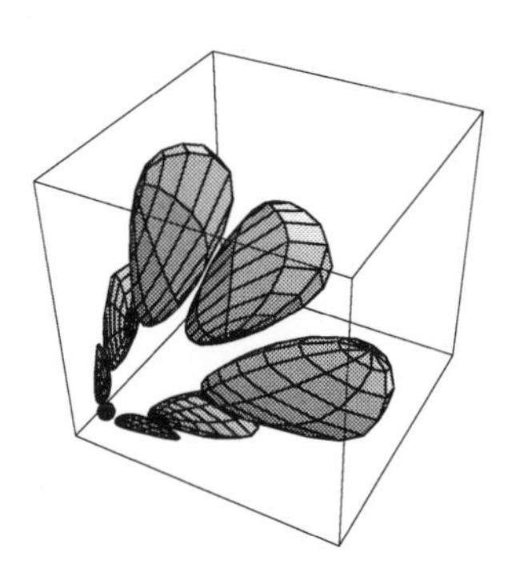

3.10.3 Gingerbreadman Equation

We now discuss an equation that generates the map known as the gingerbreadman map. The equation is

$$x_{n+1} = 1 - x_{n-1} + |x_n|. \tag{3.46}$$

The corresponding map is generated first.

```
In[10]:= GingerEq = x[n + 1] ==
            1 - x[n - 1] + Abs[x[n]];

In[11]:= GingerMap = DEToMap[GingerEq];

In[12]:= GingerMap[{x, y}]
Out[12]= {y, 1 - x + Abs[y]}
```

In the literature (for example, see [D2]) it is common to find the orbit with the initial conditions $x_0 = -0.1$, $x_1 = 0$.

```
In[13]:= ListPlot[Orbit[GingerMap,
           {-0.1, 0.}, 6000],
           PlotRange → {{-3, 8}, {-3, 8}},
           Frame → True, Axes → None];
```

The orbit seems to exhibit chaotic behavior. This is exactly the claim in the paper by R. Devaney [D2]. It is mentioned there for the first time that the plot of the orbit resembles a gingerbreadman. We reproduce this figure easily with a plot of just 6,000 points of the orbit using floating point arithmetic.

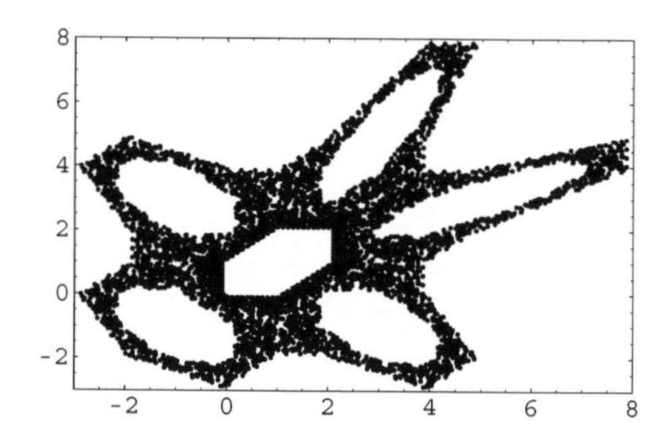

These are some of the terms in the orbit.

```
In[14]:= gingerorb = Orbit[
           GingetMap, {-0.1, 0}, 130];

In[15]:= Take[gingerorb, {1, 10}]
Out[15]= {{-0.1, 0}, {0, 1.1}, {1.1, 2.1},
         {2.1, 2.}, {2., 0.9}, {0.9, -0.1},
         {-0.1, 0.2}, {0.2, 1.3}, {1.3, 2.1},
         {2.1, 1.8}}

In[16]:= Take[gingerorb, {120, 130}]
Out[16]= {{1.5, 0.3}, {0.3, -0.2},
         {-0.2, 0.9}, {0.9, 2.1}, {2.1, 2.2},
         {2.2, 1.1}, {1.1, -0.1},
         {-0.1, -5.54294 × 10^-11},
         {-5.54294 × 10^-11, 1.1}, {1.1, 2.1},
         {2.1, 2.}}
```

A time series plot.

```
In[17]:= TimeSeriesPlot[
           GingerMap, {-0.1, 0.},
           100, PlotRange → {-3, 7}];
```

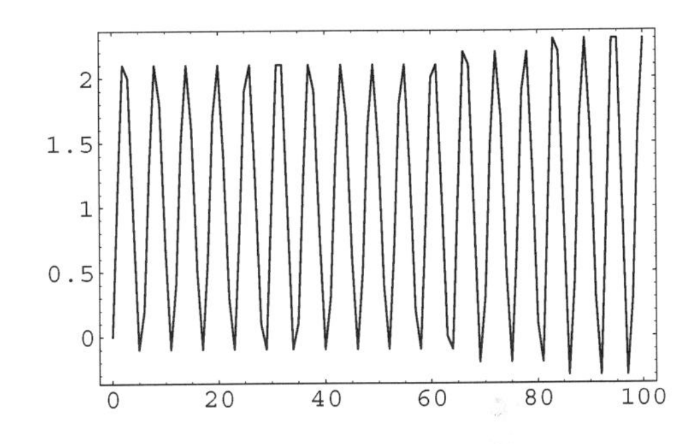

A plot of x_n versus n. The plot shows behavior approximately periodic for n not too large, with a period around 120 or so. However, chaos seems to set in for n larger than 400. This suggests the possibility of numerical error buildup.

```
In[18]:= OrbitPlot[
           GingerMap, {-0.1, 0.}, 600,
           PlotRange → {-3, 7}];
```

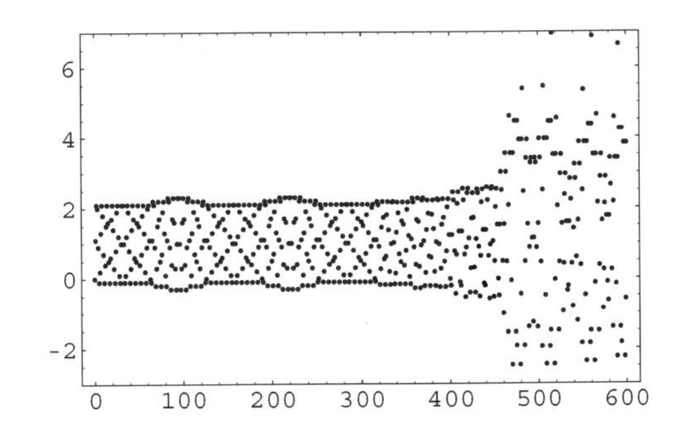

The presence of numerical error is dramatically confirmed by simply calculating the orbit in *exact arithmetic*. Here is a plot of x_n versus n.

```
In[19]:= TimeSeriesPlot[
           GingerMap, {-1/10, 0},
           600, PlotRange → {-3, 7}];
```

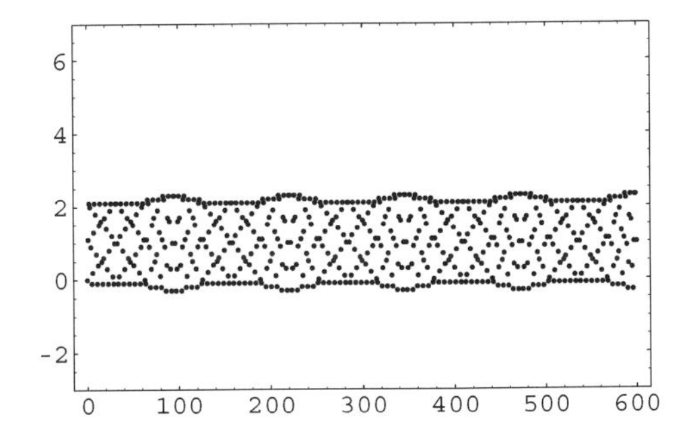

The phase plane plot of the *true* orbit is quite different from the gingerbread-man figure.

```
In[20]:= ListPlot[Orbit[GingerMap,
            {-1/10, 0}, 6000],
         PlotRange → {{-3, 8}, {-3, 8}},
         Frame → True, Axes → None];
```

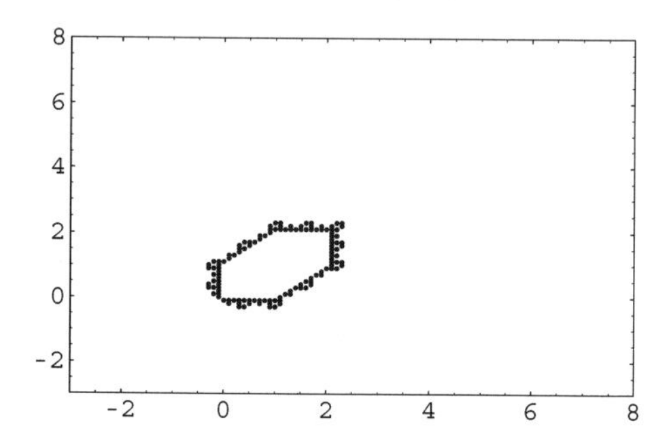

Here is close-up view of the phase plane plot.

```
In[21]:= ListPlot[Orbit[
         GingerMap, {-1/10, 0}, 6000],
         PlotRange → {{-1, 3}, {-1, 3}},
         Frame → True, Axes → None];
```

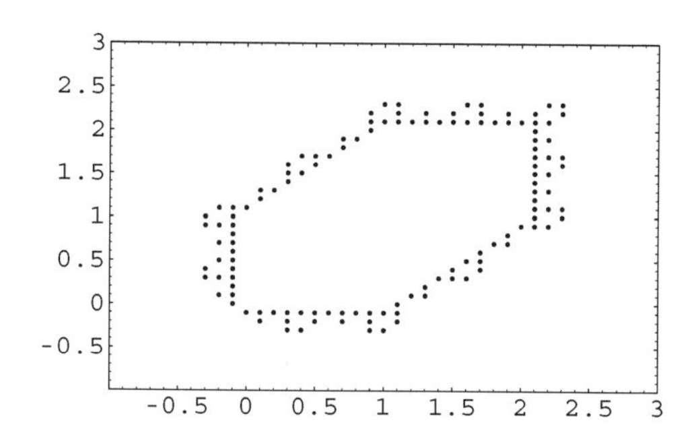

By inspecting the orbit one may verify that it is indeed periodic.

```
In[22]:= gingerorb = Orbit[GingerMap,
            {-1/10, 0}, 130];

In[23]:= Take[gingerorb, {1, 10}]
```

Out[23]= $\left\{\left\{-\frac{1}{10}, 0\right\}, \left\{0, \frac{11}{10}\right\}, \left\{\frac{11}{10}, \frac{21}{10}\right\}, \left\{\frac{21}{10}, 2\right\}, \left\{2, \frac{9}{10}\right\}, \left\{\frac{9}{10}, -\frac{1}{10}\right\}, \left\{-\frac{1}{10}, \frac{1}{5}\right\}, \left\{\frac{1}{5}, \frac{13}{10}\right\}, \left\{\frac{13}{10}, \frac{21}{10}\right\}, \left\{\frac{21}{10}, \frac{9}{5}\right\}\right\}$

The period is 125.

```
In[24]:= Take[gingerorb, {120, 130}]
```

$$Out[24]= \left\{\left\{\frac{3}{2}, \frac{3}{10}\right\}, \left\{\frac{3}{10}, -\frac{1}{5}\right\}, \left\{-\frac{1}{5}, \frac{9}{10}\right\}, \left\{\frac{9}{10}, \frac{21}{10}\right\}, \left\{\frac{21}{10}, \frac{11}{5}\right\}, \left\{\frac{11}{5}, \frac{11}{10}\right\}, \left\{\frac{11}{10}, -\frac{1}{10}\right\}, \left\{-\frac{1}{10}, 0\right\}, \left\{0, \frac{11}{10}\right\}, \left\{\frac{11}{10}, \frac{21}{10}\right\}, \left\{\frac{21}{10}, 2\right\}\right\}$$

3.10.4 Rational Equation

We now consider the rational difference equation

$$x_{n+1} = \frac{r + px_n + x_{n-1}}{qx_n + x_{n-1}}$$

where r, p, q, and the initial conditions x_{-1}, x_0, are positive, see [KL] for some theoretical results. We shall find the fixed points of the map for a particular case of this equation ($r = 1/16$, $q = 8$, $p = 1/2$) and perform a semicycle analysis of a given orbit.

First define the equation and the corresponding map.

```
In[25]:= rationeq = x[n + 1] ==
             r + p x[n] + x[n - 1]
             ---------------------;
               q x[n] + x[n - 1]

In[26]:= rationmap = DEToMap[rationeq];

In[27]:= rationmap[{x, y}]
```

$$Out[27]= \left\{y, \frac{r + x + p\,y}{x + q\,y}\right\}$$

```
In[28]:= r = 1/16.; q = 8.; p = 1/2.;
```

For the semicycle analysis, we need to find the equilibrium points of the difference equation.

```
In[29]:= Solve[rationmap[{x, y}] ==
              {x, y}, {x, y}]

Out[29]= {{x → -0.0345178, y → -0.0345178},
          {x → 0.201184, y → 0.201184}}
```

This is the positive equillibrium.

```
In[30]:= pt = x/.%[[2]];
Out[30]= 0.201184
```

The analysis of semicycles with respect to the positive equilibrium will be performed on $\gamma(\{1,2\})$. It is clear from the first few points of the orbit $\gamma(\{1,2\})$ that there is oscillation about the equilibrium.

```
In[31]:= TimeSeriesPlot[rationmap,
           {1., 2.}, 30, AxesOrigin → {0, pt},
           PlotRange → All, Axes → True]
```

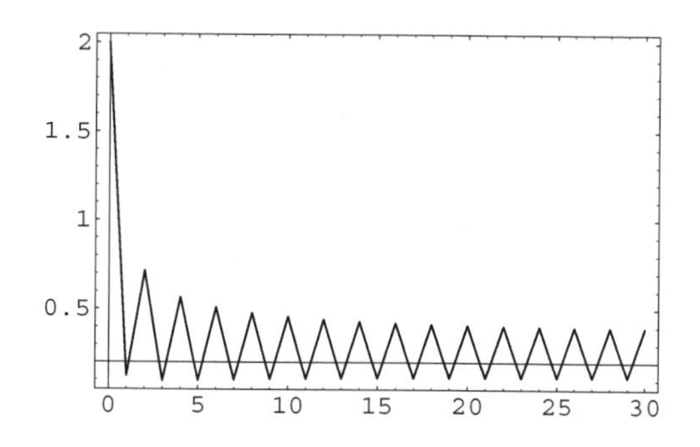

SemicycleTest gives information about the positions and maxima of semicycles.

```
In[32]:= orb = Orbit[
             rationmap, {1, 2}, 30];

In[33]:= SemicycleTest[orb, pt]
  Positions and Maxima of Positive Semicycles
  {2, {2, 2}, {4, 0.714728}, {6, 0.565565},
{8, 0.510034}, {10, 0.478331}, {12, 0.457223},
{14, 0.442145}, {16, 0.430911}, {18, 0.422292},
{20, 0.415533}, {22, 0.410145}, {24, 0.405792},
{26, 0.402238}, {28, 0.399313}, {30, 0.396888}}
  Positions and Minima of Negative Semicycles
{{3, 0.121324}, {5, 0.0926826}, {7, 0.0948551},
{9, 0.0987687}, {11, 0.102011}, {13, 0.104559},
{15, 0.106579}, {17, 0.108202}, {19, 0.109521},
{21, 0.110603}, {23, 0.111498}, {25, 0.112243},
{27, 0.112867}, {29, 0.11339}}
  Lengths of Positive Semicycles


  {2, 1, 1, 1, 1, 1, 1, 1, 1, 1, 1, 1, 1, 1, 1}
  Lengths of Negative Semicycles
 {-1, -1, -1, -1, -1, -1, -1, -1, -1, -1, -1, -1,
  - 1, -1, -1}
  Positions and Lengths of both Positive and
  Negative Semicycles
{2, -1, 1, -1, 1, -1, 1, -1, 1, -1, 1, -1, 1, -1, 1,
-1, 1, -1, 1, -1, 1, -1, 1, -1, 1, -1, 1, -1, 1, -1}
```

We see that all semicycles are of length one or two before the solution becomes equal to the equilibrium. This is observed also if other simulations are attempted. In fact, it has been proved in [GKL1] and [KL] that for all values of parameters p, q and r the semicycles of every oscillatory solution are of length one or two. This fact has been successfully used in [KL] to prove results on global stability and asymptotic behavior of solutions of the equation.

3.10.5 Conclusion

We used *Dynamica* to study semicycles as indicated next.

1. One can show with *Dynamica* that every solution of certain equations is periodic with a certain period for specific values of parameters, see Sections 2.15 and 5.5.
2. One can analyze lengths and the positions of the semicycles as well as locations of the maxima of positive semicycles and the locations of the minima of negative semicycles of given orbits.
3. The information obtained in 2. can be used to prove global asymptotic results for solutions of the difference equation (see [GKL], [KoL], [KL], [KLP1], [KLP2]).

3.11 Exercises

Exercise 3.1 Use `RSolve` to find the general solution of the following difference equations:

1. $x_{n+1} - 7x_n + 14x_{n-1} - 8x_{n-2} = 0,$
2. $x_{n+1} - 6x_n + 9x_{n-1} - 4x_{n-2} = 0,$
3. $x_{n+1} - 2x_n + x_{n-1} - 2x_{n-2} = 0,$
4. $x_{n+1} - 3x_n + 3x_{n-1} - x_{n-2} = n^2,$
5. $x_{n+1} + 3x_{n-1} + 2x_{n-3} = 0.$

Exercise 3.2
Use `RSolve` to find the solution of the following initial value problems:

1. $x_{n+1} + x_n - 4x_{n-1} - 4x_{n-2} = 0, \quad x_{-2} = 1, x_{-1} = 2, x_0 = -4,$
   $y_{n+1} - y_n - 12y_{n-1} - 4y_{n-2} + 16y_{n-3} = 2, \quad x_{-3} = 2, x_{-2} = 0, x_{-1} = 1, x_0 = -1.$

Exercise 3.3
Use `RSolve` to find the general solution of the following systems of difference equations.

1.

$$x_{n+1} = -4x_n + 6y_n - 3z_n$$

$$y_{n+1} = -x_n + 3y_n - z_n$$

$$z_{n+1} = 4x_n - 4y_n + 3z_n.$$

$$x_{n+1} = 4x_n - 6y_n + 3z_n$$

$$y_{n+1} = x_n + z_n + n$$

$$z_{n+1} = -2x_n + 2y_n + 1.$$

Exercise 3.4

Use `RSolve` to find the solution of the following initial value problems.

$$x_{n+1} = -x_n + 2y_n + 5z_n + n$$

$$y_{n+1} = x_n - z_n + 1$$

$$z_{n+1} = -3x_n + 3y_n + 7z_n$$

$$x_1 = 1, y_1 = -1, z_1 = 2.$$

Exercise 3.5

Perform a semicycle analysis for each of the following equations:

1.

$$x_{n+1} = \frac{px_n + qx_{n-1}}{1 + x_n}$$

(see [KLP1] and [KL]).

2.

$$x_{n+1} = \frac{px_n + qx_{n-1}}{1 + x_{n-1}}$$

(see [KLP2] and [KL]).

3.

$$x_{n+1} = \frac{p + qx_{n-1}}{1 + x_n + rx_{n-1}},$$

(see [KL] and [KLMR]).

4.

$$x_{n+1} = \frac{p + x_{n-1}}{qx_n + x_{n-1}},$$

(see [KKLT] and [KL]).

5.
$$x_{n+1} = x_n(A(1 - x_n) - B(Ax_{n-1}(1 - x_{n-1}) - x_n),$$

(Predator–prey equation, see [KJ] and references therein).

6.
$$x_{n+1} = -x_{n-1} + 2Ax_n + 4(1 - A)\frac{x_n^2}{1 + x_n^2},$$

(Mira's equation, see [GM]).

7.
$$x_{n+1} = A\frac{x_n^2}{(1 + x_n)x_{n-1})},$$

(May's host parasitoid equation).

8.
$$x_{n+1} = x_n \cos\theta - (y_n - x_n^2)\sin\theta$$

$$y_{n+1} = x_n \sin\theta - (y_n - x_n^2)\cos\theta,$$

(Cremona equation, see [K] and [KJ] and references therein).

Exercise 3.6
Replace x_n and y_n in the gingerbreadman equation (3.46) by $10x_n$ and $10y_n$ to obtain the difference equation

$$x_{n+1} = 10 - x_{n-1} + |x_n|.$$

Verify that $x_0 = -1$ and $x_1 = 0$ yield a periodic orbit with period 125, consisting of integers only. Produce a plot of the orbit that shows the periodic character of it.

(Integer version of gingerbreadman map).

Exercise 3.7
Use *Dynamica* function `PoincarePlot2D` to visualize that every solution of the difference equation

$$x_{n+1} = \frac{|x_{n-1}| + x_{n-1}}{2} - x_n$$

is periodic with period five. See [Sk]. Prove this fact algebraically.

Exercise 3.8 Applying `PoincarePlot2D` on the initial circle with the center at (1.2, 1.2) 17 times gives the following plot showing that every solution of

$$x_{n+1} = \frac{|x_{n-1}| - x_{n-1}}{2} - x_n \tag{3.47}$$

is periodic with period seven. See [Sk].

`PoincarePlot2D` applied on the circle with the center at (1.2, 1.2) 17 times produces only seven different shapes showing visually that every solution of equation (3.47) is periodic with period seven.

```
In[34]:= PoincarePlot2D[fper7,
            17, Center → {1.2, 1.2}];
```

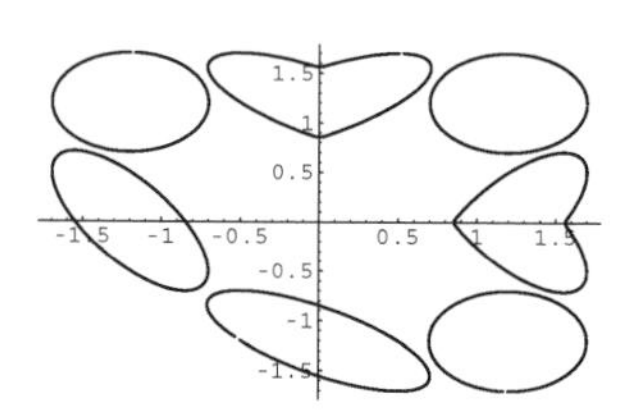

Prove this fact algebraically.

Exercise 3.9 Applying `PoincarePlot2D` on the initial circle with the center at (2, 2) fourteen times gives the following plot showing that every solution of

$$x_{n+1} = \frac{|x_{n-1}| - x_{n-1}}{2} - x_n \tag{3.48}$$

is periodic with period nine. See [Sk].

The *Dynamica* function `PoincarePlot2D` applied on the circle with the center at (2, 2) fourteen times produces only nine different shapes, thus showing visually that every solution of equation (3.48) is periodic with period nine.

```
In[35]:= PoincarePlot2D[fper9,
            14, Center → {2, 2}];
```

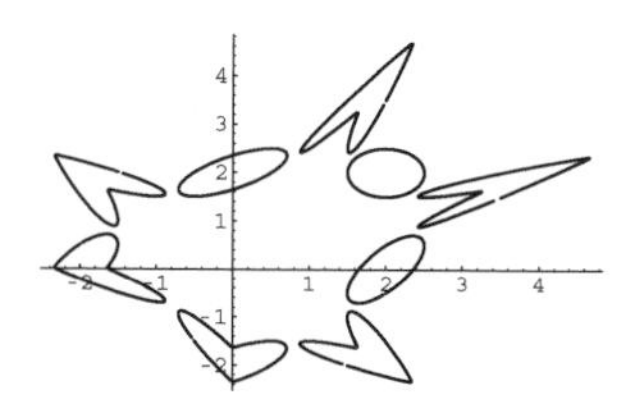

Prove this fact algebraically.

Exercise 3.10 Applying *Dynamica* function `PoincarePlot2D` on the initial circle with the center at (1.2, 2.2) fourteen times gives the following plot, thus showing that every solution of

$$x_{n+1} = \frac{(\sqrt{3} - \sqrt{2})|x_{n-1}| + (\sqrt{3} - \sqrt{2})x_{n-1}}{2} - x_n \tag{3.49}$$

is periodic with period ten. See [Sk].

`PoincarePlot2D` applied on the circle with center at $(1.2, 2.2)$ fourteen times produces only ten different shapes, thus showing visually that every solution of equation (3.49) is periodic with period ten.

```
In[36]:= PoincarePlot2D[fper10,
    14, Center → {1.2, 2.2}];
```

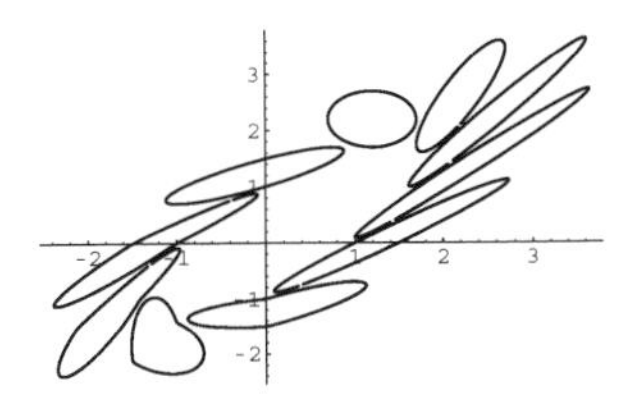

Use *Dynamica* to prove this fact algebraically.

Exercise 3.11

Use *Dynamica* function `PoincarePlot2D` to visualize that every solution of the following difference equations:

$$x_{n+1} = \frac{\sqrt{2}(|x_{n-1}| + x_{n-1})}{2} - x_n,$$

and

$$x_{n+1} = \frac{|x_{n-1}| - 3x_{n-1}}{2} - x_{n-1},$$

is periodic with the same period. See [Sk]. Prove this fact algebraically.

Chapter 4

Invariants and Related Lyapunov Functions

4.1 Introduction

In this chapter we shall address a fast developing area of discrete dynamical systems and difference equations that provides essential tools for finding the solutions of difference equations in exact form and investigating the short-term and long-term behavior of the solutions. The theory is based on the existence of expressions that remain constant or invariant along solutions of a difference equation and which reveal the behavior of the solutions of the considered equation. Naturally, this expression is called an *invariant* or a *first integral* of difference equation. The term first integral is used to emphasize the analogy with the case of differential equations. The two-dimensional version of this theory was given in Section 2.13.

Consider the system of nonlinear difference equations:

$$x_{n+1} = \frac{x_n + y_n}{2}, \quad y_{n+1} = \frac{2}{\dfrac{1}{x_n} + \dfrac{1}{y_n}}. \tag{4.1}$$

Two sequences $\{x_n\}$ and $\{y_n\}$ that satisfy (4.1) are monotone and bounded and so converge to limiting values a and b, respectively, where $a = b$; however, the equations cannot be used to find the limiting value. Instead, by using the invariant $I(x, y) = xy$, which satisfy $I(x_{n+1}, y_{n+1}) = I(x_n, y_n)$, we get $x_0 y_0 = a^2$ and so by choosing x_0 and y_0 such that $x_0 y_0 > 0$ we get an efficient numerical algorithm that converges to $\sqrt{x_0 y_0}$ and so it can be used as the algorithm of choice for computing the radicals.

Some simple looking equations may possess essentially different invariants. The simplest known example is probably the aritmetic-geometric mean iteration of Gauss and Legendre

$$x_{n+1} = \frac{x_n + y_n}{2}, \quad y_{n+1} = \sqrt{x_n y_n}, \quad n = 0, 1, \ldots \tag{4.2}$$

for approximating the following complete elliptic integral of the first kind:

$$I(a,b) = \int_0^{\pi/2} \frac{dt}{\sqrt{a^2 cos^2 t + b^2 sin^2 t}},$$

see [BB]. A straightforward calculation shows that $I(a,b)$ is an invariant of (4.2) since $I(x_{n+1}, y_{n+1}) = I(x_n, y_n)$. This in turn implies that $I(x_n, y_n) = I(x_0, y_0) = I(a,b)$ where we take $x_0 = a$, $y_0 = b$. On the other hand, it is easy to see that the sequences $\{x_n\}$ and $\{y_n\}$, are convergent, as monotone and bounded sequences, but the equations cannot be used to find the limiting values x_∞ and y_∞. Instead, by using the invariant we immediately get $I(x_\infty, y_\infty) = I(a,b)$, which shows that our iteration method converges to the elliptic integral. A similar iteration method for numerical computation of the integral

$$F(a,b) = \frac{2}{\pi} \int_0^{\pi/2} \ln(a \sin^2 t + b \cos^2 t) dt$$

is given by the following system of equations:

$$x_{n+1} = \left(\frac{x_n + y_n}{2}\right)^2, \quad y_{n+1} = x_n y_n,$$

where $x_0 = a > 0$ and $y_0 = b > 0$. Many complicated integrals in mathematical physics can be found efficiently in this way, see [BB] and [Wi].

As our final motivating example, consider Lyness' equation

$$x_{n+1} = \frac{a + x_n}{x_{n-1}}, \tag{4.3}$$

where $a > 0$ is a parameter and $x_{-1} > 0, x_0 > 0$. Recently, this equation attracted a lot of attention, see [KoL], [GJKL], and [Ku] and its stability has been proved showing that it can be transformed to an area-preserving map, see Section 2.15 and [KLTT]. Also KAM theory has been applied to provide another proof of stability, see [KLTT]. Here we will construct a Lyapunov function for this equation using the well-known invariant

$$I(x_n, x_{n-1}) = \left(1 + \frac{1}{x_n}\right)\left(1 + \frac{1}{x_{n-1}}\right)(a + x_n + x_{n-1}).$$

In fact, a Lyapunov function $V(x,y)$ is given by

$$V(x,y) = I(x,y) - I(p,p) = I(x,y) - \frac{(p+1)^3}{p},$$

where

$$p = \frac{1 \pm \sqrt{1 + 4a}}{2}$$

is the equilibrium of this equation.

Thus, the invariants are used to find the value of the equilibrium, to show boundedness of the solutions, to prove stability, etc. Now we give the formal definitions of invariants.

DEFINITION 4.1 *Consider the difference equation*

$$\mathbf{x}_{n+1} = \mathbf{f}(\mathbf{x}_n), \quad n = 0, 1, \ldots, \tag{4.4}$$

where $\mathbf{x}_n$ is in R^k and $\mathbf{f} : D \to D$ is continuous, where $D \subset R^k$. We call a nonconstant continuous function $I : R^k \to R$ an invariant for the system (4.4) if

$$I(\mathbf{x}_{n+1}) = I(\mathbf{f}(\mathbf{x}_n)) = I(\mathbf{x}_n), \quad \textit{for every } n = 0, 1, \ldots .$$

4.2 Invariants for Linear Equations and Systems

First, we shall consider the existence of invariants for systems of linear difference equations. The basic result in this direction has been established recently in [BH]:

THEOREM 4.1
Let

$$\mathbf{x}_{n+1} = B\mathbf{x}_n, \quad n = 0, 1, \ldots, \tag{4.5}$$

be a linear difference equation where B is a $k \times k$ real matrix and $\mathbf{x} \in R^k$. Then (4.5) possesses a continuous invariant $I : R^k \to R$ if and only if one of the following two conditions holds:

(1) B has an eigenvalue $|\lambda| = 1$.

(2) B has eigenvalues λ_1 and λ_2 with $|\lambda_1| > 1$ and $0 < |\lambda_2| < 1$.

In other words, neither all eigenvalues of B lie inside of the unit disk nor all eigenvalues of B lie outside of the unit disk.

Notice that both of eigenvalues of (4.3) are on the unit disk.

REMARK 4.1 The requirement that the invariant is continuous is essential. If this requirement was dropped, then any linear system (4.5) would possess a nontrivial invariant.

As an example, consider the system

$$x_{n+1} = 2x_n, \quad y_{n+1} = 4y_n. \tag{4.6}$$

Theorem 4.1 implies that this system does not possess any real continuous invariant. However, if we do not require continuity, the following is an invariant:

$$I(x,y) = \begin{cases} \frac{y}{x^2}, & if\ x \neq 0; \\ 0, & if x = 0. \end{cases}$$

The construction of a discontinuous invariant of system (4.6) can be extended to any linear system. In fact, we can show that any homeomorphism of R^k possesses an invariant. Namely, each orbit of any homeomorphism of R^k is countable. So, by cardinality theory, there are precisely as many orbits as there are real numbers, i.e., there is a bijection from the set of orbits to the set of real numbers. Any such correspondence is a (discontinuous) invariant. See [BH] and [TB]. ▮

4.3 Invariants and Corresponding Lyapunov Functions for Nonlinear Systems

In the case of nonlinear systems we present the following result, which establishes a connection between invariants and Lyapunov functions and stability of equilibrium points. This result has been used extensively to prove the stability of several difference equations, see [Ku].

THEOREM 4.2 (Discrete Dirichlet Theorem)
Consider the difference equation (4.4) where $\mathbf{x}_n$ *is in* R^k *and* $\mathbf{f} : D \to D$ *is continuous, where* $D \subset R^k$*. Suppose that* $I : R^k \to R$ *is a continuous invariant of (4.4). If* I *attains an isolated local minimum or maximum value at the equilibrium point* $\bar{\mathbf{x}}$ *of this system, then there exists a Lyapunov function equal to*

$$\pm(I(\mathbf{x}) - I(\bar{\mathbf{x}}))$$

and so the equilibrium $\bar{\mathbf{x}}$ *is stable.*

Theorem 4.2 is a discrete analogue of the Dirichlet theorem, see [SM], p. 208.

Theorem 4.2 shows that whenever certain conditions are satisfied, there exists a Lyapunov function that is actually a special invariant of the equation. As is well known, see [GM] and [Mi], an invariant is not uniquely determined but rather if one invariant is given then any appropriate function (continuous, analytic) of this invariant is an invariant itself. Furthermore, the difference equation may have several "independent" invariants (e.g., two invariants neither of which can be derived from the other through homeomorphic changes of the variable), see [BB].

REMARK 4.2 An important question that could be raised here is what are the implications of Theorem 4.2 on the linearized stability theory for the difference equation. The long-standing conjecture is that the difference equation

$$\mathbf{x}_{n+1} = \mathbf{f}(\mathbf{x}_n)$$

has an invariant if and only if the corresponding linearized equation

$$\mathbf{y}_{n+1} = D_{\mathbf{f}}(\bar{\mathbf{x}})\,\mathbf{y}_n$$

has an invariant, which in turn is equivalent to the fact that the matrix $D_{\mathbf{f}}(\bar{\mathbf{x}})$ does not have all eigenvalues inside or outside the unit disk. ∎

4.3.1 Applications

In this section, we present several equations with known invariants. By using Theorem 4.2, we obtain a Lyapunov function and prove the stability of the corresponding equilibrium points. We note that the positive definiteness of the Hessian at the critical point is verified by checking that all the principal minors are positive. After presenting the hand calculations we proceed to perform similar calculations by using *Dynamica*.

Example 4.1 The generalized Lyness' equation (see [FJL], [GJKL], and [KoL]) has the form:

$$x_{n+1} = \frac{ax_n + b}{(cx_n + d)x_{n-1}},$$

where a, b, c, d are positive parameters and $x_{-1} > 0, x_0 > 0$. This equation has an invariant:

$$I(x_n, x_{n-1}) = \frac{ab}{x_{n-1}x_n} + (a^2 + bd)\left(\frac{1}{x_n} + \frac{1}{x_{n-1}}\right)$$

$$+ad\left(\frac{x_{n-1}}{x_n} + \frac{x_n}{x_{n-1}}\right) + (ac + d^2)(x_n + x_{n-1}) + cdx_{n-1}x_n.$$

The equilibrium point p satisfies $cp^3 + dp^2 - ap - b = 0$. Obviously, this equation has at least one positive solution and we will investigate the stability of this equilibrium. In order to investigate the stability of the negative equilibrium point, we should first prove the existence of the solutions in some neighborhood of such solutions, which may be a formidable task.

The necessary conditions for the invariant to have the extremum give:

$$\frac{\partial I}{\partial x} \equiv -\frac{ab}{x^2 y} + cdy - \frac{a^2 + bd}{x^2} + ad\left(\frac{1}{y} - \frac{y}{x^2}\right) + ac + d^2 = 0,$$

and

$$\frac{\partial I}{\partial y} \equiv -\frac{ab}{xy^2} + cdx - \frac{a^2 + bd}{y^2} + ad\left(\frac{1}{x} - \frac{x}{y^2}\right) + ac + d^2 = 0,$$

which after some manipulations lead to $x = y$ and to the equation

$$cdx^4 + (ac + d^2)x^3 - (a^2 + bd)x - ab = (dx + a)(cx^3 + dx^2 - ax - b) = 0.$$

This shows that the critical points are exactly the equilibrium points.

Now, we check the Hessian at a positive critical point (p_+, p_+) which will be denoted simply as (p, p). We have,

$$H = \begin{pmatrix} A & B \\ B & A \end{pmatrix},$$

where

$$A = \frac{\partial^2 I}{\partial x^2}(p, p) = \frac{2}{p^4}[adp^2 + (a^2 + bd)p + ab] > 0,$$

and

$$B = \frac{\partial^2 I}{\partial x \partial y}(p, p) = \frac{cdp^4 - 2adp^2 + ab}{p^4}.$$

Now,

$$det\ H = A^2 - B^2 = (A - B)(A + B) = \frac{1}{p^8}[cdp^4 + 2(a^2 + bd)p + 3ab]E,$$

where by using the equilibrium equation

$$E = -cdp^4 + 4adp^2 + 2(a^2 + bd)p + ab = d^2p^3 + 3adp^2 + 2a^2p + ab > 0.$$

This shows that the Hessian H is positive definite at the point (p, p), which implies that the invariant I attains a minimum at (p, p). Thus,

$$\min\{I(x, y) : (x, y) \in D\} = I(p, p),$$

and so by Theorem 4.2

$$V(x, y) = I(x, y) - I(p, p),$$

which shows that the equilibrium point p is stable. ∎

Example 4.2 Todd's equation (see [KoL] and [Ku]),

$$x_{n+1} = \frac{a + x_n + x_{n-1}}{x_{n-2}},$$

where $a > 0$ is a parameter, and $x_1 > 0$, $x_0 > 0$, $x_1 > 0$, has an invariant:

$$I(x_n, x_{n-1}, x_{n-2}) = \left(1 + \frac{1}{x_n}\right)\left(1 + \frac{1}{x_{n-1}}\right)\left(1 + \frac{1}{x_{n-2}}\right)(a + x_n + x_{n-1} + x_{n-2}).$$

The equilibrium points $p = 1 \pm \sqrt{1+a}$ satisfy the equation $p^2 - 2p - a = 0$.

The necessary conditions for the extremum give:

$$\frac{\partial I}{\partial x} \equiv \left(1 + \frac{1}{y}\right)\left(1 + \frac{1}{z}\right)\left(1 - \frac{a+y+z}{x^2}\right) = 0,$$

$$\frac{\partial I}{\partial y} \equiv \left(1 + \frac{1}{x}\right)\left(1 + \frac{1}{z}\right)\left(1 - \frac{a+x+z}{y^2}\right) = 0,$$

and

$$\frac{\partial I}{\partial z} \equiv \left(1 + \frac{1}{x}\right)\left(1 + \frac{1}{y}\right)\left(1 - \frac{a+x+y}{z^2}\right) = 0.$$

This leads to $x = y = z$, and to the equation $x^2 - 2x - a = 0$, which shows that the critical points are exactly the equilibrium points. Evaluating the Hessian at a positive critical point (p, p, p) where $p = 1 + \sqrt{1+a}$ yields

$$H = \begin{pmatrix} A & B & B \\ B & A & B \\ B & B & A \end{pmatrix},$$

where

$$A = \frac{\partial^2 I}{\partial x^2}(p, p) = \frac{\partial^2 I}{\partial y^2}(p, p) = \frac{\partial^2 I}{\partial z^2}(p, p) = 2\frac{(p+1)^2(a+2p)}{p^5} > 0,$$

and

$$B = \frac{\partial^2 I}{\partial x \partial y}(p, p) = \frac{\partial^2 I}{\partial y \partial z}(p, p) = \frac{\partial^2 I}{\partial x \partial z}(p, p) = \frac{(p+1)(a+p-2p^2)}{p^5}.$$

We have

$$\det H_1 = A > 0,$$

$$\det H_2 = \det\begin{pmatrix} A & B \\ B & A \end{pmatrix} = \frac{(p+1)^2(a+(2a+3)p+6p^2)(3a+(2a+5)p+2p^2)}{p^{10}} > 0,$$

and

$$\det H = det\begin{pmatrix} A & B & B \\ B & A & B \\ B & B & A \end{pmatrix} = (A-B)^2(A+2B)$$

$$= 2\frac{(p+1)^3(2a+(a+3)p)(a+(2a+3)p+6p^2)}{p^{15}} > 0,$$

which shows that the invariant I attains a minimum at (p, p, p). Thus,

$$\min\{I(x, y, z) : (x, y, z) \in D\} = I(p, p, p) = \frac{(p+1)^3(a+3p)}{p^3},$$

and so by Theorem 4.2

$$V(x, y, z) = I(x, y, z) - \frac{(p+1)^3(a+3p)}{p^3}.$$

This shows that p is stable.

See Section 5.5 for the complete study of Todd's equation. ∎

Example 4.3 The $k+2$-th-order Lyness' equation (see [KoL] and [Ku]) has the form

$$x_{n+1} = \frac{a + x_n + x_{n-1} + \ldots + x_{n-k}}{x_{n-k-1}},$$

where $a > 0$ is a parameter, and $x\_k > 0, \ldots, x\_1 > 0$. An invariant of this equation is given by:

$$I(x_n, \ldots, x_{n-k-1}) = \left(1 + \frac{1}{x_n}\right)\left(1 + \frac{1}{x_{n-1}}\right)\ldots\left(1 + \frac{1}{x_{n-k-1}}\right)(a + x_n + x_{n-1} + \ldots + x_{n-k-1}).$$

The equilibrium points of this equation are given by $p = \frac{k+1 \pm \sqrt{(k+1)^2+4a}}{2}$ and satisfy the equation $p^2 - (k+1)p - a = 0$. The necessary conditions for the extremum give:

$$\frac{\partial I}{\partial x_{n-i}} \equiv \prod_{j=0, j\neq i}^{k+1}\left(1 + \frac{1}{x_{n-j}}\right)\left(1 - \frac{a + \sum_{j=0, j\neq i}^{k+1} x_{n-j}}{x_{n-i}^2}\right) = 0, \quad i = 1, \ldots, k+1,$$

which gives

$$x_{n-i}^2 - a - \sum_{j=0, j\neq i}^{k+1} x_{n-j} = 0, \quad i = 1, \ldots, k+1. \tag{4.7}$$

Subtracting the consecutive equations in (4.7) we get

$$x_n = x_{n-1} = \ldots = x_{n-k-1} = p,$$

which shows that the critical points are exactly the equilibrium points $(p, \ldots, p)$.

Now, we check the Hessian at a positive critical point $(p_+, \ldots, p_+)$, which for simplicity will be denoted by $(p, \ldots, p)$. Computing the second-order derivatives of I with respect to x_{n-i} we get

$$\frac{\partial^2 I}{\partial x_{n-i}^2} = \frac{2}{x_{n-i}^3}\left(a + \sum_{j=0, j\neq i}^{k+1} x_{n-j}\right)\prod_{j=0, j\neq i}^{k+1}\left(1 + \frac{1}{x_{n-j}}\right),$$

and

$$\frac{\partial^2 I}{\partial x_{n-i}\partial x_{n-j}} = \prod_{p=0, p\neq i,j}^{k+1}\left(1 + \frac{1}{x_{n-p}}\right)\left(\frac{a + \sum_{p=0, p\neq i,j}^{k+1} x_{n-p}}{x_{n-i}^2 x_{n-j}^2} - \frac{1}{x_{n-i}^2} - \frac{1}{x_{n-j}^2}\right),$$

for $i = 1, \ldots, k+1$. Calculating the second derivatives at the equilibrium point $(p, p, \ldots, p)$, we get

$$A = \frac{\partial^2 I}{\partial x_{n-i}^2}(p, p, \ldots, p) = 2\frac{(p+1)^{k+1}(a + (k+1)p)}{p^{k+4}} > 0,$$

and

$$B = \frac{\partial^2 I}{\partial x_{n-i}\partial x_{n-j}}(p, p, \ldots, p) = \frac{(p+1)^k(a+kp-2p^2)}{p^{k+4}}.$$

Thus, the Hessian takes the form

$$H = \begin{pmatrix} A & B & \ldots & B \\ B & A & \ldots & B \\ \vdots & \vdots & \ddots & \vdots \\ B & B & \ldots & A \end{pmatrix}.$$

Now, the principal minors of the Hessian are:

$$\det H_1 = A > 0,$$

$$\det H_m = det\begin{pmatrix} A & B & \ldots & B \\ B & A & \ldots & B \\ \vdots & \vdots & \ddots & \vdots \\ B & B & \ldots & A \end{pmatrix} = (A-B)^{m+1}[A+(m-1)B]$$

$$= K[2(k+2)p^2+(k+2+a)p+a]^{m+1}[(2(k-m)+4)p^2+(2a+(m+1)k+2)p+(m+1)a] > 0,$$

where

$$K = \frac{(p+1)^{(m+2)k}}{p^{(k+4)(m+2)}}, \quad k = 1, \ldots, m,$$

where H_m is m-th order principal minor of the Hessian H.

This shows that the invariant I attains a minimum at $(p, \ldots, p)$. Thus,

$$\min\{I(x_n, x_{n-1}, \ldots, x_{n-k-1}) : (x_n, x_{n-1}, \ldots, x_{n-k-1}) \in D\}$$

$$= I(p, p, \ldots, p) = \frac{(p+1)^{k+2}(a+(k+2)p)}{p^{k+2}}$$

and so, by Theorem 4.2,

$$V(x_n, x_{n-1}, \ldots, x_{n-k-1}) = I(x_n, x_{n-1}, \ldots, x_{n-k-1}) - \frac{(p+1)^{k+2}(a+(k+2)p)}{p^{k+2}},$$

shows that p is stable.

Similarly we can establish the stability of the second equilibrium point $(p\_, \ldots, p\_)$ in some neighborhood of this point where the solution exists. The existence of the solution in this case may be very hard to establish.

Obviously, for $k = 0$ and $k = 1$ we get Lyness' and Todd's equations, respectively. ∎

4.4 Invariants of a Special Class of Difference Equations

In this section we present the explicit formulas for the class of the rational difference equations and their invariants. The formulas are based on recent research results from [QRT1] and [QRT2]. Our main result is based on the following lemmas:

LEMMA 4.1
Consider the vectors $\mathbf{x} = (x^2, x, 1)^T$, *and* $\mathbf{z} = (z^2, z, 1)^T$, *and the vector-valued function*

$$\mathbf{f}(y) = (f_1(y), f_2(y), f_3(y))^T, \quad y \in R,$$

where f_1, f_2, f_3 *are the given functions and*

$$z \neq x, \quad f_2(y) - xf_3(y) \neq 0. \tag{4.8}$$

Then, the mixed product of $\mathbf{x}$, $\mathbf{z}$ *and* $\mathbf{f}(y)$ *is zero, that is,*

$$\mathbf{x} \times \mathbf{f}(y) \cdot \mathbf{z} = 0,$$

if and only if

$$z = \frac{f_1(y) - xf_2(y)}{f_2(y) - xf_3(y)}. \tag{4.9}$$

The proof follows from the simple identity

$$0 = \mathbf{x} \times \mathbf{f(y)} \cdot \mathbf{z} = f_1(y)(z - x) + f_2(y)(x^2 - z^2) + f_3(y)(xz^2 - zx^2)$$
$$= (z - x)\left(f_1(y) - f_2(y)(x + z) + f_3(y)xz\right).$$

If (4.8) holds, we conclude that the last equality is equivalent to (4.9).

REMARK 4.3 Taking in (4.9) $z = x_{n+1}$, $y = x_n$, and $x = x_{n-1}$ we get the following difference equation

$$x_{n+1} = \frac{f_1(x_n) - x_{n-1}f_2(x_n)}{f_2(x_n) - x_{n-1}f_3(x_n)}, \tag{4.10}$$

while if we take $z = x_{n+1}$, $y = x_{n-1}$, and $x = x_n$ we obtain the difference equation

$$x_{n+1} = \frac{f_1(x_{n-1}) - x_nf_2(x_{n-1})}{f_2(x_{n-1}) - x_nf_3(x_{n-1})}. \tag{4.11}$$

∎

The next result establishes a necessary connection between difference equations of the types (4.10), and (4.11) and their invariants.

LEMMA 4.2

Suppose that $\mathbf{x} \neq \mathbf{0}$, $\mathbf{f}(\mathbf{y}) = A_1\mathbf{y} \times A_2\mathbf{y}$ *in Lemma 4.1, where*

$$\mathbf{y} = (y^2, y, 1)^T,$$

where f *is a given function, and* A_i, $i = 1, 2$ *are given matrices*

$$A_i = \begin{pmatrix} a_i & b_i & c_i \\ d_i & e_i & f_i \\ g_i & h_i & k_i \end{pmatrix}, i = 1, 2.$$

Then, for every vector $\mathbf{x}$, *such that* $(\mathbf{A_2y}) \cdot \mathbf{x} \neq 0$, *the expression*

$$\frac{(\mathbf{A_1y}) \cdot \mathbf{x}}{(\mathbf{A_2y}) \cdot \mathbf{x}}, \tag{4.12}$$

is independent of $\mathbf{x}$, *that is,*

$$\frac{(\mathbf{A_1y}) \cdot \mathbf{x}}{(\mathbf{A_2y}) \cdot \mathbf{x}} = \frac{(\mathbf{A_1y}) \cdot \mathbf{z}}{(\mathbf{A_2y}) \cdot \mathbf{z}}, \tag{4.13}$$

for every $\mathbf{z}$, *for which* $(\mathbf{A_2y}) \cdot \mathbf{z} \neq 0$.

Using the fact that $(\mathbf{a} \times \mathbf{b}) \times \mathbf{c} = (\mathbf{a} \cdot \mathbf{c})\mathbf{b} - (\mathbf{a} \cdot \mathbf{b})\mathbf{c}$, we get

$$\begin{aligned} 0 = \mathbf{x} \times \mathbf{f}(\mathbf{y}) \cdot \mathbf{z} = \mathbf{x} \cdot (\mathbf{f}(\mathbf{y}) \times \mathbf{z}) &= \mathbf{x} \cdot [(\mathbf{A_1y} \times \mathbf{A_2y}) \times \mathbf{z}] \\ &= -\mathbf{x} \cdot [(\mathbf{z} \cdot \mathbf{A_2y})\mathbf{A_1y} - (\mathbf{z} \cdot \mathbf{A_1y})\mathbf{A_2y}], \end{aligned}$$

which, by $\mathbf{x} \neq \mathbf{0}$, is equivalent to

$$(\mathbf{z} \cdot \mathbf{A_2y})(\mathbf{A_1y} \cdot \mathbf{x}) = (\mathbf{z} \cdot \mathbf{A_1y})(\mathbf{A_2y} \cdot \mathbf{x}),$$

which, in view of the conditions of the lemma, implies the condition (4.13).

The expression (4.12) is exactly an invariant and (4.13) is the corresponding condition for the invariance of this expression. In fact, taking in (4.13), $y = x_n$, and $x = x_{n-1}$ we get a polynomial-like invariant:

$$\begin{aligned} I(x_n, x_{n-1}) &= (a_1 + Ka_2)x_{n-1}^2x_n^2 + (b_1 + Kb_2)x_nx_{n-1}^2 + (d_1 + Kd_2)x_n^2x_{n-1} \\ &+ (c_1 + Kc_2)x_{n-1}^2 + (e_1 + Ke_2)x_{n-1}x_n + (g_1 + Kg_2)x_n^2 \\ &+ (f_1 + Kf_2)x_{n-1} + (h_1 + Kh_2)x_n + k_1 + Kk_2, \end{aligned} \tag{4.14}$$

where K is a constant with its value determined by the initial points x_0, and x_1 as

$$K = \frac{(\mathbf{A_1y_0}) \cdot \mathbf{x_0}}{(\mathbf{A_2y_0}) \cdot \mathbf{x_0}},$$

where $\mathbf{y_0} = (x_1^2, x_1, 1)$, and $\mathbf{x_0} = (x_0^2, x_0, 1)$. Solving (4.14) for K, we get an invariant. The expression (4.14) is exactly the invariant of equation

$$x_{n+1} = \frac{f_1(x_n) - x_{n-1} f_2(x_n)}{f_2(x_n) - x_{n-1} f_3(x_n)}. \tag{4.15}$$

Taking in (4.13), $y = x_{n-1}$, and $x = x_n$, we get a polynomial-like invariant:

$$\begin{aligned} I(x_n, x_{n-1}) &= (a_1 + Ka_2)x_{n-1}^2 x_n{}^2 + (b_1 + Kb_2)x_n^2 x_{n-1} + (d_1 + Kd_2)x_n x_{n-1}^2 \\ &+ (c_1 + Kc_2)x_n^2 + (e_1 + Ke_2)x_{n-1}x_n + (g_1 + Kg_2)x_{n-1}^2 \\ &+ (f_1 + Kf_2)x_n + (h_1 + Kh_2)x_{n-1} + k_1 + Kk_2, \end{aligned} \tag{4.16}$$

where K is a constant with the value determined by the initial points x_0 and x_1 in a similar way as above. The expression (4.16) is an invariant of the equation

$$x_{n+1} = \frac{f_1(x_{n-1}) - x_n f_2(x_{n-1})}{f_2(x_{n-1}) - x_n f_3(x_{n-1})}. \tag{4.17}$$

These cases has been obtained in [QRT1] and [QRT2], and used in several papers referred to in [RQ] and [HBQC]. Further simplifications can be obtained by assuming that the matrices $\mathbf{A_i}, i = 1, 2$ are symmetric, that is, $\mathbf{A_i^T} = \mathbf{A_i}, i = 1, 2$ which reduces the invariants (4.14) and (4.16) to the same form

$$\begin{aligned} I(x_n, x_{n-1}) &= (a_1 + Ka_2)x_{n-1}^2 x_n^2 + (b_1 + Kb_2)(x_n^2 x_{n-1} + x_n x_{n-1}^2) \\ &+ (c_1 + Kc_2)(x_n^2 + x_{n-1}^2) + (e_1 + Ke_2)x_{n-1}x_n \\ &+ (f_1 + Kf_2)(x_n + x_{n-1}) + k_1 + Kk_2. \end{aligned} \tag{4.18}$$

Some of these results can be formulated as a theorem.

THEOREM 4.3

The expression (4.14) is an invariant of the difference equation (4.15). Similarly, the expression (4.16) is an invariant of the difference equation (4.15).

Some general results on the existence of the rational invariants for rational difference equations and systems of such equations have been obtained recently in [PS2], [Sc1], and [Sc2].

4.4.1 Applications

In this part, we shall use Theorem 4.3 to find the invariants of several well-known equations such as Lyness' equation, May's host parasitoid model, soliton model, as well as some new types of equations. We will also combine some of these results with Theorem 4.2 to obtain a Lyapunov function and prove the stability of their equilibria.

Example 4.4 The Gumovski-Mira equation (see [GM] and [La]) is

$$x_{n+1} = \frac{2x_n}{A^2 - x_n^2} - x_{n-1},$$

where a is a positive parameter. This equation is a special case of the class of equations that can be obtained from (4.9) by taking that $f_3(y) \equiv 0$, that is, the equations of the type

$$x_{n+1} = \frac{f_1(x_n)}{f_2(x_n)} - x_{n-1}. \tag{4.19}$$

Assuming that $f_1(y)$ and $f_2(y)$ are polynomials of the above type, and taking that $a_1 = b_1 = c_1 = e_1 = f_1 = 0$, and dropping the indices in other coefficients, we get the following equation

$$x_{n+1} = \frac{bx_n^2 + ex_n + f}{ax_n^2 + bx_n + c} - x_{n-1}, \tag{4.20}$$

with an invariant

$$Ka_2x_{n-1}^2x_n^2 + Kb_2x_nx_{n-1}^2 + (d_1 + Kd_2)x_n^2x_{n-1} + Kc_2x_{n-1}^2$$

$$+Ke_2x_{n-1}x_n + (g_1 + Kg_2)x_n^2 + Kf_2x_{n-1} + (h_1 + Kh_2)x_n + k_1 + Kk_2 = 0,$$

or in symmetric form

$$Ka_2x_{n-1}^2x_n^2 + Kb_2(x_n^2x_{n-1} + x_nx_{n-1}^2) + Kc_2(x_n^2 + x_{n-1}^2)$$

$$+Ke_2x_{n-1}x_n + Kf_2(x_n + x_{n-1}) + k_1 + Kk_2 = 0,$$

or

$$a_2x_{n-1}^2x_n^2 + b_2(x_n^2x_{n-1} + x_nx_{n-1}^2) + c_2(x_n^2 + x_{n-1}^2) + e_2x_{n-1}x_n + f_2(x_n + x_{n-1}) = \widetilde{K}, \tag{4.21}$$

where $\widetilde{K}$ is the constant determined by the initial values x_0, and x_1.

Taking in (4.20) and (4.21) $b = f = 0$, and $a = -1, c = A^2, a = 1$ we get the Gumovski-Mira equation with an invariant of the form:

$$I(x_n, x_{n-1}) = x_n^2x_{n-1}^2 - A^2(x_n^2 + x_{n-1}^2) + 2x_nx_{n-1}.$$

∎

Example 4.5 The equation

$$x_{n+1} = \frac{a + bx_n + cx_n^2}{(c + dx_n + ex_n^2)x_{n-1}}, \tag{4.22}$$

where a, b, c, d, e are positive parameters, can be obtained by taking $a_1 = b_1 = c_1 = f_1 = k_1 = 0$, and $a_2 = e, b_2 = d, c_2 = c, f_2 = b, k_2 = a$ in equation (4.15)), see [GKL]. The corresponding invariant (4.14) results in an invariant of the form (it is solved for K)

$$I(x_n, x_{n-1}) = \frac{a}{x_n x_{n-1}} + b\left(\frac{1}{x_n} + \frac{1}{x_{n-1}}\right) + c\left(\frac{x_{n-1}}{x_n} + \frac{x_n}{x_{n-1}}\right) + d(x_n + x_{n-1}) + e x_n x_{n-1}. \quad (4.23)$$

The equilibrium p of this equation satisfies $ep^4 + dp^3 - bp - a = 0$. The necessary conditions for the extremum in Theorem 4.2 give:

$$\frac{\partial I}{\partial x} \equiv -\frac{a}{x^2 y} - \frac{b}{x^2} + c\left(\frac{1}{y} - \frac{y}{x^2}\right) + d + ey = 0,$$

and

$$\frac{\partial I}{\partial x} \equiv -\frac{a}{xy^2} - \frac{b}{y^2} + c\left(\frac{1}{x} - \frac{x}{y^2}\right) + d + ex = 0,$$

which after some manipulations lead to $x = y$, and to the equation

$$ex^4 + dx^3 - bx - a = 0.$$

This shows that the critical points are exactly equilibrium points. Now, we check the Hessian at the positive critical point (p_+, p_+), which will be denoted simply as (p, p):

$$H = \begin{pmatrix} A & B \\ B & A \end{pmatrix},$$

where

$$A = \frac{\partial^2 I}{\partial x^2}(p, p) = \frac{2}{p^4}(a + bp + cp^2) > 0,$$

and

$$B = \frac{\partial^2 I}{\partial x \partial y}(p, p) = \frac{ep^4 - 2cp^2 + a}{p^4}.$$

Now,

$$det\ H = A^2 - B^2 = (A - B)(A + B) = \frac{1}{p^7}(dp^2 + 4cp + b)(ep^4 + 2bp + 3a) > 0,$$

which shows that the Hessian H is positive definite at the point (p, p), which means that the invariant I attains a minimum at (p, p). Thus,

$$\min\{I(x, y) : (x, y) \in D\} = I(p, p),$$

and so by our theorem $V(x, y) = I(x, y) - I(p, p)$ which shows that p is stable. The Lyapunov function is given by:

$$V(x, y) = I(x, y) - I(p, p) = I(x, y) - \frac{dp^3 + 2cp^2 + 3bp + 2a}{p^2}.$$

In the special case where $a_2 = 0, b_2 = c_2 = 1, k_2 = a, f_2 = a + 1$, we obtain Lyness' equation

$$x_{n+1} = \frac{a + x_n}{x_{n-1}},$$

with an invariant of the form

$$I(x_n, x_{n-1}) = \left(1 + \frac{1}{x_n}\right)\left(1 + \frac{1}{x_{n-1}}\right)(a + x_n + x_{n-1}).$$

In the special case where $a_2 = f_2 = k_2 = 0, b_2 = c_2 = 1$, we get a special case of May's host-parazitoid model

$$x_{n+1} = \frac{x_n^2}{(1 + x_n)x_{n-1}}$$

which has the invariant

$$I(x_n, x_{n-1}) = x_n + x_{n-1} + \frac{x_n}{x_{n-1}} + \frac{x_{n-1}}{x_n}.$$

In the special case $c = d = 0$, and $a = -d, b = d^2, e = 1$, we get the well-known Yang-Baxter's equation, see [QRT1] and [QRT2]

$$x_{n+1} = \frac{d^2 x_n - d}{x_{n-1} x_n^2},$$

which has the invariant given by

$$I(x_n, x_{n-1}) \equiv x_n^2 x_{n-1}^2 + K x_n x_{n-1} + d^2(x_n + x_{n-1}) - d = 0.$$

∎

Example 4.6 (see [QRT1] and [QRT2])
Taking in (4.20) and (4.21) $b = f = 0$ and $e = \omega + 2, a = 1/2, c = 1$, we get the soliton's equilibrium equation from [QRT1] and [QRT2]

$$x_{n+1} = \frac{(\omega + 2)x_n}{1 + x_n^2/2} - x_{n-1},$$

with an invariant of the form

$$I(x_n, x_{n-1}) = x_n^2 x_{n-1}^2 + 2(x_n^2 + x_{n-1}^2) - 2(\omega + 2)x_n x_{n-1}.$$

The equilibrium p of this equation satisfies $p^2 - \omega = 0$, that is, $p = \pm\sqrt{\omega}$, if we additionally assume that $\omega \geq 0$. Checking all neccessary conditions of Theorem 4.2 we can show that the invariant I attains a minimum at (p, p). Thus,

$$\min\{I(x, y) : (x, y) \in D\} = I(p, p) = -\omega^2,$$

and so by our theorem, $V(x,y) = I(x,y) + \omega^2$, which shows that p is stable. ▮

Example 4.7 Taking in (4.15) and (4.14) $f_1(x) = x + c - c^2$, $f_2(x) = -cx$ and $f_3(x) = -x^2$, we get the following equation:

$$x_{n+1} = \frac{cx_n x_{n-1} + x_n + c - c^2}{x_n(x_n x_{n-1} - c)},$$

with an invariant of the form:

$$I(x_n, x_{n-1}) = (1 + x_{n-1})(1 + x_n)\left(1 + \frac{1}{x_n x_{n-1} - c}\right).$$

This equation was considered by C. Schinas and I. Schinas, and the stability was established by using the KAM theory.

The equilibrium p of this equation satisfies $p^4 - 2cp^2 - p + c^2 - c = 0$, or

$$(p^2 - c)^2 = p + c.$$

It is not hard to see that this equation has a positive root p_+, which will be denoted as p in what follows. The above equilibrium equation implies that $p^2 > c$. We will be interested in the stability of this root. The necessary conditions for the extremum give:

$$\frac{\partial I}{\partial x} \equiv (1 + y)(1 + \frac{1}{xy - c} - \frac{y + xy}{(xy - c)^2}) = 0,$$

and

$$\frac{\partial I}{\partial x} \equiv (1 + x)(1 + \frac{1}{xy - c} - \frac{x + xy}{(xy - c)^2}) = 0,$$

which lead to $x = y$, and to the above equilibrium equation. Thus, $x = y = p$. This shows that the critical points are exactly the equilibrium points. The Hessian at p has the form

$$H = \begin{pmatrix} A & B \\ B & A \end{pmatrix},$$

where

$$A = \frac{\partial^2 I}{\partial x^2}(p, p) = \frac{2p(1 + p)}{p^2 - c} > 0,$$

and

$$B = \frac{\partial^2 I}{\partial x \partial y}(p, p) = \frac{(p^2 - c)(p + c) + (c + 1)p^2 + 4cp + c - c^2}{(p^2 - c)(p + c)}.$$

Now,

$$det\ H = A^2 - B^2 = \frac{(p + 1)(p^2 - c) + 2c^2}{(p^2 - c)(p + c)} > 0,$$

which shows that the invariant I attains a minimum at (p, p). Thus,

$$\min\{I(x, y) : (x, y) \in D\} = I(p, p),$$

and so by Theorem 4.2, $V(x, y) = I(x, y) - I(p, p)$, which shows that p is stable. ∎

4.5 *Dynamica* Session on Invariants

4.5.1 Introduction

In this section we present a *Dynamica* session on invariants. The results given in this chapter so far point to the importance of developing methods for finding invariants and CAS (Computer Algebra System) implementation. In this regard it is important to observe that if I is an invariant then also any continuous function of it is also an invariant, see [GM] and [Mi]. In particular, if an equation possesses a polynomial invariant, then every power of it is an invariant itself. This simple fact has the consequence on our strategy of choosing the powers in the expansion of prospective invariant.

Essentially, our method for finding invariants is the method of undetermined coefficients where one can choose the basis functions. The default basis consists of powers with integer exponents, both positive and negative but it can be easily changed to any other basis. In view of the above remark about the nonuniqueness of an invariant one should start with low powers in our expansion and if they do not lead to the solution proceed with the next higher power. Using higher-order terms may require a substantial amount of time and should not be tried as a first choice.

We shall use *Dynamica* to find invariants of the form

$$I(x, y, \ldots, z) = \sum_{i,j,\ldots,k=-1}^{1} a_{i,j,\ldots,k} x^i y^j \cdots z^k.$$

`AllDE` is a list of several difference equations that are known to have interesting properties and applications.

```
In[1]:= << Dynamica`

In[2]:= ?AllDE
A list of difference equations
in the following order:
{Lyness,Todd,Lyness2,Lyness3,YangBaxter,
GumovskiMira,KdV,KdV2,Ladas,Mira,Lyness4,May}
```

Here is the contents of AllDE.

```
In[3]:= AllDE
```

$$Out[3]= \{x[1+n] == \frac{A+x[n]}{x[-1+n]},$$
$$x[1+n] == \frac{A+x[-1+n]+x[n]}{x[-2+n]},$$
$$x[1+n] == \frac{A+x[-2+n]+x[-1+n]+x[n]}{x[-3+n]},$$
$$x[1+n] == \frac{A2+A1\,x[n]}{x[-1+n]\,(A3+x[n])},$$
$$x[1+n] == \frac{-A+A^2\,x[n]}{x[-1+n]\,x[n]^2},$$
$$x[1+n] == -x[-1+n]+\frac{2\,A\,x[n]}{1+x[n]^2},$$
$$x[1+n] == -x[-1+n]+\frac{-A+x[n]}{2\,B^2\,x[n]^2},$$
$$x[1+n] == -x[-1+n]+\frac{-\frac{1}{2}+A+x[n]^2}{x[n]-x[n]^2},$$
$$x[1+n] == \frac{A1+A2\,x[n]+A3\,x[n]^2}{x[-1+n]\,(B1+B2\,x[n]+B3\,x[n]^2)},$$
$$x[1+n] == -x[-1+n]+2\,A\,x[n]+\frac{4\,(1-A)\,x[n]^2}{1+x[n]^2},$$
$$x[1+n] == \frac{A-A^2+x[n]+A\,x[-1+n]\,x[n]}{x[n]\,(-A+x[-1+n]\,x[n])},$$
$$x[1+n] == \frac{A\,x[n]^2}{x[-1+n]\,(1+x[n])}\}$$

To recall an equation, just type its name.

```
In[4]:= Lyness
```

$$Out[4]= x[1+n] == \frac{A+x[n]}{x[-1+n]}$$

4.5.2 An Invariant for Lyness-Type Equation

With the function RationalInvariant we find an invariant for $x_{n+1} = \frac{px_n+qx_{n-1}}{p+q}$

```
In[5]:= RationalInvariant[
            x[n + 1] == (p x[n] + q x[n - 1])/(p + q)]
Solve :: "svars": Equations may not give
solutions for all "solve" variables.
```

$$Out[5]= c[1]+c[3]\,x[-1+n]+\frac{(p+q)\,c[3]\,x[n]}{q}$$

Choosing c[1] to be 0 and c[3] to be q we obtain a specific invariant.

```
In[6]:= %/.{c[1] → 0, c[3] → q}
Out[6]= q x[-1 + n] + (p + q) x[n]
```

Here we specify what powers to use for the invariant.

```
In[7]:= RationalInvariant[x[n + 1] ==
          (2 x[n] - x[n - 1]) x[n - 1]    x[n]
          ---------------------------- + ----,
                   2 x[n]                  2
        BasisRange ->
           {{x[n], 0, 2}, {x[n - 1], 0, 2}}]

Solve :: "svars": Equations may not give
solutions for all "solve" variables.
```

$Out[7]= c[1] - \frac{1}{2} c[5]\, x[-1+n]^2 + c[5]\, x[-1+n]\, x[n]$

Choosing `c[1]` to be 0 and `c[5]` to be 2 we obtain a specific invariant:

```
In[8]:= %/.{c[1] -> 0, c[5] -> 2}
```

$Out[8]= -x[-1+n]^2 + 2\, x[-1+n]\, x[n]$

An attempt to produce an invariant for Lyness' equation [GKL] yields an expression that depends on two parameters, `c[1]` and `c[7]`, which represent arbitrary constants.

```
In[9]:= RationalInvariant[
                    Ax[n] + B
        x[n + 1] == ------------]
                    x[n]x[n - 1]
Solve :: "svars" : Equations may not give
solutions for all "solve" variables.
```

$$Out[9]= c[1] + \frac{A\, c[7]}{x[-1+n]} + c[7]\, x[-1+n] + \frac{A\, c[7]}{x[n]} + \frac{B\, c[7]}{x[-1+n]\, x[n]} + c[7]\, x[n]$$

The answer can be normalized by setting `c[1]` to be 0 and `c[7]` to be 1.

```
In[10]:= Simplify[%/.c[1] -> 0, c[7] -> 1]
```

$$Out[10]= x[-1+n] + \frac{A + \frac{B}{x[n]}}{x[-1+n]} + \frac{A}{x[n]} + x[n]$$

To make effective use of an invariant it is sometime necessary to have it in factored form. Here no factorization is returned by *Mathematica*.

```
In[11]:= Factor[%]
```

$$Out[11]= \frac{B + A\, x[-1+n] + A\, x[n] + x[-1+n]^2\, x[n] + x[-1+n]\, x[n]^2}{x[-1+n]\, x[n]}$$

4.5.3 Specifying a Basis for the Invariant, Part I

Let us recall Gumovski-Mira's equation

```
In[12]:= GumovskiMira
```

$$Out[12]= x[1+n] == -x[-1+n] + \frac{2\, A\, x[n]}{1 + x[n]^2}$$

We see that the function `RationalInvariant` applied to Gumovski-Mira's equation produces a trivial invariant.

```
In[13]:= RationalInvariant[GumovskiMira]

Solve    ::    "svars": Equations may not give
solutions for all "solve" variables.

Out[13]= c[1]
```

More can be done. When necessary, we may change the basis in order to have a better shot at finding an invariant. Here we specify the invariant to be a sum of powers of `x[n]` and `x[n-1]`, with exponents ranging from `-1` to 4.

```
In[14]:= RationalInvariant[GumovskiMira,
             BasisRange → { {x[n], -1, 4},
                  {x[n - 1], -1, 4}} ]
Solve    ::    "svars": Equations may not give
solutions for all ''solve" variables.
```

$$\text{Out[14]= } c[1] + c[19]\, x[-1+n]^2 + \frac{1}{2} c[34]\, x[-1+n]^4 - 2 A\, c[19]\, x[-1+n]\, x[n] - 2 A\, c[34]\, x[-1+n]^3\, x[n] + c[19]\, x[n]^2 + (c[19] + c[34] + 2 A^2 c[34]) \times x[-1+n]^2\, x[n]^2 + c[34]\, x[-1+n]^4\, x[n]^2 - 2 A\, c[34]\, x[-1+n]\, x[n]^3 - 2 A\, c[34]\, x[-1+n]^3\, x[n]^3 + \frac{1}{2} c[34]\, x[n]^4 + c[34]\, x[-1+n]^2\, x[n]^4 + \frac{1}{2} c[34]\, x[-1+n]^4\, x[n]^4$$

We have three arbitrary constants in the previous output. To keep the degree of invariant as low as possible, choose `c[34]=0`. A similar effect can be achieved by requiring that exponents of the model have a smaller range.

```
In[15]:= RationalInvariant[GumovskiMira,
             BasisRange → { {x[n], -1, 2},
             {x[n - 1], -1, 2}}]
Solve    ::    "svars": Equations may not give
solutions for all "solve" variables.
```

$$\text{Out[15]= } c[1] + c[13]\, x[-1+n]^2 - 2 A\, c[13]\, x[-1+n]\, x[n] + c[13]\, x[n]^2 + c[13]\, x[-1+n]^2\, x[n]^2$$

Setting `c[13]=1` and `c[1]=0`, we get a simple invariant.

```
In[16]:= %/.{c[13] → 1, c[1] → 0}
```

$$\text{Out[16]= } x[-1+n]^2 - 2 A\, x[-1+n]\, x[n] + x[n]^2 + x[-1+n]^2\, x[n]^2$$

4.5.4 Specifying a Basis for the Invariant, Part II

Consider the difference equation $x_{n+1} = \frac{A+x_{n-1}}{x_n x_{n-1}-1}$.

$$\text{In[17]:= eqn = } x[n+1] == \frac{A + x[n-1]}{x[n]\, x[n-1] - 1}$$

We now generate a basis consisting of powers of x[n], x[n-1], and of x[n]x[n-1]-1.

```
In[18]:= RationalInvariant[eqn,
            BasisRange →
            {{x[n], -1, 1}, {x[n - 1], -1, 1},
            {x[n] x[n - 1] - 1, -1, 1}}]
Solve :: "svars": Equations may not give
solutions for all "solve" variables.
```

Out[18]= $c[1] + c[2]\,x[-1+n] + c[2]\,x[n] + \frac{c[2]\,x[-1+n]\,x[n]}{A} + \frac{c[2] + A^2 c[2] - A\,c[18]}{A\,(-1 + x[-1+n]\,x[n])} + \frac{c[2]\,x[-1+n]}{-1 + x[-1+n]\,x[n]} + \frac{c[2]\,x[n]}{-1 + x[-1+n]\,x[n]} + \frac{c[18]\,x[-1+n]\,x[n]}{-1 + x[-1+n]\,x[n]}$

```
In[19]:= Simplify[% /. c[2] → 1]
```

Out[19]= $\frac{1}{A\,(-1 + x[-1+n]\,x[n])}(1 + A^2 - A\,(c[1] + c[18]) + x[-1+n]^2\,x[n]\,(A + x[n]) + x[-1+n]\,x[n]\,(-1 + A\,(c[1] + c[18]) + A\,x[n]))$

A factored form is possible with this choice of constants (communicated by E. Janowski).

```
In[20]:= Simplify[%/.
           {c[1] → A, c[18] → 1/A}]
```

Out[20]= $\frac{x[-1+n]\,(A + x[-1+n])\,x[n]\,(A + x[n])}{A\,(-1 + x[-1+n]\,x[n])}$

4.6 *Dynamica* Session on Lyapunov Functions

In this section we shall see how to use *Dynamica* to find the Lyapunov function when an invariant is known. The technique is useful for a large class of nonlinear difference equations.

Let us consider the soliton's equilibrium equation, see [QRT1] and [QRT2].

```
In[1]:= << Dynamica`
```

In[2]:= soliton := $x[n+1] == \frac{2\,(A+2)\,x[n]}{2 + x[n]^2} - x[n-1]$;

The first attempt to find an invariant is not successful.

```
In[3]:= RationalInvariant[soliton]
Solve :: "svars" : Equations may not give
solutions for all "solve" variables.
Out[3]= c[1]
```

As we have seen earlier, we may change the basis for our invariant.

```
In[4]:= RationalInvariant[soliton,
          BasisRange → {{x[n], -2, 2},
            {x[n - 1], -2, 2}}]
Solve   ::   "svars": Equations may not give
solutions for all "solve" variables.
```

$Out[4]= c[1] + c[21]\, x[-1+n]^2 - (2+A)\, c[21]\, x[-1+n]\, x[n] + c[21]\, x[n]^2 + \frac{1}{2} c[21]\, x[-1+n]^2 x[n]^2$

By choosing c[1] to be 0 and c[21] to be 2 we obtain a simple expression for the invariant.

```
In[5]:= %/.{c[1] → 0, c[21] → 2}
```

$Out[5]= 2\, x[-1+n]^2 - 2\,(2+A)\, x[-1+n]\, x[n] + 2\, x[n]^2 + x[-1+n]^2 x[n]^2$

```
In[6]:= fsoliton = %/.
            {x[n] → x, x[n - 1] → y}
```

In order to find the equilibrium points we need the corresponding map.

$Out[6]= 2\, x^2 - 2\,(2+A)\, x\, y + 2\, y^2 + x^2 y^2$

```
In[7]:= Solitonmap := DEToMap[soliton];

In[8]:= Solitonmap[{x, y}]
```

$Out[8]= \left\{y, -x + \frac{2\,(2+A)\, y}{2+y^2}\right\}$

There are three fixed poins.

```
In[9]:= Solve[Solitonmap[{x, y}] == {x, y},
            {x, y}]
```

$Out[9]= \left\{\{x \to 0, y \to 0\}, \left\{x \to -\sqrt{A}, y \to -\sqrt{A}\right\}, \left\{x \to \sqrt{A}, y \to \sqrt{A}\right\}\right\}$

Here we name the fixed points fp[1], fp[2], and fp[3].

```
In[10]:= {fp[1], fp[2], fp[3]} = {x, y} /. %
```

$Out[10]= \left\{\{0, 0\}, \left\{-\sqrt{A}, -\sqrt{A}\right\}, \left\{\sqrt{A}, \sqrt{A}\right\}\right\}$

It is clear that for all values of the parameter $A > 0$, the hessian of the invariant function has positive principal minors at `fp[2]` and `fp[3]`, hence the invariant function has a strict local minimum at these points.

```
In[11]:= FixedPointTest[
          fsoliton[{x,y}],
          Solitonmap[x,y],
          {x,y}];
The point number 1 is {0, 0}
Principal Minors of the Hessian for this point:
 d1[1] = 4
 d2[1] = -4 A (4 + A)
--------------------------
```

The point number 2 is $\{-\sqrt{A}, -\sqrt{A}\}$

```
Principal Minors of the Hessian for this point:
 d1[2] = 2 (2 + A)
 d2[2] = 32 A
--------------------------
```

The point number 3 is $\{\sqrt{A}, \sqrt{A}\}$

```
Principal Minors of the Hessian for this point:
 d1[3] = 2 (2 + A)
 d2[3] = 32 A
--------------------------
```

This is a Lyapunov function at $(-\sqrt{A}, -\sqrt{A})$.

```
In[12]:= lyap2[{x_, y_}] =
          fsoliton[{x,y}] - fsoliton[fp[2]]
```

Out[12]= $A^2 - 4\,x\,y - 2\,A\,x\,y + 2\,y^2 + x^2\,(2 + y^2)$

This is a Lyapunov function at $(\sqrt{A}, \sqrt{A})$.

```
In[13]:= lyap3[{x_, y_}] =
          fsoliton[{x,y}] - fsoliton[fp[3]]
```

Out[13]= $A^2 - 4\,x\,y - 2\,A\,x\,y + 2\,y^2 + x^2\,(2 + y^2)$

This is a plot of the Lyapunov function for a specific value of the parameter `A`.

```
In[14]:= Plot3D[lyap2[{x,y}]/.A → 2.,
              {x, -3, 3.}, {y, -3, 3.}]
```

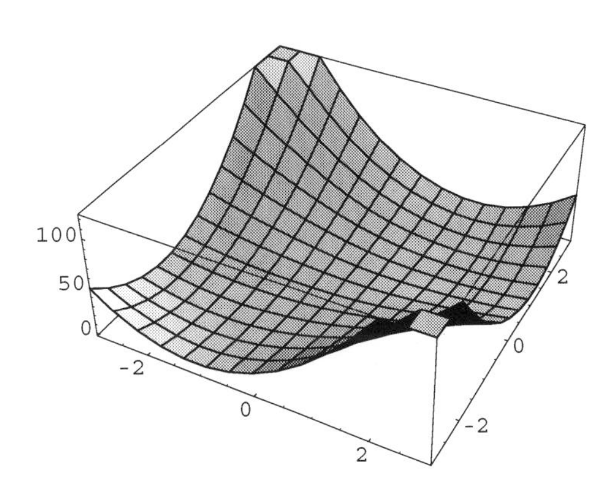

Another informative plot is a contour plot.

```
In[15]:= ContourPlot[lyap2[{x,y}]/.A→2.,
           {x,-3,3.},{y,-3,3.},
           Contours→31,
           PlotPoints→30,
           ContourShading→False]
```

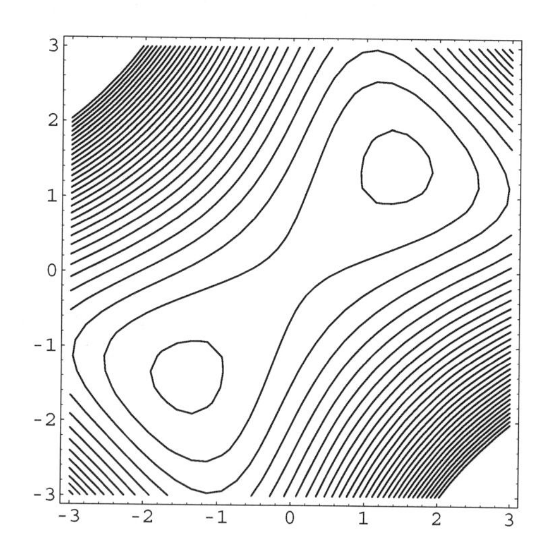

4.7 Invariance under Lie Group Transformations

In this section we consider some specific invariants that remain invariant under certain one-parameter Lie group transformations (see [Ar], [O]). The theory of Lie group transformations for difference equations, though still in its infancy, is experiencing rapid development, see [Ar], [HBQC], [QRT1], [QRT2], [QS], and [SQ]. The potential applications of this theory are mainly in the area of systematic methods for finding the exact solutions of difference equations and their invariants. There is already substantial progress toward this objective in the case of first-order difference equations and some encouraging results for higher-order difference equations.

First, we will present some specific results for the first-order autonomous difference equation:

$$x_{n+1} = f(x_n), \quad n = 0, 1, \ldots, \tag{4.24}$$

where the function f is analytic on its domain $D \subset R$ and $x_0 \in D$.

DEFINITION 4.2 *Let f be an analytic function on its domain $D \subset R$. An analytic function $H : D \to R$ is said to be a global invariant for f provided*

$$H(f(x)) = H(x),$$

for all $x \in D$.

The next result is a partial answer to the conjecture mentioned in Remark 4.2 in the one-dimensional case, see [Ar].

THEOREM 4.4

Assume the function f is analytic on its domain and that f has at least one equilibrium point $\bar{x}$. If any equilibrium point is hyperbolic, then the global invariant is trivial, that is, the identically constant function.

Notice that when the equilibrium point $\bar{x}$ is hyperbolic, then, according to Theorem 4.1, the corresponding linearized equation around $\bar{x}$ does not have a nontrivial invariant.

Consider the one-parameter Lie group of transformations (see [O] and [Ma]):

$$X = x + \epsilon\psi(x) + O(\epsilon^2), \tag{4.25}$$

$$Y = y + \epsilon\psi(y) + O(\epsilon^2) \tag{4.26}$$

for some analytic function ψ on the domain of f. Invariance of the relation $y = f(x)$ or the function f under the Lie group of transformations (4.25) and (4.26) means that $Y = f(X)$.

The next result gives the necessary and sufficient condition for a function f to be invariant under the considered Lie group of transformations, see [Ar] and [Ma].

THEOREM 4.5

The function f is invariant under the Lie group of transformations (4.25) and (4.26) if and only if the function ψ satisfies the functional equation

$$\psi(y) = \psi(f(x)) = f'(x)\psi(x). \tag{4.27}$$

The next result gives an explicit formula for an invariant of a difference equation (4.24) given that an associated function is invariant under the considered Lie group of transformations, see [Ar] and [Ma].

THEOREM 4.6

The function

$$H(x) = \int \frac{1}{\psi(x)} dx \tag{4.28}$$

where the function $\psi(x) \neq 0$ on the domain of f is given in Theorem 4.5, is a global invariant for $y = f(x)$ if and only if the function f is invariant under the Lie group of transformations (4.25) and (4.26).

For the sake of convenience we write the invariant H and the corresponding function ψ in the more convenient forms as

$$H(x) = \int (x - f(x))\phi(x)dx,$$

where $\phi(x)$ is defined with

$$\psi(x) = \frac{1}{(x - f(x))\phi(x)}.$$

In this case, H is a global invariant for $y = f(x)$, provided that there exists an analytic function ϕ that satisfies the condition:

$$(x - f(x))\phi(x) = f'(x)(x - f(x))\phi(f(x)),$$

for every x in domain of f. An immediate consequence of Theorem 4.5 is that if the difference equation (4.24) allows a global invariant, then all its equilibrium points are nonhyperbolic. As we emphasized before, we believe that this result holds for an autonomous difference equation of any order.

In summary, assume that difference equation (4.24) has at least one equilibrium point. If any of the equilibrium points are hyperbolic, then the global invariant is trivial, that is, identically constant. Nonconstant global invariants exist if and only if the difference equation (4.24) is invariant under a Lie group of transformations (4.25) and (4.26). Finally, if the difference equation (4.24) allows a global invariant, then all its equilibrium points are nonhyperbolic.

4.8 Exercises

Exercise 4.1 Check that each of the difference equations below has a nontrivial invariant:

1. $2x_{n+1} - 5x_n + 2x_{n-1} = 0$,
2. $x_{n+1} - x_n - x_{n-1} - x_{n-2} = 0$,
3. $x_{n+1} - 6x_n + 9x_{n-1} - 6x_{n-2} + 8x_{n-3} = 0$,
4. $x_{n+k+2} + x_{n+k} - x_{n+3} + x_{n+2} - x_{n+1} + x_n = 0$, where $k > 1$ is an integer.

For each of the above equations find a nontrivial invariant using *Dynamica*'s `RationalInvariant` function.

Exercise 4.2 Check that each of the systems below has a nontrivial invariant. Then use *Dynamica*'s `RationalInvariant` function to find any nontrivial invariant.

1.
$$x_{n+1} = -7x_n + \tfrac{5}{2}y_n$$
$$y_{n+1} = 9x_n - \tfrac{5}{2}y_n$$

2.
$$x_{n+1} = 3x_n + 2y_n$$
$$y_{n+1} = -4x_n - 3y_n,$$

3.
$$x_{n+1} = \tfrac{1}{2}x_n + \tfrac{\sqrt{3}}{2}y_n$$
$$y_{n+1} = -\tfrac{\sqrt{3}}{2}x_n + \tfrac{1}{2}y_n,$$

4.
$$x_{n+1} = -3x_n + 5y_n - 2z_n$$
$$y_{n+1} = x_n + y_n + z_n$$
$$z_{n+1} = 6x_n - 6y_n + 5z_n.$$

Exercise 4.3 Find all values of the parameter a such that the following equations have a nontrivial invariant:

1. $x_{n+1} - ax_n + x_{n-1} = 0,$
2. $x_{n+1} - ax_n + x_{n-2} = 0,$
3. $x_{n+k} - ax_{n+k-1} + x_n = 0, \quad n = 0, 1, \ldots$ where $k > 1$ is an integer.
4. $x_{n+k} - ax_{n+k-1} + x_{n+1} - ax_n = 0, \quad n = 0, 1, \ldots$ where $k > 1$ is an integer.

Exercise 4.4
Find all values of the parameters $a, b, c,$ and d such that the following system has a nontrivial invariant:

$$x_{n+1} = ax_n + by_n$$
$$y_{n+1} = cx_n + dy_n.$$

Find a nontrivial invariant for the obtained values of parameters.

Exercise 4.5
Find all values of the parameter a such that the following systems have a nontrivial invariant:

1.
$$\begin{aligned} x_{n+1} &= (6-a)x_n + (2a-6)y_n + (3-a)z_n \\ y_{n+1} &= (2-a)x_n + (2a-2)y_n + (2-a)z_n \\ z_{n+1} &= -2x_n + 2y_n + z_n, \end{aligned}$$

2.
$$\begin{aligned} x_{n+1} &= (4-a)x_n + (2a-4)y_n + (2-a)z_n \\ y_{n+1} &= x_n + (a-1)y_n + z_n \\ z_{n+1} &= (2a-2)x_n + (2-2a)y_n + 2az_n. \end{aligned}$$

Find the nontrivial invariants for all the values of parameter a that admit such invariants.

Exercise 4.6 Show that the difference equation

$$x_{n+1} = -x_{n-1} + \frac{bx_n^2 + dx_{n-1}}{ax_n^2 + bx_n + c}$$

admits an invariant of the form:

$$I(x, y) = ax^2y^2 + b(x^2y + xy^2) + c(x^2 + y^2) + dxy.$$

First, use hand calculation and then use *Dynamica*. Try to apply Theorem 4.2 to prove the stability of the feasible equilibrium point. Use *Dynamica* whenever possible.

Exercise 4.7

(a) For which difference equations in the `AllDE` list does the `RationalInvariant` function produce an invariant?

(b) Whenever an invariant is produced, try to obtain the simplest nontrivial invariant possible by choosing appropriate values for the constants.

Exercise 4.8 Show that the difference equation

$$x_{n+1} = \frac{x_n + x_{n-1} + A}{(x_n x_{n-1} + B)x_{n-2}}$$

admits an invariant of the form:

$$I(x, y, z) = \left(B + \frac{1}{x}\right)\left(B + \frac{1}{y}\right)\left(B + \frac{1}{z}\right)(AB + xyz + B(x + y + z)).$$

First, use hand calculation and then use *Dynamica*. Try to apply Theorem 4.2 to prove the stability of the feasible equilibrium point. Use *Dynamica* whenever possible.

Exercise 4.9 For all the difference equations in the `AllDE` list for which an invariant is produced, try to obtain the corresponding Lyapunov function by using the method based on Theorem 4.2. Use *Dynamica* whenever possible.

Exercise 4.10 Find an invariant for the system

$$x_{n+1} = \frac{ay_n}{1 + x_n^2}$$

$$y_{n+1} = \frac{bx_n}{1 + y_n^2},$$

where a and b are parameters (see [E2] and [HK], p. 462). Use this invariant to find the corresponding Lyapunov function and using it study the stability of all equilibrium points. Use *Dynamica* if possible.

(*Hint*: Reduce the above system to a second-order difference equation.)

Exercise 4.11 Show that the difference equation

$$x_{n+1} = \frac{\max\{x_n, a\}}{x_{n-1}}, \quad n = 0, 1, \dots$$

(max version of Lyness' equation) possesses the invariant

$$I(x, y) = \max\left\{1, \frac{1}{x}\right\} \max\left\{1, \frac{1}{y}\right\} \max\{a, x, y\}$$

(see [L2]).

Exercise 4.12 Check if the difference equation

$$x_{n+1} = \frac{\min\{x_n, a\}}{x_{n-1}}, \quad n = 0, 1, \dots$$

(min version of Lyness' equation) possesses the invariant

$$I(x, y) = \frac{\min\{a, x, y, xy\} \min\{a, x, y\}}{xy}.$$

Use hand calculation (see [L2]).

Exercise 4.13 Show that the difference equation

$$x_{n+1} = \frac{\max\{x_n, x_{n-1}, a\}}{x_{n-2}}, \qquad n = 0, 1, \ldots$$

(max version of Todd's equation) possesses the invariant (see [L2])

$$I(x, y, z) = \max\left\{1, \frac{1}{x}\right\} \max\left\{1, \frac{1}{y}\right\} \max\left\{1, \frac{1}{z}\right\} \max\{a, x, y, z\}.$$

Exercise 4.14 Difference equations with variable coefficients may also possess invariants. Show that the difference equation

$$x_{n+1} = \frac{a_n x_n + b_n}{x_{n-1}}, \qquad n = 0, 1, \ldots, \qquad x_1 = c > 0, x_0 = d > 0,$$

where $\{a_n\}$ and $\{b_n\}$ are nonnegative periodic sequences each of period two with $a_0 + b_0 > 0$ and $a_1 + b_1 > 0$ (periodic version of Lyness' equation) possesses the invariant (see [JLV] and [Lop1]

$$\begin{aligned}
I(x_n, x_{n-1}) = {} & a_{n+1} b_n x_{n-1} + a_n b_{n+1} x_n + \frac{a_{n+1}(a_n^2 b_{n+1} + b_n^2)}{x_{n-1}} \\
& + \frac{a_n(a_{n+1}^2 b_n + b_{n+1}^2)}{x_n} + \left(\frac{x_{n-1}}{x_n} b_{n+1} + \frac{x_n}{x_{n-1}} b_n\right) a_{n+1} a_n \\
& + \frac{a_{n+1} a_n b_{n+1} b_n)}{x_n x_{n-1}}.
\end{aligned}$$

Prove this fact by hand calculation.

Chapter 5

Dynamics of Three-Dimensional Dynamical Systems

5.1 Introduction

In this chapter we investigate and visualize the dynamics of three-dimensional dynamical systems of the form

$$\begin{aligned} x_{n+1} &= f(x_n, y_n, z_n) \\ y_{n+1} &= g(x_n, y_n, z_n) \\ z_{n+1} &= h(x_n, y_n, z_n) \end{aligned} \tag{5.1}$$

where f, g, and h are given functions.

We begin by presenting theoretical results in this section which are mainly an application of the general theory presented in Chapter 3.

Section 5.2 is a *Dynamica* session featuring third order difference equations. In Section 5.3 we present two results for dissipative systems. In Section 5.4 we present a *computer-assisted proof* of local asymptotic stability of period-two solutions for a particular difference equation. Section 5.5 is a *Dynamica* session on Todd's difference equation. The chapter ends with a collection of research projects in biology in Section 5.6.

Along with system (5.1) we consider the corresponding vector map $\mathbf{F} = (f, g, h)$. An equilibrium point of (5.1) is a point $(\bar{x}, \bar{y}, \bar{z})$ that satisfies

$$\begin{aligned} \bar{x} &= f(\bar{x}, \bar{y}, \bar{z}) \\ \bar{y} &= g(\bar{x}, \bar{y}, \bar{z}) \\ r &= h(\bar{x}, \bar{y}, \bar{z}). \end{aligned} \tag{5.2}$$

The point $(\bar{x}, \bar{y}, \bar{z})$ is also called a fixed point of the vector map $\mathbf{F}$. A periodic point (r, s, t) of period m is an equilibrium point of the m-th iterate $\mathbf{F}^m$ of the map $\mathbf{F}$. Basic definitions of stability of periodic points were given in Chapter 3.

The linearized system at the equilibrium point $(\bar{x}, \bar{y}, \bar{z})$ has the form

$$\begin{aligned} x_{n+1} &= f_x x_n + f_y y_n + f_z z_n \\ y_{n+1} &= g_x x_n + g_y y_n + g_z z_n \\ z_{n+1} &= h_x x_n + h_y y_n + h_z z_n, \end{aligned} \tag{5.3}$$

where all partial derivatives are evaluated at the equilibrium point $(\bar{x}, \bar{y}, \bar{z})$. The characteristic equation of this system is given by

$$\det(D_{\mathbf{f}}(\bar{x}, \bar{y}, \bar{z}) - \lambda I) = 0,$$

which is a complicated cubic equation that we do not display here. By using the Schur-Cohn criterion for the cubic equation given in Section 3.4, one may obtain explicit conditions for the local asymptotic stability and instability of the equilibrium point. In the case of the corresponding third order difference equation

$$x_{n+1} = f(x_n, x_{n-1}, x_{n-2}), \quad n = 0, 1, \ldots, \tag{5.4}$$

these conditions are simple enough to display explicitly.

Let

$$r = \frac{\partial f}{\partial u}(\bar{x}, \bar{x}, \bar{x}) \quad \text{and} \quad s = \frac{\partial f}{\partial v}(\bar{x}, \bar{x}, \bar{x}) \quad t = \frac{\partial f}{\partial w}(\bar{x}, \bar{x}, \bar{x})$$

denote the partial derivatives of $f(u, v, w)$ evaluated at an equilibrium $\bar{x}$ of (5.4). Then the equation

$$y_{n+1} = r y_n + s y_{n-1} + t y_{n-2}, \quad n = 0, 1, \ldots \tag{5.5}$$

is called the **linearized equation** associated with (3.43) about the equilibrium point $\bar{x}$. In this case, the linearized stability theorem has the following form.

THEOREM 5.1

(a) If all the roots of the cubic equation

$$\lambda^3 - r\lambda^2 - s\lambda - t = 0 \tag{5.6}$$

lie inside of open unit disk $||\lambda|| < 1$, then the equilibrium $\bar{x}$ of (5.4) is locally asymptotically stable.

(b) If at least one of the roots λ of (5.6) lies outside the closed unit disk, that is, $|\lambda|| > 1$, then the equilibrium $\bar{x}$ of (5.4) is unstable.

(c) *A necessary and sufficient condition for all roots* λ *of (5.6) to lie inside the open unit disk, that is,* $|\lambda| < 1$, *is*

$$\begin{aligned} |r + t| &< 1 - s, \\ |r - 3t| &< 3 - s, \qquad (5.7) \\ t^2 - s - rt &< 1. \end{aligned}$$

5.2 *Dynamica* Session on Third Order Difference Equations

In this section we perform an analysis of various difference equations of third order.

Begin by defining the ACT map, which stands for Arneodo, Coullet, and Tresser map, see [ACT] and [K].

```
In[1]:= << Dynamica`

In[2]:= ACTMap[{x_, y_, z_}] :=
            {ax - b(y - z),
             bx + a(y - z),
             cx - dx^3 + ez; }

In[3]:= ACTMap[{x, y, z}]
Out[3]= {ax-b(y-z), bx+a(y-z), cx-dx^3+ez}
```

There are three fixed points.

```
In[4]:= fp = {x, y, z} /. Solve[
           ACTMap[{x, y, z}] == {x, y, z},
              {x, y, z}];

In[5]:= Length[fp]
Out[5]= 3
```

This is one of the fixed points

```
In[6]:= fp1 = fp[[1]]
Out[6]= {0, 0, 0}
```

This is the Jacobian of the map at the fixed point fp1.

```
In[7]:= ACTJac = Jac[ACTMap, fp1];

In[8]:= MatrixForm[ACTJac]
```

$$Out[8]= \begin{pmatrix} a & -b & b \\ b & a & -a \\ c & 0 & e \end{pmatrix}$$

Calculation of eigenvalues in symbolic form is possible for small matrices.

```
In[9]:= evals = Eigenvalues[ACTJac];
```

Some of the parameters may be assigned values.

```
In[10]:= a = 0.6; b = 0.5; d = 1; e = 1;

In[11]:= fp//Chop
Out[11]= {{0, 0, 0},
          {-1.41421 √0.5 c,
           -0.0282843 √0.5 c,
           -1.15966 √0.5 c},
          {1.41421 √0.5 c,
           0.0282843 √0.5 c,
           1.15966 √0.5 c}}
```

In this case, the equilibrium point fp1 is a saddle point with a one-dimensional local unstable manifold and a two-dimensional local stable manifold.

```
In[12]:= c = 0.3

In[13]:= Eigenvalues[ACTJac]
Out[13]= {1.27197, 0.464017 + 0.514063 i,
          0.464017 - 0.514063 i}

In[14]:= Abs[%]
Out[14]= {1.27197, 0.692512, 0.692512}
```

5.2.1 Numerical Explorations

These are the first 10 terms of the orbit $\gamma((0.1, 0.2, 0.3))$ for the ACT map with parameter values $a = 0.6$, $b = 0.5$, $c = 0.3$, $d = 1$, $e = 1$.

```
In[15]:= orb1 = Take[Orbit[ACTMap,
              {0.1, 0.2, 0.3}, 1000],
              {1, 10}]
Out[15]= {{0.1, 0.2, 0.3},
          {0.11, -0.01, 0.329},
          {0.2355, -0.1484, 0.360669},
          {0.395835, -0.187691, 0.418258},
          {0.540475, -0.165652, 0.474987},
          {0.644605, -0.114146, 0.479249},
          {0.683461, -0.0337348, 0.404787},
          {0.629338, 0.078617, 0.290568},
          {0.483578, 0.187498, 0.230111},
          {0.311453, 0.216222, 0.2621}}
```

This is a phase portrait of an orbit for the parameter value $c = 0.38$ for which there is a complicated behavior of the orbit. Further simulations and visualizations are needed to reveal the possibly chaotic character of the solutions of this equation. See [ACT] for some results.

```
In[16]:= c = 0.38;

In[17]:= orb2 = Orbit[ACTMap,
            {0.2, 0.1, 0.1}, 4000];

In[18]:= PhasePortrait3D[orb1]
```

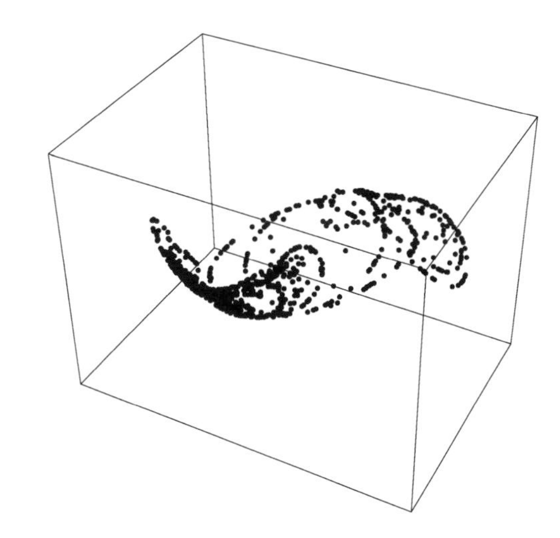

We now consider maxima and minima of positive and negative semicycles of solutions of the equation

$$x_{n+1} = \frac{A + x_{n-2}}{x_n} \tag{5.8}$$

where A is a positive parameter and the initial conditions are all positive. The definitions of semicycles and terminology are given in Section 3.9.

The equation and corresponding map are generated.

```
In[19]:= eqnA := x[n + 1] == (A + x[n - 2])/x[n];

In[20]:= mapA := DEToMap[eqnA];

In[21]:= mapA[{x, y, z}]
Out[21]= {y, z, (A + x)/z}
```

Switching back to the equation may be done by using `MapToDE` function:

```
In[22]:= MapToDE[mapA, 3]
Out[22]= x[1 + n] == (A + x[-2 + n])/x[n]
```

There are two equilibrium points.

```
In[23]:= fpA = {x, y, z}/.Solve[
              mapA[{x, y, z}] == {x, y, z}
                , {x, y, z}]
```

$$Out[23]= \left\{\left\{\frac{1}{2}\left(1-\sqrt{1+4A}\right), \frac{1}{2}\left(1-\sqrt{1+4A}\right), \frac{1}{2}\left(1-\sqrt{1+4A}\right)\right\}, \left\{\frac{1}{2}\left(1+\sqrt{1+4A}\right), \frac{1}{2}\left(1+\sqrt{1+4A}\right), \frac{1}{2}\left(1+\sqrt{1+4A}\right)\right\}\right\}$$

These are approximate values of the fixed points when $A = 1$.

```
In[24]:= A = 1;
Out[24]= {{-0.618034, -0.618034, -0.618034},
              {1.61803, 1.61803, 1.61803}}
```

Based on simulations such as the one shown here it seems that either the solution is periodic or it is eventually periodic, with period 5.

```
In[25]:= TimeSeriesPlot[mapA,
              {1., 2., 2.}, 40];
```

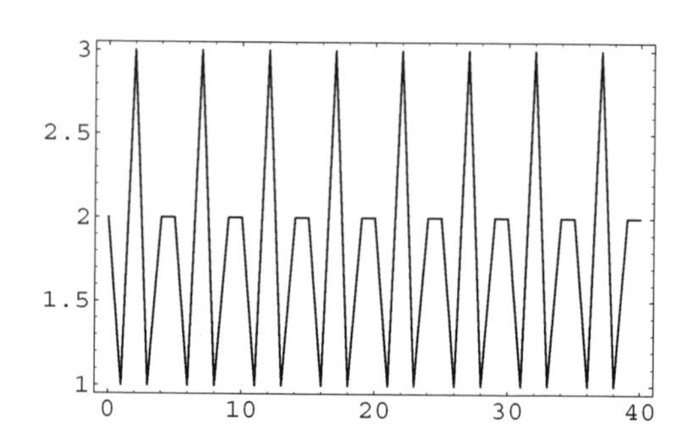

A point plot also suggests periodicity.

```
In[26]:= OrbitPlot[mapA, {1., 2., 2.}, 40];
```

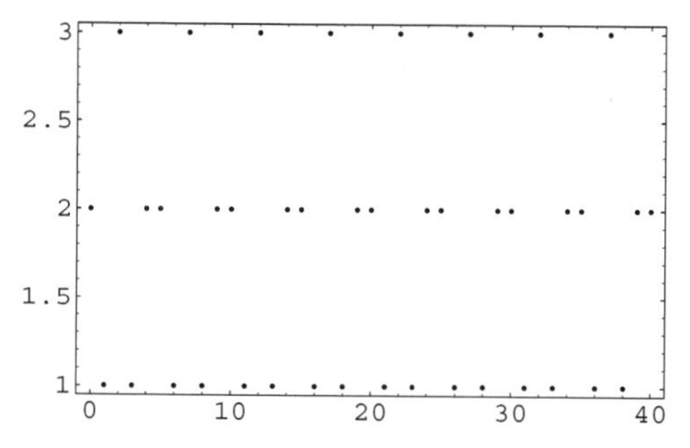

The function `SemicycleTest` may be used to get more precise information on semicycles. See [KL] for some general results about the semicycles of second order difference equations and [KoL] for some results about the semicycles of difference equations of orders three and higher.

The complexity of the behavior of equation (5.8) for different values of parameter A can be seen from the bifurcation diagrams.

The bifurcation plot displayed here shows that the solution of equation (5.8) has complicated behavior for the specified range of values of the parameter A. We note that the theory for this equation has not been developed yet.

```
In[27]:= BifurcationPlotND[mapA,
         {A, 0.1, 2.1}, {1.2, 1., 4.},
         PlotRange → {-1, 7},
         FirstIt → 300,
         Iterates → 150,
         Steps → 300];
```

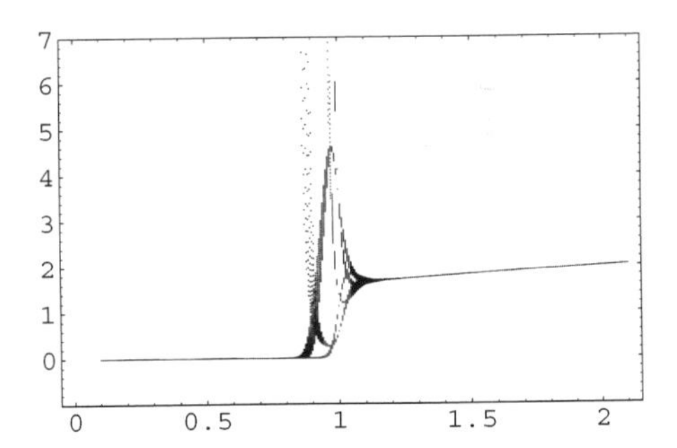

5.3 Dissipative Difference Equation of Third Order

5.3.1 Introduction

In this section we study the dissipative difference equation

$$x_{n+1} = f(x_n, x_{n-1}, x_{n-2}) \tag{5.9}$$

and the corresponding map.

We begin by presenting recently proven results that can be used to investigate the asymptotic behavior of these equations and in particular to prove global attractivity of the equilibrium. These results give sufficient conditions for all solutions to converge to the equilibrium, and they have been successfully applied to the case of the third order rational difference equation of the form

$$x_{n+1} = \frac{\alpha + \beta x_n + \gamma x_{n-1} + \delta x_{n-2}}{A + Bx_n + Cx_{n-1} + Dx_n - 2} \qquad n = 0, 1, \dots . \tag{5.10}$$

Here the parameters α, β, γ, δ, A, B, C, D, are nonnegative real numbers and the initial conditions x_{-2}, x_{-1}, x_0 are arbitrary nonnegative real numbers, such that the solution exists for all $n \geq 0$.

The regions where the function $f(x, y, z)$ is monotonic in its arguments play a fundamental role in these results. In this regard there are eight possible scenarios depending on whether $f(x, y, z)$ is nondecreasing in all three arguments, or nonincreasing in two arguments and nondecreasing in the third, or nonincreasing in one and nondecreasing in the other two, or nondecreasing in all three. The proofs of these results can be found in [KL]. We now present two of the results with corresponding applications.

THEOREM 5.2
Let $[a, b]$ be an interval of real numbers, and let

$$f : [a, b]^3 \to [a, b]$$

be a continuous function satisfying the following properties:

a. $f(x, y, z)$ is nondecreasing in each of its arguments;

b. The equation

$$f(x, x, x) = x,$$

has a unique solution in the interval $[a, b]$.

Then (5.9) has a unique equilibrium $\bar{x} \in [a, b]$ and every solution of (5.9) converges to $\bar{x}$.

Example 5.1 Consider the equation

$$x_{n+1} = \sqrt{x_n} + \sqrt{x_{n-1}} + \sqrt{x_{n-2}}, \quad n = 0, 1, \dots, \tag{5.11}$$

where the initial values x_{-2}, x_{-1} and x_0 are nonnegative numbers and such that their sum is positive,

$$x_{-2} + x_{-1} + x_0 > 0.$$

Obviously, equation (5.11) has a unique positive equilibrium $\bar{x} = 9$. The function $f(x, y, z) = \sqrt{x} + \sqrt{y} + \sqrt{z}$ is increasing in all three variables and any interval $[L, U]$, where $U > 9 > L > 0$ is an invariant interval. Indeed, if $x, y, z \in [L, U]$ then

$$L < 3\sqrt{L} = f(L, L, L) < f(x, y, z) < f(U, U, U) < U,$$

which proves that $[L, U]$ is an invariant interval. In addition the positive equilibrium is unique in this interval, and so all the conditions of Theorem 5.2 are satisfied. Thus all solutions with the given initial conditions converge to the positive equilibrium.

A similar result holds for a more general equation of the form:

$$x_{n+1} = a\sqrt{x_n} + b\sqrt{x_{n-1}} + c\sqrt{x_{n-2}}, \quad n = 0, 1, \dots,$$

where $a, b,$ and c are positive parameters and the initial values satisfy the same condition as above. ∎

THEOREM 5.3
Let $[a, b]$ be an interval of real numbers and let

$$f : [a, b]^3 \to [a, b]$$

be a continuous function satisfying the following properties:

a. *$f(x, y, z)$ is nonincreasing in all three variables x, y, z in $[a, b]$;*

b. *Whenever $(m, M) \in [a, b] \times [a, b]$ is a solution of the system*

$$\begin{aligned} M &= f(m, m, m) \\ m &= f(M, M, M) \end{aligned} \tag{5.12}$$

then necessarily $m = M$.

Then (5.9) has a unique equilibrium $\bar{x} \in [a, b]$ and every solution of (5.9) converges to $\bar{x}$.

Example 5.2 Consider the equation

$$x_{n+1} = \frac{1}{1 + x_n} + \frac{1}{1 + x_{n-1}} + \frac{1}{1 + x_{n-2}}, \quad n = 0, 1, \dots, \tag{5.13}$$

where the initial values are nonnegative numbers. This equation has a unique positive equilibrium $\bar{x} = -\frac{1}{2} + \frac{1}{2}\sqrt{13} \approx 1.3028$. The function

$$f(x, y, z) = \frac{1}{1 + x} + \frac{1}{1 + y} + \frac{1}{1 + z}$$

is decreasing in all three variables and an invariant interval is $[\frac{3}{4}, 3]$. Now the system (5.12) takes the form:

$$M = \frac{3}{1 + m}, \quad m = \frac{3}{1 + M},$$

and implies that $M = m$. Thus all the conditions of Theorem 5.3 are satisfied, hence all solutions converge to the positive equilibrium $\bar{x}$. ∎

5.4 *Dynamica* Session on Local Asymptotic Stability of Period-Two Solutions

5.4.1 Introduction

In this section we determine conditions on the parameters of the equation

$$y_{n+1} = \frac{y_{n-1} + py_n}{y_{n-1} + qy_{n-2}} \tag{5.14}$$

under which there are period-two positive solutions

$$\phi, \psi, \phi, \psi, \ldots \tag{5.15}$$

where $\phi \neq \psi$. Throughout this section we assume that both p and q are positive, and that $p \neq q$. The example and results of this section are from [KK].

5.4.2 A Quadratic Equation Satisfied by Period-Two Solutions

If (5.14) has a periodic solution (5.15), we may substitute ϕ and ψ into (5.14) to obtain these equations.

```
In[1]:= << Dynamica`

In[2]:= eq1 = ψ == (ψ + p ϕ)/(ψ + q ϕ);

In[3]:= eq2 = ϕ == (ϕ + p ψ)/(ϕ + q ψ);
```

Solving for ϕ or ψ gives formulas with radicals. Instead, we eliminate one of the variables.

```
In[4]:= eq3 = Eliminate[{eq1, eq2}, ϕ]
Out[4]= p^3 - 2 p^2 q ψ +
          p (-1 + ψ + q ψ - q ψ^2 + q^2 ψ^2)
        == ψ (1 - q - 2 ψ + q ψ + q^2 ψ + ψ^2 - q^2 ψ^2)
```

We factor the difference of the two sides of eq3. This helps in the analysis.

```
In[5]:= factors =
          Factor[eq3[[1]] - eq3[[2]]]
Out[5]= (1 + p - ψ - q ψ) (-p + p^2 - ψ + p ψ +
          q ψ - p q ψ + ψ^2 - q ψ^2)
```

Setting the first factor equal to zero gives, after some calculations, $\phi = \psi$, which is a case we are ruling out. Thus we do not need to consider further the first factor.

```
In[6]:= ans1 = Solve[factors[[1]] == 0, ψ]
Out[6]= {{ψ → -(-1 - p)/(1 + q)}}
In[7]:= eq1 /. ans1 //Simplify
Out[7]= {(1 + p)/(1 + q) == (1 + p + p ϕ + p q ϕ)/(1 + p + q ϕ + q^2 ϕ)}
In[8]:= Solve[%, ϕ]
Out[8]= {{ϕ → (1 + p)/(1 + q)}}
```

The second factor is quadratic in ψ.

```
In[9]:= Collect[factors[[2]], ψ]
Out[9]= -p + p^2 + (-1 + p + q - p q) ψ + (1 - q) ψ^2
```

Note that by symmetry, a similar equation is satisfied by ϕ. Also, note that if $q = 1$, then necessarily we have $p = 1$, and so $p = q$, which we are not allowing. Hence we may assume that $q \neq 1$.

5.4.3 Main Results

We may sum up our calculations in Subsection 5.4.2 in the following statement.

PROPOSITION 5.1
Let p, q, ϕ, ψ be positive, with $p \neq q$ and $\phi \neq \psi$. Then, the sequence

$$\phi, \psi, \phi, \psi, \ldots$$

is a solution to (5.14) if and only if $q \neq 1$ and the quadratic equation

$$z^2 + (p-1)z - \frac{(p-1)p}{q-1} = 0 \tag{5.16}$$

has exactly two solutions $z = \phi$ and $z = \psi$ which are positive and distinct.

Moreover, in either case, we have

$$\psi + \phi = 1 - p \tag{5.17}$$

$$\phi\psi = -\frac{(p-1)p}{q-1} \tag{5.18}$$

A quadratic polynomial $z^2 + bz + c$ has positive, distinct roots if and only if $b < 0$, $c > 0$, and $b^2 > 4c$. Thus we have the following corollary.

COROLLARY 5.1
In either one of the two cases of Proposition 5.1, the following inequalities hold:

$$1 < q, \tag{5.19}$$

$$0 < p < 1, \tag{5.20}$$

and

$$1 + 3p + pq < q. \tag{5.21}$$

Conversely, if (5.19), (5.20) and (5.21) hold, then equation (5.14) has a unique positive period-two solution whose first two terms ϕ, and ψ, are solutions to equation (5.16).

We shall see in Subsections 5.4.4 through 5.4.7 that period-two solutions are indeed locally asymptotically stable. For convenience we state the result here.

PROPOSITION 5.2

Positive, period-two solutions to equation (5.14) are locally asymptotically stable.

5.4.4 Conditions for Linearized Local Stability

We now turn to the study of the linearized stability of period-two solutions to (5.14). We do this by applying the Schur-Conn criterion to the characteristic polynomial of the Jacobian of the second iterate of the map of equation (5.14).

We begin by assigning a name to the equation.

```
In[10]:= pqeqn = x[n + 1] ==
              (x[n - 1] + p x[n - 2]) / (x[n - 1] + q x[n - 2])
Out[10]= x[1 + n] == (p x[-2 + n] + x[-1 + n]) / (q x[-2 + n] + x[-1 + n])
```

This is the map associated with (5.14).

```
In[11]:= pqmap = DEToMap[pqeqn];

In[12]:= pqmap[{x, y, z}]
Out[12]= {y, z, (p x + y) / (q x + y)}
```

The second iterate of the map is given by this expression.

```
In[13]:= pqmap2[{x_, y_, z_}] =
              pqmap[pqmap[{x, y, z}]]
Out[13]= {z, (p x + y) / (q x + y), (p y + z) / (q y + z)}
```

Here we evaluate the Jacobian of the second iterate of the map at $\{\phi, \psi, \phi\}$.

```
In[14]:= J = Jac[pqmap2, {φ, ψ, φ}];

In[15]:= MatrixForm[Simplify[J2]]
```

$$\text{Out[15]=} \begin{pmatrix} 0 & 0 & 1 \\ \frac{(p-q)\,\psi}{(q\,\phi+\psi)^2} & \frac{(-p+q)\,\phi}{(q\,\phi+\psi)^2} & 0 \\ 0 & \frac{(p-q)\,\phi}{(\phi+q\,\psi)^2} & \frac{(-p+q)\,\psi}{(\phi+q\,\psi)^2} \end{pmatrix}$$

Period-two solutions to (5.14) are fixed points of the second iterate of pqmap. For local asymptotic stability of the period two solutions, we require that the eigenvalues of J be located inside the unit circle.

To find the locations of the eigenvalues, we compute the characteristic polynomial of J, and then apply the Schur-Conn criterion.

```
In[16]:= charpol = Collect[
            CharacteristicPolynomial[J, t], t]
```

$$\text{Out[16]=} -t^3 + t^2\left(-\frac{p\,\phi}{(q\,\phi+\psi)^2} - \frac{\psi}{(q\,\phi+\psi)^2} + \frac{1}{q\,\phi+\psi} + \frac{-\phi-p\,\psi}{(\phi+q\,\psi)^2} + \frac{1}{\phi+q\,\psi}\right) - \left(\frac{p\,q\,\phi}{(q\,\phi+\psi)^2} + \frac{q\,\psi}{(q\,\phi+\psi)^2} - \frac{p}{q\,\phi+\psi}\right)\left(-\frac{q\,(\phi+p\,\psi)}{(\phi+q\,\psi)^2} + \frac{p}{\phi+q\,\psi}\right) + t\left(-\frac{p\,\phi\,(\phi+p\,\psi)}{(q\,\phi+\psi)^2\,(\phi+q\,\psi)^2} - \frac{\psi\,(\phi+p\,\psi)}{(q\,\phi+\psi)^2\,(\phi+q\,\psi)^2} + \frac{\phi+p\,\psi}{(q\,\phi+\psi)\,(\phi+q\,\psi)^2} + \frac{p\,\phi}{(q\,\phi+\psi)^2\,(\phi+q\,\psi)} + \frac{\psi}{(q\,\phi+\psi)^2\,(\phi+q\,\psi)} - \frac{1}{(q\,\phi+\psi)\,(\phi+q\,\psi)}\right)$$

Note that the characteristic polynomial of J has the form

$$p(t) = -t^3 - At^2 + Bt + B,$$

where A and B are defined as follows:

```
In[17]:= A = Factor[
            Coefficient[-charpol, t^2]]
```

$$\text{Out[17]=} \frac{(p-q)\,(\phi+\psi)\,(\phi^2-\phi\,\psi+2\,q\,\phi\,\psi+q^2\,\phi\,\psi+\psi^2)}{(q\,\phi+\psi)^2\,(\phi+q\,\psi)^2}$$

```
In[18]:= B = Factor[
            Coefficient[-charpol, t]]
```

$$\text{Out[18]=} \frac{(p-q)^2\,\phi\,\psi}{(q\,\phi+\psi)^2\,(\phi+q\,\psi)^2}$$

When specialized to our setting, the Schur-Conn criterion states that the roots of $p(t)$ are located inside the unit circle if and only if the following inequalities hold:

$$|B - A| < 1 + B \tag{5.22}$$

$$|3B + A| < 3 + B \tag{5.23}$$

$$B^2 + B + AB < 1 \tag{5.24}$$

Note that conditions (5.22) and (5.23) may be written as follows:

$$0 \leq A + 1 \tag{5.25}$$

$$0 < 2B + 1 - A \tag{5.26}$$

$$0 < 3 - 2B - A \tag{5.27}$$

$$0 < A + 4B + 3 \tag{5.28}$$

Our task is to find what values of p and q ensure that inequalities (5.24) through (5.28) hold.

Rather than treating inequalities (5.24) through (5.28) with the expressions we have for A and B, here we shall follow [KK] to obtain in Subsection 5.4.5 formulas for A and B that depend on p and q only. Then, in Subsection 5.4.6 we shall prove that A and B satisfy specific bounds that, as we shall see in Subsection 5.4.7, guarantee the inequalities (5.24) to (5.28).

5.4.5 Formulas for A and B in Terms of p, q

The formulas for A and B given in Subsection 5.4.4 depend on ϕ and ψ. In this subsection we derive equivalent expressions for A and B that depend on p and q only.

By using equations (5.17) and (5.18), it is possible to eliminate ψ and ϕ from the numerator of A.

```
In[19]:= numA = Simplify[Numerator[A] //.
            {ϕ + ψ → 1 - p,
             ϕ ψ → - (p - 1) p/(q - 1),
             ϕ^2 + ψ^2 → (1 - p)^2 +
                2 (p - 1) p
                ----------- }]
                  q - 1
Out[19]= (-1 + p)^2 (p - q) (1 + p (2 + q))
```

The denominator of A may also be written as a rational expression in p and q.

```
In[20]:= denA = Simplify[Expand[
           Denominator[A]]] //.
          {ϕ + ψ → 1 - p,
                     (p - 1) p
           ϕ ψ → - ----------- ,
                       q - 1
           ϕ^2 + ψ^2 →
                         2 (p - 1) p
              (1 - p)^2 + ----------- }
                            q - 1
         (1 - p) p
Out[20]= --------- +
          -1 + q
                      2 (-1 + p) p        (1 - p) p q^2  2
         ((1 - p)^2 + ------------) q + -------------)
                        -1 + q              -1 + q

In[21]:= Simplify[Expand[denA]]
Out[21]= (-1 + p)^2 (p - q)^2
```

We may proceed in a similar manner to obtain a formula for B in terms of p and q.

```
In[22]:= numB = Simplify[Numerator[B] //.
              {φψ → -(p - 1) p/(q - 1)}]
```

$$\text{Out[22]=} -\frac{(-1+p)\,p\,(p-q)^2}{-1+q}$$

```
In[23]:= denB = denA;
```

The expressions for A and B in terms of p and q can now be produced.

```
In[24]:= Apq = Factor[numA/denA]
```

$$\text{Out[24]=} \frac{1+2p+pq}{p-q}$$

```
In[25]:= Bpq = Factor[numB/denB]
```

$$\text{Out[25]=} -\frac{p}{(-1+p)\,(-1+q)}$$

5.4.6 Bounds for A and B

In this section we prove that A and B have certain bounds. We shall see that such bounds immediately yield inequalities (5.24) through (5.28).

LEMMA 5.1

The quantities A and B satisfy

$$-1 < A < 1 \quad \textit{and} \quad 0 < B < \frac{1}{4}. \tag{5.29}$$

We present a proof assisted with *Mathematica.*

Since $p - q < 0$, the inequality $0 < A + 1$ is seen to be equivalent to

$$0 > 1 + 3p - q + pq, \tag{5.30}$$

which is precisely inequality (5.21).

```
In[26]:= Factor[Apq + 1]
```

$$\text{Out[26]=} \frac{1+3p-q+pq}{p-q}$$

To prove that $A < 1$, consider the expression $A - 1$. Clearly, since $p - q < 0$, we have that $A - 1 < 0$.

```
In[27]:= Factor[Apq - 1]
```

$$\text{Out[27]=} \frac{(1+p)\,(1+q)}{p-q}$$

That $0 < B$ is clear from the fact that $p > 0$, $p-1 < 0$ and $q - 1 > 0$.

```
In[28]:= Bpq
```

$$\text{Out[28]=} -\frac{p}{(-1+p)\,(-1+q)}$$

It remains to show that $B < \frac{1}{4}$. But this is a consequence of inequality 5.21 in Proposition 5.1. Indeed, $B - \frac{1}{4}$ is obviously negative.

```
In[29]:= Factor[Bpq - 1/4]
Out[29]= -(1 + 3 p - q + p q)/(4 (-1 + p) (-1 + q))
```

5.4.7 Completion of the Proof of Stability

It is easy to see that the inequalities (5.24) through (5.28) are a direct consequence of Lemma 5.1. This completes the proof of stability of period two solutions of equation (5.14).

5.5 *Dynamica* Session on Todd's Difference Equation

In this section we study Todd's equation

$$x_{n+1} = \frac{A + x_n + x_{n-1}}{x_{n-2}} \tag{5.31}$$

and the corresponding map.

Todd's equation may be considered as a $3D$-version of Lyness' equation which was investigated in Section 2.15. Todd's equation appears for the first time in [To], where it was established that for $A = 1$, every solution is periodic with period 8.

There are few results known for this equation (mentioned in Subsection 5.5.1) and many open problems (some of them are mentioned in Subsection 5.5.1). More about equation (5.31) can be found in [KoL], pp. 133–153, [KLR], and [Ku].

Todd's map is available in *Dynamica*.

```
In[1]:= << Dynamica`
"Dynamica Version 1.0"

In[2]:= ToddMap[{x, y, z}]
Out[2]= {y, z, (A + y + z)/x}
```

There are two solutions to the equation satisfied by fixed points.

```
In[3]:= fp = {x, y, z} /. Solve[
          ToddMap[{x, y, z}] == {x, y, z},
          {x, y, z}]
Out[3]= {{1 - √(1 + A), 1 - √(1 + A), 1 - √(1 + A)},
          {1 + √(1 + A), 1 + √(1 + A), 1 + √(1 + A)}}
```

Clearly in the case $A < -1$ there are no fixed points. When $A = -1$, both solutions obtained above are the same. Note that $\{0, 0, 0\}$ is not allowed as a fixed point. Thus in the case $A = 0$ there is only one fixed point, $(2, 2, 2)$. Then, we have the following cases.

a. $A < -1$: There are no fixed points.

b. $A = -1$: There is only one fixed point, namely, $(1, 1, 1)$

c. $A = 0$. There is only one fixed point, namely, $(2, 2, 2)$.

d. $A \in (-1, 0) \cup (0, \infty)$: There are two fixed points, namely, $(1 - \sqrt{1 + A}, 1 - \sqrt{1 + A}, 1 - \sqrt{1 + A})$ and $(1 + \sqrt{1 + A}, 1 + \sqrt{1 + A}, 1 + \sqrt{1 + A}\})$.

This calculation shows that the equilibrium point `fp[[1]]` is a nonhyperbolic point for all values of parameter A. This is an indication of possibly complicated dynamics.

```
In[4]:= {x, y, z} = fp[[1]];

In[5]:= Eigenvalues[toddjac]

Out[5]= { 1,

  3  A + 3 √(1 + A)   √(14  17A  3A² + 14 √(1 + A) + 10A √(1 + A))
  ------------------------------------------------------------------ ,
                      2 (2  A + 2 √(1 + A))

  3  A + 3 √(1 + A) + √(14  17A  3A² + 14 √(1 + A) + 10A √(1 + A))
  ------------------------------------------------------------------ }
                      2 (2  A + 2 √(1 + A))
```

In the case of the second equilibrium point `fp[[2]]` also have a nonhyperbolic point for all values of A.

```
In[6]:= {x, y, z} = fp[[2]];

In[7]:= Eigenvalues[toddjac]

Out[7]= { 1,

  3 + A + 3 √(1 + A)   √(14  17A  3A²  14 √(1 + A)  10A √(1 + A))
  ----------------------------------------------------------------- ,
                      2 (2 + A + 2 √(1 + A))

  3 + A + 3 √(1 + A) + √(14  17A  3A²  14 √(1 + A)  10A √(1 + A))
  ----------------------------------------------------------------- }
                      2 (2 + A + 2 √(1 + A))
```

In the case $A = -1$ we have that all eigenvalues of the Jacobian are on the unit circle.

```
In[8]:= A = -1;

In[9]:= fp
Out[9]= {{1, 1, 1}, {1, 1, 1}}

In[10]:= {x, y, z} = fp[[1]];

In[11]:= Eigenvalues[toddjac]
Out[11]= {-1, 1, 1}
```

Recall that when $A = 0$, the only fixed point is $(2,2,2)$. In this case all eigenvalues of the Jacobian are on the unit circle.

```
In[12]:= A = 0;

In[13]:= fp
Out[13]= {{0, 0, 0}, {2, 2, 2}}

In[14]:= {x, y, z} = fp[[2]];

In[15]:= Eigenvalues[toddjac]
```

$$Out[15]= \left\{-1, \frac{1}{4}\left(3 - i\sqrt{7}\right), \frac{1}{4}\left(3 + i\sqrt{7}\right)\right\}$$

```
In[16]:= Abs[%]
Out[16]= {1, 1, 1}
```

We are now interested in establishing some facts about the location of eigenvalues `e1`, `e2`, and `e3` of the Jacobian at the first fixed point `fp[[1]]`. It is clear that, depending on the value of A, `e2` and `e3` are both real numbers or a complex conjugate pair. Thus we must analyze the sign of the discriminant.

Here the discriminant is given the name `disc1`.

```
In[17]:= disc1 =
            -14 - 17 A - 3 A^2 + 14 Sqrt[1 + A] + 10 A Sqrt[1 + A];
```

$$-14 - 17A - 3A^2 + 14\sqrt{1+A} + 10A\sqrt{1+A}$$

The plot shows for which values of A the sign of `disc1` is positive. In this case there are two distinct real eigenvalues of the Hessian at the fixed point.
The value of A for which the discriminant is zero is of interest, as this is where roots evolve from real values to complex values or vice versa.

```
In[18]:= Plot[disc1, {A, 0, 1.5},
            PlotRange → All];
```

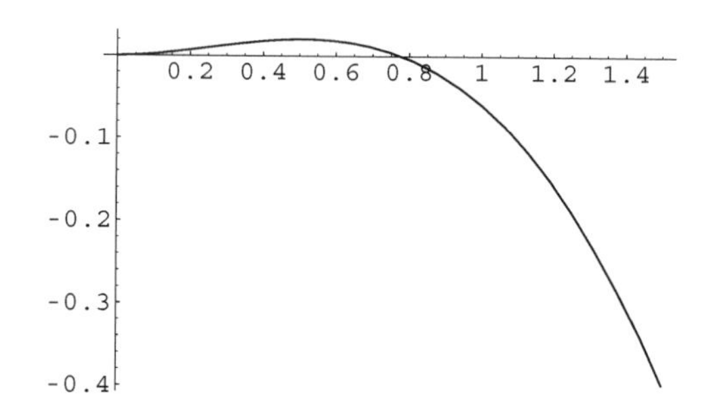

The discriminant is zero at these values.

```
In[19]:= Solve[disc1 == 0, A]
```

$$Out[19]= \left\{\{A \to -1\}, \{A \to 0\}, \left\{A \to \frac{7}{9}\right\}\right\}$$

We now determine the eigenvalues of the Jacobian when A approaches 0, $\frac{7}{9}$ or ∞.

```
In[20]:= Limit[{e1, e2, e3}, A → 0]
Out[20]= {-1, 0, -∞}

In[21]:= Limit[{e1, e2, e3}, A → ∞]
```

$$Out[21]= \left\{-1, \frac{1}{2}\left(1 + i\sqrt{3}\right), \frac{1}{2}\left(1 - i\sqrt{3}\right)\right\}$$

```
In[22]:= {e1, e2, e3} /. A → 7/9
Out[22]= {-1, -1, -1}
```

This "root locus plot," together with the previous calculation, indicates that there are essentially three cases when $A > 0$. These are:

- if $0 < A < \frac{7}{9}$, then all roots are real with `e3<e1<e3<0` and `e1` = -1.
- if $A = \frac{7}{9}$, then `e1` = `e2` = `e3` = -1
- if $A > \frac{7}{9}$, then `e2` and `e3` are two complex conjugate roots on the unit circle, and `e1` = -1.

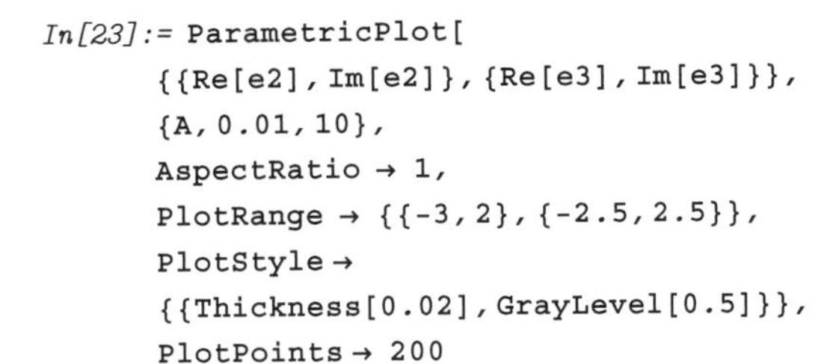

```
In[23]:= ParametricPlot[
          {{Re[e2], Im[e2]}, {Re[e3], Im[e3]}},
          {A, 0.01, 10},
          AspectRatio → 1,
          PlotRange → {{-3, 2}, {-2.5, 2.5}},
          PlotStyle →
          {{Thickness[0.02], GrayLevel[0.5]}},
          PlotPoints → 200
```

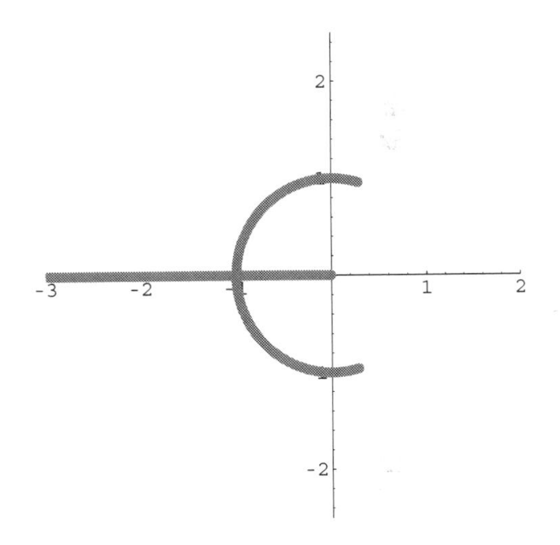

We may confirm that `e2` is on the unit circle with this algebraic calculation. Note that this implies that `e3` is also on the unit circle, as `e3` is the complex conjugate of `e2`.

```
In[24]:= abs1 = Simplify[
  ((-3 + -A + 3 √(1 + A))^2 -
  (-14 - 17 A - 3 A^2 + 14 √(1 + A) + 10 A √(1 + A)))/
      (2 (-2 - A + 2 √(1 + A)))^2]

Out[24]= 1
```

We have succeeded in using *Dynamica* to classify the fixed point `fp[[1]]` for all values of the parameter A. We leave it to the reader to classify `fp[[2]]`. Many second and third order difference equations may be treated with explicit symbolic computation, but there are serious limitations for higher order equations.

Now we shall perform numerical explorations and visualization of orbits for particular values of A

```
In[25]:= Clear[A, x, y, z]

In[26]:= A = 3;
```

A point plot of an orbit.

```
In[27]:= orb = Orbit[ToddMap,
                {1., 2., 3.}, 100];

In[28]:= TimeSeriesPlot[ToddMap,
                {1., 2., 3.}, 100]
```

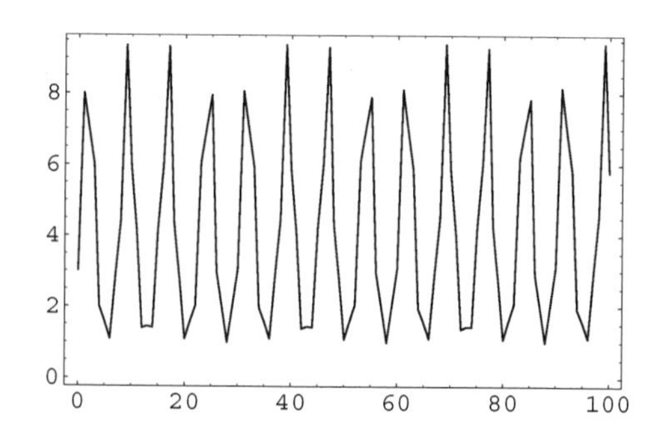

A time series plot of an orbit.

```
In[29]:= OrbitPlot[ToddMap, {1., 2., 3.},
           100]
```

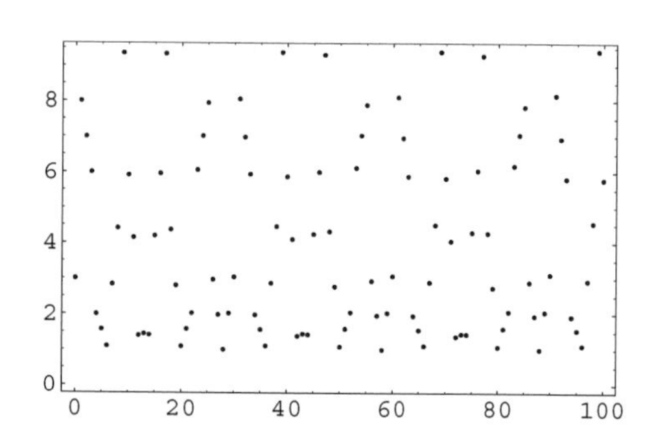

A phase portrait plot of an orbit.

```
In[30]:= orb = Orbit[ToddMap,
              {1., 2., 3.}, 1000];

In[31]:= PhasePortrait3D[{orb},
              FrameTicks → True]
```

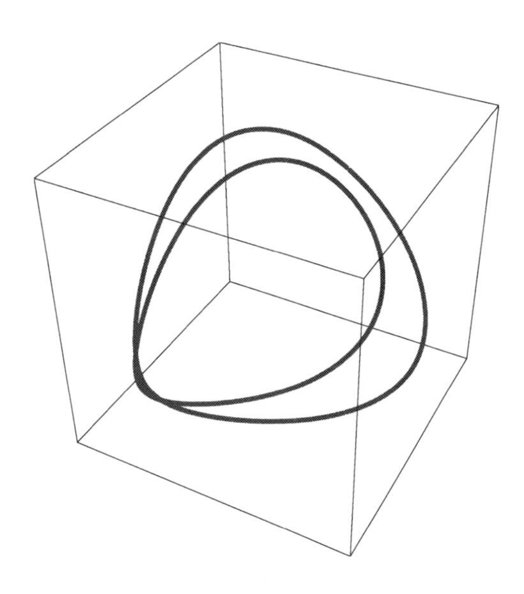

There is a known invariant for Todd's difference equation. We may find it with *Dynamica*'s `RationalInvariant` function.

```
In[32]:= Clear[A, x, y, z]

In[33]:= RationalInvariant[Todd]
Solve :: svars : Equations may not give
solutions for all "solve" variables.
```

$$\text{Out[33]= } c[1] + \frac{(2+A)\,c[23]}{x[-2+n]} + c[23]\,x[-2+n] + \frac{(2+A)\,c[23]}{x[-1+n]} + \frac{(1+A)\,c[23]}{x[-2+n]\,x[-1+n]} + \frac{c[23]\,x[-2+n]}{x[-1+n]} + c[23]\,x[-1+n] + \frac{c[23]\,x[-1+n]}{x[-2+n]} + \frac{(2+A)\,c[23]}{x[n]} + \frac{(1+A)\,c[23]}{x[-2+n]\,x[n]} + \frac{c[23]\,x[-2+n]}{x[n]} + \frac{(1+A)\,c[23]}{x[-1+n]\,x[n]} + \frac{A\,c[23]}{x[-2+n]\,x[-1+n]\,x[n]} + \frac{c[23]\,x[-2+n]}{x[-1+n]\,x[n]} + \frac{c[23]\,x[-1+n]}{x[n]} + \frac{c[23]\,x[-1+n]}{x[-2+n]\,x[n]} + c[23]\,x[n] + \frac{c[23]\,x[n]}{x[-2+n]} + \frac{c[23]\,x[n]}{x[-1+n]} + \frac{c[23]\,x[n]}{x[-2+n]\,x[-1+n]}$$

Since c[23] appears in all the monomials, it may be set to 1. In addition, we set c[1] = A+3, which, as we shall see later, is helpful for factoring the invariant. This was discovered in [FJL]. Also, we change notation so simplify the way the expression looks.

```
In[34]:= invtodd = %/.
           {c[23] → 1, c[1] → A + 3,
            x[n] → x, x[n - 1] → y, x[n - 2] → z}
```

$$\text{Out[34]= } 3+A+\frac{2+A}{x}+x+\frac{2+A}{y}+\frac{1+A}{x\,y}+\frac{x}{y}+y+\frac{y}{x}+\frac{2+A}{z}+\frac{1+A}{x\,z}+\frac{x}{z}+\frac{1+A}{y\,z}+\frac{A}{x\,y\,z}+\frac{x}{y\,z}+\frac{y}{z}+\frac{y}{x\,z}+z+\frac{z}{x}+\frac{z}{y}+\frac{z}{x\,y}$$

Here is the factorization of the invariant.

```
In[35]:= Factor[invtodd]
```

$$\text{Out[35]= } \frac{(1+x)\,(1+y)\,(1+z)\,(A+x+y+z)}{x\,y\,z}$$

It is not hard to conclude from the factored form of the invariant that positive solutions are bounded. This is a typical use of the invariant to prove qualitative statements on solutions.

Here we consider the special case $A = 2$, for which we try to determine graphically whether the invariant we found has a minimum at fp[[2]]. We shall need the value of the invariant function at fp[[2]].

```
In[36]:= A = 2.;

In[37]:= {x, y, z} = fp[[2]];

In[38]:= invtodd
Out[38]= 25.9906
Out[38]= Clear[x, y, z];
```

After loading *Mathematica*'s ContourPlot3D, we plot several levels together with fp[[2]].

```
In[39]:= << Graphics`ContourPlot3D`

In[40]:= Show[
          {Graphics3D[{PointSize[0.05],
              Point[fp[[2]]]}],
            ContourPlot3D[invtodd,
          {x, 0.1, 2.7305}, {y, 0.1, 6},
          {z, 0.1, 6},
          Contours → {26.2, 26.5, 27., 28.,
                29., 30.}, Axes → True,
            PlotPoints → {3, 8},
            DisplayFunction → Identity]},
        DisplayFunction → $DisplayFunction
```

Note that the range of the variables in the plot was chosen to show how level surfaces nest with `fp[[2]]` at the center. This suggests that a minimum is attained by the invariant function at `fp2`.

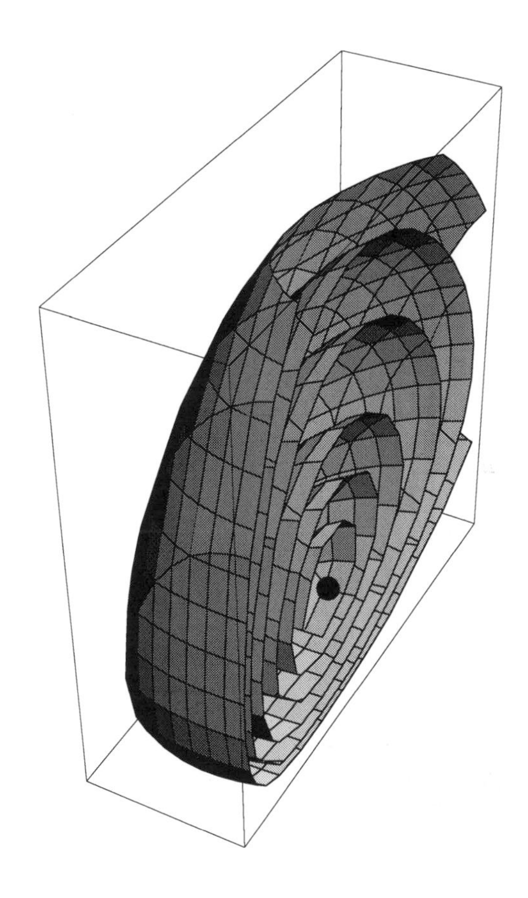

We now use *Dynamica*'s `FixedPointTest` function to determine the nature of the fixed points for different parameter values.

```
In[41]:= Clear[A, x, y, z]

In[42]:= FixedPointTest[invtodd,
           ToddMap[{x, y, z}], {x, y, z}]
```

Obviously, all principal minors `d1[2]`, `d2[2]`, and `d3[2]` are positive and so the invariant attains its minimum value at the second equilibrium point.

The point number 1 is $\left\{1 \quad \sqrt{1+A}, 1 \quad \sqrt{1+A}, 1 \quad \sqrt{1+A}\right\}$

Principal Minors of the Hessian for this point:

$$d1[1] = \frac{2A^2 + 36\left(\quad 1+\sqrt{1+A}\right) + 6A\left(\quad 5+2\sqrt{1+A}\right)}{\left(\quad 1+\sqrt{1+A}\right)^5}$$

$$d2[1] =$$

$$\frac{1}{\left(\quad 1+\sqrt{1+A}\right)^{10}}\left(4A^4 + A^2\left(2008 \quad 854\sqrt{1+A}\right) + A^3\left(263 \quad 48\sqrt{1+A}\right)\right.$$

$$\left. 2430\left(\quad 1+\sqrt{1+A}\right) \quad 243A\left(\quad 17+12\sqrt{1+A}\right)\right)$$

$$d3[1] = \frac{1}{\left(-1+\sqrt{1+A}\right)^{15}} \left(2\left(-4A^6 + 72900\left(-1+\sqrt{1+A}\right) + 3A^5\left(-203+24\sqrt{1+A}\right) + 3645A\left(-53+43\sqrt{1+A}\right) + 243A^2\left(-752+467\sqrt{1+A}\right) + A^4\left(-12048+3191\sqrt{1+A}\right) + A^3\left(-73894+32189\sqrt{1+A}\right)\right)\right)$$

The point number 2 is $\left\{1+\sqrt{1+A},\ 1+\sqrt{1+A},\ 1+\sqrt{1+A}\right\}$

Principal Minors of the Hessian for this point:

$$d1[2] = \frac{2\left(A^2 + 18\left(1+\sqrt{1+A}\right) + 3A\left(5+2\sqrt{1+A}\right)\right)}{\left(1+\sqrt{1+A}\right)^5}$$

$$d2[2] = \frac{1}{\left(1+\sqrt{1+A}\right)^{10}}\left(4A^4 + 2430\left(1+\sqrt{1+A}\right) + 243A\left(17+12\sqrt{1+A}\right) + A^3\left(263+48\sqrt{1+A}\right) + 2A^2\left(1004+427\sqrt{1+A}\right)\right)$$

$$d3[2] = \frac{1}{\left(1+\sqrt{1+A}\right)^{15}}\left(2\left(4A^6 + 72900\left(1+\sqrt{1+A}\right) + 3A^5\left(203+24\sqrt{1+A}\right) + 3645A\left(53+43\sqrt{1+A}\right) + 243A^2\left(752+467\sqrt{1+A}\right) + A^4\left(12048+3191\sqrt{1+A}\right) + A^3\left(73894+32189\sqrt{1+A}\right)\right)\right)$$

This is the second fixed point, and a corresponding Lyapunov function.

```
In[43]:= {x0, y0, z0} = fplist[[2]]
```

Out[43]= $\left\{1+\sqrt{1+A},\ 1+\sqrt{1+A},\ 1+\sqrt{1+A}\right\}$

```
In[44]:= Lyap = invartodd - (invartodd /.
              {x → x0, y → y0 , z → z0 })
```

Out[44]= $$-\frac{\left(2+\sqrt{1+A}\right)^3\left(3+A+3\sqrt{1+A}\right)}{\left(1+\sqrt{1+A}\right)^3} + \frac{(1+x)\ (1+y)\ (1+z)\ (A+x+y+z)}{x\,y\,z}$$

The signs of `d1[1]`, `d2[1]`, and `d3[1]` must be analyzed to determine if the equilibrium `fp[[1]]` is a point of local minimum or local maximum or neither. This is left as an exercise.

We now proceed to use *Dynamica* to simulate the semicycle analysis of solutions. In particular we find maxima and minima of positive and negative semicycles of solutions.

We begin by taking initial conditions to be arbitrary constants. Here are the first eight terms of the orbit $\gamma(a, b, c)$ when $A = 1$. Clearly, every solution is periodic with period eight.

```
In[45]:= A = 1;

In[46]:= Simplify[Orbit[
             ToddMap, {a, b, c}, 8]]
```

$$Out[46]= \Big\{\{a, b, c\}, \Big\{b, c, \frac{1+b+c}{a}\Big\}, \Big\{c, \frac{1+b+c}{a}, \frac{1+a+b+c+ac}{ab}\Big\}, \Big\{\frac{1+b+c}{a}, \frac{1+a+b+c+ac}{ab}, \frac{(1+a+b)(1+b+c)}{abc}\Big\}, \Big\{\frac{1+a+b+c+ac}{ab}, \frac{(1+a+b)(1+b+c)}{abc}, \frac{1+a+b+c+ac}{bc}\Big\}, \Big\{\frac{(1+a+b)(1+b+c)}{abc}, \frac{1+a+b+c+ac}{bc}, \frac{1+a+b}{c}\Big\}, \Big\{\frac{1+a+b+c+ac}{bc}, \frac{1+a+b}{c}, a\Big\}, \Big\{\frac{1+a+b}{c}, a, b\Big\}, \{a, b, c\}\Big\}$$

This is a simulation for the case $A = 3$.

```
In[47]:= A = 3;

In[48]:= TimeSeriesPlot[ToddMap,
           {2., 1.7, 1.}, 70,
           AxesOrigin → {0, fp[[2, 1]]},
           Axes → True]
```

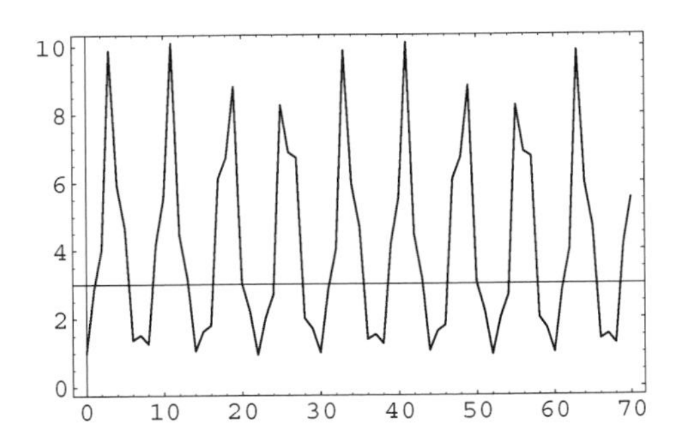

Semicycles with respect to the equilibrium fp[[2]] are found with SemicycleTest

```
In[49]:= SemicycleTest[
        Transpose[Orbit[ToddMap,
            {2.,1.7,1.},70]][[2]],
            fp[[2,1]]]
```

```
Positions and Maxima of Positive Semicycles
5, {5, 9.87941}, {13, 10.094}, {21, 8.81177},
{27, 8.27019}, {35, 9.869}, {43, 10.1},
{51, 8.83078}, {57, 8.24915}, {65, 9.85843}
Positions and Minima of Negative Semicycles
{2, 1.}, {8, 1.37677}, {10, 1.26402}, {16, 1.04926},
{24, 0.93463}, {32, 0.998387}, {38, 1.38153},
{40, 1.26007}, {46, 1.05145}, {54, 0.934889},
{62, 0.996795}, {68, 1.38631}, {70, 1.25615}
Lengths of Positive Semicycles
{4, 5, 4, 3, 4, 5, 4, 3, 4, 1}
Lengths of Negative Semicycles
{-3, -3, -3, -4, -4, -3, -3, -4, -4, -3}
Positions and Lengths of both Positive and Negative Semicycles
{-3, 4, -3, 5, -3, 4, -4, 3, -4, 4, -3, 5, -3, 4, -4,
3, -4, 4, -3, 1}
```

Another simulation yields the following time series, which looks complicated and may be chaotic. Chaotic behavior of solutions of Todd's equation and some parameter ranges has not been established theoretically.

```
In[50]:= A = 8.;

In[51]:= TimeSeriesPlot[ToddMap,
        {2.,4.7,1.},100]
```

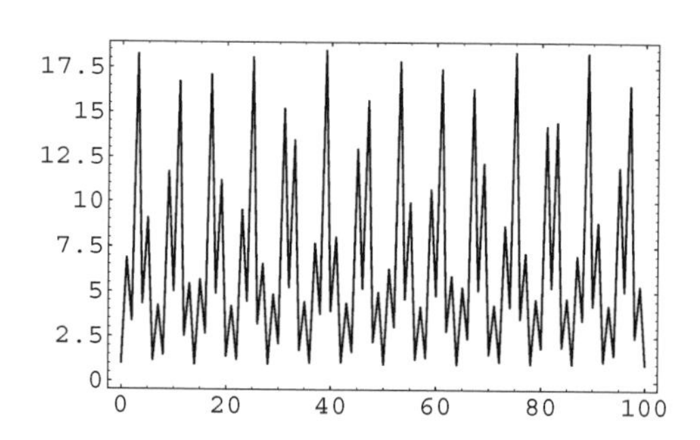

If many simulations of this kind are performed, one will find out that the semicycles of all solutions after the first semicycle are of length $1, 2, 3, 4$, or 5 and that the maxima of positive semicycles and minima of negative semicycles have special locations. See [KoL], pp. 141–143 for a proof. This fact can be used to obtain further global properties of the solutions of this equation.

The complexity of behavior of Todd's equation for different values of parameter A can be seen from the bifurcation diagrams discussed in Chapters 1 and 2.

This is a bifurcation diagram for Todd's equation.

```
In[52]:= Clear[A];

In[53]:= BifurcationPlotND[ToddMap,
           {A, 0.1, 6.1}, {1.2, 1., 4.},
           FirstIt → 400,
           Iterates → 150,
           Steps → 300];
```

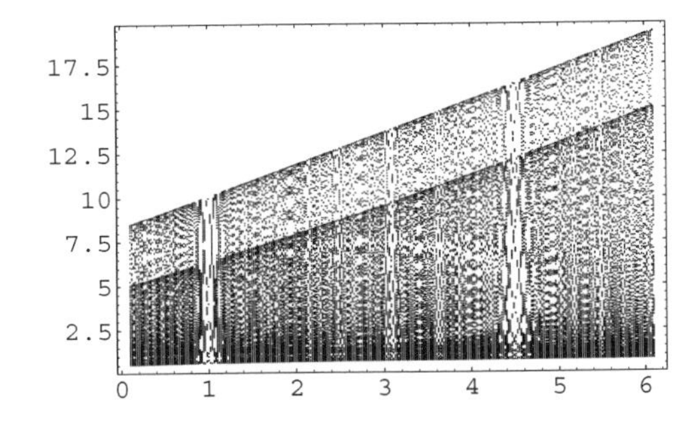

Zooming around the value $A = 1$, we see that we have a period eight solution for this value of parameter. However, the dynamics of this equation for most of the other values of parameter is not known.

```
In[54]:= Clear[A];

In[55]:= BifurcationPlotND[ToddMap,
           {A, 0.998, 1.002}, {1.2, 1., 4.},
           FirstIt → 400,
           Iterates → 150,
           Steps → 300];
```

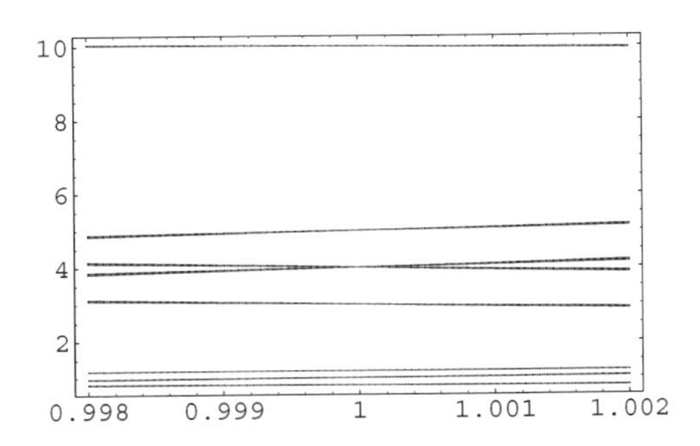

5.5.1 Summary of Results and Open Problems.

In this section we give an overview of the results about Todd's equation that were obtained so far. We assume that the parameter A and the initial values are positive.

1. Both equilibrium points of Todd's equation are nonhyperbolic equilib-

rium points.

2. Todd's equation possesses an invariant (established in [KLR]).

3. The positive equilibrium of Todd's equation is stable (established in [Ku]).

4. Every positive solution of Todd's equation, with $A = 1$, is periodic with period 8 (established in [To]).

5. Every positive nontrivial solution of Todd's equation is strictly oscillatory about the positive equilibrium. Furthermore, every semicycle of a nontrivial solution contains at most five terms (established in [KLR]).

The following research problems are suggested in part by our work in this chapter.

1. Find all possible periods of solutions of Todd's equation for different values of parameter A.

2. Show that no nontrivial solution of Todd's equation has a limit.

3. Find the values of A for which the solution of Todd's equation is chaotic.

4. Find an invariant set in negative octant $\{(x, y, z) : x < 0, y < 0, z < 0\}$ that contains the equilibrium point.

5. Investigate the existence of solutions and the stability of the equilibrium points for negative values of parameters and initial conditions.

5.6 Biology Applications Projects

Problem 1.

The discrete-time *SIR* epidemic model divides the population into three subgroups: susceptibles (S), infectives (I), and removed or isolated (R). The system of difference equations has the form

$$\begin{aligned} S_{n+1} &= S_n\left(1 - \alpha\tfrac{\Delta t}{N}I_n\right) \\ I_{n+1} &= I_n\left(1 - \gamma\Delta t + \alpha\tfrac{\Delta t}{N}S_n\right) \\ R_{n+1} &= R_n + \gamma\Delta t I_n, \end{aligned} \tag{5.32}$$

with positive initial conditions $S_0 > 0, I_0 > 0$, and $R_0 \geq 0$ satisfying $S_0+I_0+R_0 = N$, where $\gamma > 0$ is the probability that one infective will be removed from the infection process during a unit time interval (relative removal rate). Unlike the *SI* model (2.107), individuals in the *SIR* model recover from the disease and become permanently immune (*R* subgroup). *SIR* epidemic models were considered in several papers, see [Al] and [AB] and references therein. The parameter $\alpha > 0$ is the contact rate as in the *SI* model (2.107). Obviously, the sum of susceptibles, infectives, and removed remains constant throughout time and equals the total population, i.e.,

$$S_{n+1} + I_{n+1} + R_{n+1} = S_n + I_n + S_n = \ldots = S_0 + I_0 + R_0, \quad n = 0, 1, \ldots. \tag{5.33}$$

If we assume that $\max\{\alpha\Delta t, \alpha\Delta t\} \leq 1$, then solutions to (5.32) are positive.

The global behavior of system (5.32) is determined by the value of the reproductive rate $R = \frac{S_0\alpha}{N\gamma}$.

Perform the following analysis for this system:

1. Find equilibrium points.
2. Check the linear stability of the equilibrium points.
3. Discuss the global behavior of the solutions of (5.32) when $R \leq 1$ and $R > 1$.
4. Use *Dynamica* to visualize trajectories and plot bifurcation diagrams.

Problem 2.

The discrete-time *SIR* epidemic model divides the population into three subgroups: susceptibles (S), infectives (I), and removed or isolated (R). The system of difference equations has the form

$$\begin{aligned} S_{n+1} &= (1 - \lambda_n)S_n + \beta(N - S_n) \\ I_{n+1} &= (1 - (\gamma + \beta))I_n + \lambda_n S_n \\ R_{n+1} &= (1 - \beta)R_n + \gamma I_n, \end{aligned} \tag{5.34}$$

with positive initial conditions $S_0 > 0, I_0 > 0$, and $R_0 \geq 0$ satisfying $S_0 + I_0 + R_0 = N$, where $\alpha > 0, \beta > 0$ and $\gamma > 0$. The force of infection λ_n has the same meaning as in the *SIS* model (2.111), and satisfies some conditions given in [AB]. This model is a generalization of (5.32). The global behavior of system (5.34) is determined by the value of the basic reproductive number $R_0 = \frac{\alpha+\beta}{\gamma}$. This number is defined as the average number of secondary infections caused by one infected individuals during his/her infectious period in an entirely susceptible population, see [AB].

Perform the following analysis for this system:

1. Find equilibrium points.

2. Check the linear stability of the equilibrium points.

3. Discuss global behavior of solutions of (5.34) when $R_0 \leq 1$. Show that every solution satisfies

$$(S_n, I_n, R_n) \to (N, 0, 0) \quad \text{as } n \to \infty.$$

4. Discuss global behavior of solutions of (5.34) when $R_0 > 1$. Show that every solution satisfies

$$(S_n, I_n, R_n) \to (\bar{S}, \bar{I}, \bar{R}) \quad \text{as } n \to \infty,$$

   where $\bar{S} > 0, \bar{I} > 0$, and $\bar{R} > 0$.

5. Find conditions for system (5.34) to have period-two and period-three solutions.

6. Use *Dynamica* to visualize trajectories and plot bifurcation diagrams.

5.7 Exercises

Use *Dynamica* to analyze the following equations and systems:

Exercise 5.1 $x_{n+1} = A(1 - x_n)x_{n-1} - Bx_{n-2}$, $n = 0, 1, \ldots$ for different values of parameters A and B.

Exercise 5.2 $x_{n+1} = \frac{A+x_n}{B+x_{n-1}+x_{n-2}}$, $n = 0, 1, \ldots$ for different values of parameters A and B.

Exercise 5.3 Euler discretization of Lorenz' system (here s, r, and b are parameters).

$$x_{n+1} = (1 - s)x_n + sy_n$$

$$y_{n+1} = rx_n - x_n z_n$$

$$z_{n+1} = x_n y_n + (1 - b)z_n.$$

Exercise 5.4 Euler discretization of Rossler's system (here A is a parameter).

$$x_{n+1} = x_n - y_n + z_n$$

$$y_{n+1} = x_n + 1.2y_n$$

$$z_{n+1} = 0.2 + (x_n - a)z_n.$$

Exercise 5.5
Use *Dynamica* to analyze the equation

$$x_{n+1} = \frac{\alpha + \gamma x_{n-1}}{Cx_{n-1} + Dx_{n-2}},$$

where all coefficients are nonnegative and the initial conditions are positive. In the case when all coefficients are positive, show that there exists a unique period-two solution. Use *Mathematica* to prove that the period-two solution is locally asymptotically stable. See [EGL].

Exercise 5.6
Use *Dynamica* to analyze the equation

$$x_{n+1} = \frac{\alpha + \delta x_{n-2}}{A + Bx_n},$$

where all coefficients are nonnegative and the initial conditions are positive.

Exercise 5.7
Use *Dynamica* to analyze the equation

$$x_{n+1} = \frac{\beta x_n + \gamma x_{n-1}}{A + Dx_{n-2}},$$

where all coefficients are nonnegative and the initial conditions are positive. Find the initial conditions that generate unbounded solutions.

Exercise 5.8
Use *Dynamica* to analyze the equation

$$x_{n+1} = \frac{\alpha + \gamma x_{n-1}}{A + Dx_{n-2}},$$

where all coefficients are nonnegative and the initial conditions are positive. Find the condition that guarantees the existence of a period-two solution. Find initial conditions that generate unbounded solutions. See [AKL].

Chapter 6

Fractals Generated by Iterated Function Systems

6.1 Introduction

The goal in this chapter is to explore iterated function systems and their attractors. The word **fractal** was coined from the Latin word *fractus* by B. Mandelbrot in 1977 [Mt]. The meaning of the corresponding Latin verb *frangere* is "to break." Intuitively, a fractal is a geometrical object that consists of a motif repeating itself on a reducing scale, i.e., self-similar. Before giving a formal definition we present some simple examples of fractals.

Example 6.1 (The Middle Third Cantor Set) The Cantor set was introduced in Section 1.8.1, where we saw that it has the property of self-similarity. We also saw that the Cantor set is in some sense "large" (uncountable), and in another sense is small (it's length is zero).

There are other ways of measuring the size of sets such as the box dimension introduced in Definition 2.13. The *Middle Third Cantor Set* C is made up of two equal line segments scaled by a factor 1/3. Thus $N(r) = 2, r = 1/3$ in the formula

$$\text{BoxDimension}(C) = D_B(C) = \lim_{r\to 0} \frac{\ln N(r)}{\ln(1/r)},$$

which gives

$$D_B(C) = \frac{\ln 2}{\ln 3} \approx 0.6309....$$

Another way to measure the size of a set is the geometric or topological dimension. Here we think about the topological dimension as the minimal number of coordinates needed to locate points of the set. For example, the Cantor set has geometric dimension zero since it does not contain any line segments. A formal definition will be given in Section 6.2. ∎

Example 6.2 (The Sierpinski Triangle) The Sierpinski triangle, introduced by W. Sierpinski in 1916 [S], may be considered as a two-dimensional version of the Cantor set. The construction is as follows: start with an equilateral triangle with sides of length a which is subdivided into four identical triangles obtained by connecting the midpoints of the three sides of the triangle. Then, remove the middle triangle leaving three smaller triangles with sides of length 1/2. The same procedure is repeated on each of the three triangles to obtain 9 equilateral triangles with sides of length $(\frac{1}{2})^2$. Continue with this process so that in the n-th step 3^n equilateral triangles are produced with sides of length $(\frac{1}{2})^n$. If these steps are continued indefinitely, one obtains the Sierpinski triangle as the limit. Note that the Sierpinski triangle is self-similar since every part of it is similar to the whole set.

Let us calculate the sum of the areas of the triangles that have been removed in the construction of the Sierpinski triangle. Let A be the area of the original triangle. In the first step $\frac{1}{4}A$ is removed, and in the second step three triangles are removed, each with area $\left(\frac{1}{4}\right)^2 A$, and so on. The sum of the areas of the removed triangles is

$$\frac{1}{4}A + 3\left(\frac{1}{4}\right)^2 A + 3^2\left(\frac{1}{4}\right)^3 A + \cdots = A.$$

Thus, the Sierpinski triangle has a zero area.

Let us measure the box dimension of the Sierpinski triangle. This set is made up of three equal triangles scaled by a factor 1/2. Hence $N(r) = 3, r = 1/2$ in the formula

$$\text{BoxDimension}(S) = D_B(S) = \lim_{r \to 0} \frac{\ln N(r)}{\ln(1/r)}$$

which gives

$$D_B(S) = \frac{\ln 3}{\ln 2} \approx 1.585....$$

▮

Example 6.3 (The von Koch Curve) The von Koch Curve was introduced by von Koch in 1904 [Kc]. This set is obtained by removing the middle third of a segment of length L and replacing it with two sides of an equilateral triangle. The length of the resulting object is $\frac{4}{3}L$. In the second step the same procedure is repeated with each of the four line segments. The resulting geometrical object has 16 line segments, each of length $\left(\frac{1}{3}\right)^2 L$. The total length is $\left(\frac{4}{3}\right)^2 L$. The limiting curve obtained when this process is continued indefinitely is the von Koch curve. Note that the von Koch curve is self-similar since every part of it is similar to the whole set and the length of this curve is ∞ since the length of the curve at the n-th step is $\left(\frac{4}{3}\right)^n L$.

To obtain the box dimension, note that the von Koch curve consists of four equal line segments scaled by a factor 1/3. Hence $N(r) = 4, r = 1/3$ in the formula

$$\text{BoxDimension}(S) = D_B(S) = \lim_{r \to 0} \frac{\ln N(r)}{\ln(1/r)}$$

which gives

$$D_B(S) = \frac{\ln 4}{\ln 3} \approx 1.26\ldots.$$

∎

6.2 Basic Definitions and Results

In this section we introduce the definition of a fractal. In Section 6.1, we presented examples of sets that have a noninteger box dimension. In addition, all the sets considered in Section 6.1 share the property of self-similarity. Self-similarity is not characteristic of fractals as many nonfractals, such as lines, planes, etc. have the same property. Moreover there are fractals that do not have the property of self-similarity. According to Mandelbrot, a fractal is a set whose fractal dimension is strictly greater than its topological dimension. Now, we present formal definitions of different types of dimensions.

DEFINITION 6.1 *A set A has a* **topological dimension** 0 *if every point in A has a neighborhood whose boundary does not intersect A. If $k > 0$, a set A has* **topological dimension** k *if every point in A has a neighborhood whose boundary intersects A in a set of topological dimension $k-1$, and k is the least positive integer for which this holds. The topological dimension of a set A is denoted by $D_t(A)$.*

According to Definition 6.1, the topological dimension of the Cantor set C is 0, as it does not contain any line segments. Obviously, the set of rational numbers and the set of irrational numbers are both of topological dimension zero. Thus, the set of real numbers whose topological dimension is 1 is the union of two sets whose topological dimensions are 0.

Next, we define the box dimension of an arbitrary set.

DEFINITION 6.2 *If $a_1, a_2, \ldots, a_k$ are elements of R, the set*

$$\left\{(x_1, x_2, \ldots, x_k) \,:\, a_i \le x_i \le a_i + d, \quad i = 1, \ldots, k \quad \textit{for some } r > 0.\right\}$$

*is a k-**dimensional box** of side length r. Let $N(r)$ be the smallest number of k-dimensional boxes of side length r needed to cover a set $A \subset R^k$. Then the box dimension of A is given by*

$$D_B(A) \;=\; \lim_{r\to\infty} \frac{\ln N(r)}{\ln\left(\frac{1}{r}\right)}. \tag{6.1}$$

It can be shown [F] that for any set A in R^k,

$$D_t(A) \leq D_B(A). \tag{6.2}$$

For most of well-known geometric objects, such as lines, disks, cubes, etc., $D_t(A) = D_B(A)$. Sets for which $D_t(A)$ is strictly less than $D_B(A)$ are called fractals.

DEFINITION 6.3 *A set $A \subset R^k$ is called a* **fractal** *if*

$$D_t(A) < D_B(A).$$

As we have seen already, $0 = D_t(C) < D_B(C) = \frac{\ln 2}{\ln 3}$, for the Cantor set C.

Now, we shall define a class of fractals, which can be easily constructed by the recursive process from an original set, with self-similarity properties that can be utilized to calculate their box dimension. Intuitively, such a class does not seem to be "large," but there is a surprising result that shows that practically any set can be approximated, in some sense, with one or more sets from this class. To make these statements precise we introduce some definitions and give some basic results in the next subsection.

6.3 Iterated Function Systems

In this section we give the mathematical foundation of the special class of fractals called the iterated function system, which was explored in great details by Barnsley [B]. Most of the work will be done in the two-dimensional case R^2 and it will have a straightforward extension to the n-dimensional case R^n. We begin by recalling basic definitions and results concerning linear transformation on the plane.

We say that a map $\mathbf{L} : R^2 \to R^2$ is a **linear transformation** if

$$\mathbf{L}(\alpha \mathbf{v}_1 + \beta \mathbf{v}_2) = \alpha \mathbf{L}(\mathbf{v}_1) + \beta \mathbf{L}(\mathbf{v}_2), \quad \text{for all } \mathbf{v}_1, \mathbf{v}_2 \in R^2, \quad \text{and all } \alpha, \beta \in R. \tag{6.3}$$

As is well known, every linear transformation $\mathbf{L} : R^2 \to R^2$ may be represented by a 2×2 matrix A, i.e.,

$$\mathbf{L}\begin{pmatrix} x \\ y \end{pmatrix} = A\begin{pmatrix} x \\ y \end{pmatrix} = \begin{pmatrix} a & b \\ c & d \end{pmatrix}\begin{pmatrix} x \\ y \end{pmatrix} = \begin{pmatrix} ax + by \\ cx + dy \end{pmatrix}$$

Here are simple geometric facts about the effect of the linear transformation. First, we rewrite the matrix A in the form

$$A = \begin{pmatrix} a & b \\ c & d \end{pmatrix} = \begin{pmatrix} r\cos\theta & -s\sin\phi \\ r\sin\theta & s\cos\phi \end{pmatrix}, \tag{6.4}$$

where

$$r = \sqrt{a^2 + c^2} \quad \cos\theta = \tfrac{a}{r}$$

$$s = \sqrt{b^2 + d^2} \quad \cos\phi = -\tfrac{b}{s}.$$

We now look at some special cases.

Example 6.4

(1) $r = a, s = b, \theta = \phi = 0$. In this case the matrix A and the transformation L take the form

$$A = \begin{pmatrix} a & 0 \\ 0 & b \end{pmatrix}, \qquad L\begin{pmatrix} x \\ y \end{pmatrix} = \begin{pmatrix} ax \\ by \end{pmatrix}.$$

Geometrically, this transformation is a scaling. Precisely, it is a contraction (a dilation) in the x direction if $0 < a < 1$ $(a > 1)$ and a contraction (the dilation) in the y direction if $0 < b < 1$ $(b > 1)$. If any of these two coefficients are negative, then in addition to contraction or dilation we also have a reflection with respect to the x and/or y axis.

(2) $r = s = 1, \theta = \phi$. In this case the matrix A and the transformation L take the form

$$A = \begin{pmatrix} \cos\theta & -\sin\theta \\ \sin\theta & \cos\theta \end{pmatrix}, \qquad L\begin{pmatrix} x \\ y \end{pmatrix} = \begin{pmatrix} x\cos\theta - y\sin\theta \\ x\sin\theta + y\cos\theta \end{pmatrix}.$$

Geometrically, this transformation is a counterclockwise rotation by an angle θ.

(3) $r = s, \theta = \phi$. In this case the matrix A and the transformation L take the form

$$A = r\begin{pmatrix} \cos\theta & -\sin\theta \\ \sin\theta & \cos\theta \end{pmatrix}, \qquad L\begin{pmatrix} x \\ y \end{pmatrix} = r\begin{pmatrix} x\cos\theta - y\sin\theta \\ x\sin\theta + y\cos\theta \end{pmatrix}.$$

Geometrically, this transformation is a counterclockwise rotation by an angle θ and a scaling by a factor r.

∎

Based on Example 6.4 we see that a linear transformation in the plane may be formed by a combination of scalings, rotations, and reflections. It turns out that these transformations are the only building blocks of linear transformations [B].

Given a 2×2 matrix A and a vector $(x, y)^T$, the map $\mathbf{F} : R^2 \to R^2$ such that

$$\mathbf{F}\begin{pmatrix} x \\ y \end{pmatrix} = A\begin{pmatrix} x \\ y \end{pmatrix} + \begin{pmatrix} t_1 \\ t_2 \end{pmatrix} = \begin{pmatrix} ax + by + t_1 \\ cx + dy + t_2 \end{pmatrix}. \tag{6.5}$$

is an **affine** linear transformation.

Thus an affine linear transformation is simply a sum of a linear transformation A and a translation determined by the vector $(x, y)^T$.

For the following definition we need the fact that the plane R^2 is a metric space with metric

$$d(\mathbf{x}, \mathbf{y}) = \|\mathbf{x} - \mathbf{y}\|$$

DEFINITION 6.4 *A map* $\mathbf{F} : A \to A$, $A \subset R^2$ *is said to be a* **contraction** *if there exists* $0 < c < 1$ *such that:*

$$d(\mathbf{F}(\mathbf{x}), \mathbf{F}(\mathbf{y})) \le cd(\mathbf{x}), \mathbf{y}), \quad \textit{for all} \mathbf{x}, \mathbf{y} \in A. \tag{6.6}$$

The constant c is called **contraction constant** *or* **contraction factor**.

Now we have all the ingredients needed to define the iterated function system. Let us recall that in each of Examples 6.1 to 6.3 the limiting set was defined by using 2, 3, and 4 contractions, respectively. Let H be the collection of all closed and bounded subsets of R^2. Let

$F_1, F_2, \ldots, F_m$ be a family of contractions defined on R^2. Then the function F defined by

$$F(A) = F_1(A) \cup F_2(A) \cup \ldots \cup F_m(A), \quad A \in H \tag{6.7}$$

is called **union of the functions** (or **Hutchinson operator** of) $F_1, F_2, \ldots, F_m$ [Hu]. If the sets $F_1(A), F_2(A), \ldots, F_m(A)$ are pairwise disjoint (with the possible exception of the boundaries) and $A = F(A)$, then A is said to be **self-similar**.

Finally, we define the notion of iterated function system (IFS).

DEFINITION 6.5 *Let* $F_1, F_2, \ldots, F_m$ *be a family of contractions defined on* R^2 *and A a closed bounded subset of* R^2. *Then the system* $\{A, F\} = \{A, \cup_{i=1}^m F_i\}$ *is called an* **iterated function system** *(IFS).*

In the sequel, we will show that the union map or Hutchinson operator is a contraction map on an appropriate space with respect to a specific distance that will be introduced. Then we will use the powerful Banach contraction principle to show that the sequence of iterates of such mapping converges to the unique fixed point irrespective of the initial point of this mapping. In other words $F^n(S) \to A_F$ for every S in the appropriate space and $F(A_F) = A_F$. Since A_F is a set, the relation $F(A_F) = A_F$ means that A_F is an **invariant set** or **attractor set**. Before defining basic notions let us first revisit examples 6.1 to 6.3.

Example 6.5 Set $F_1(x) = \frac{1}{3}x$ and $F_2(x) = \frac{1}{3}x + \frac{2}{3}$. Then iterates of the IFS $A = \{[0, 1], F_1 \cup F_2\}$ generate the Middle Third Cantor Set C as a limit: $F^n(A) \to C$ Obviously $F(C) = C$. An interesting new feature is that the attractor set C does not depend on the initial set. So instead of $[0, 1]$ we could use $[0, 1/2]$

or the set of rationals or even a single point. *Dynamica* may be used to visualize this phenomenon. Consequently, the Middle Third Cantor Set is completely determined by two contractions F_1 and F_2, that is, by the 2×2 table of the coefficients shown in Table 6.1. Table 6.1 can be considered as the code for the Middle Third Cantor Set C. Thus, a very complex set such as the Cantor set C can be coded with a small number of data. This fact has important implications in the construction of one of the most powerful image compression methods know as the *fractal image compression method*, see [BHu]. ∎

Table 6.1 Coefficients of contractions of the Middle Third Cantor Set

F_1	$\frac{1}{3}$	0
F_2	$\frac{1}{3}$	$\frac{2}{3}$

Example 6.6 Set

$$F_1\begin{pmatrix} x \\ y \end{pmatrix} = \begin{pmatrix} \frac{1}{2} & 0 \\ 0 & \frac{1}{2} \end{pmatrix}\begin{pmatrix} x \\ y \end{pmatrix}$$

$$F_2\begin{pmatrix} x \\ y \end{pmatrix} = \begin{pmatrix} \frac{1}{2} & 0 \\ 0 & \frac{1}{2} \end{pmatrix}\begin{pmatrix} x \\ y \end{pmatrix} + \begin{pmatrix} \frac{1}{2} \\ 0 \end{pmatrix} \tag{6.8}$$

$$F_3\begin{pmatrix} x \\ y \end{pmatrix} = \begin{pmatrix} \frac{1}{2} & 0 \\ 0 & \frac{1}{2} \end{pmatrix}\begin{pmatrix} x \\ y \end{pmatrix} + \begin{pmatrix} \frac{1}{4} \\ \frac{\sqrt{3}}{4} \end{pmatrix},$$

and consider the IFS $\{\triangle, F = F_1 \cup F_2 \cup F_3\}$, where $\triangle$ is a solid equilateral triangle. In the limiting process, this IFS generates the Sierpinski triangle $S_\triangle$. More precisely,

$$\lim_{n\to\infty} F^n(\triangle) = S_\triangle.$$

Obviously, $F(S_\triangle) = S_\triangle$. In view of the above-mentioned Banach contraction principle the limiting attractor set $S_\triangle$ is independent of the initial set, which is in our case $\triangle$. Thus, the initial set can be any point or segment or square, etc. Consequently, the Sierpinski triangle $S_\triangle$ is generated by three contractions F_1, F_2 and F_3, all being the affine transformations, and can be coded as the list of numbers shown in Table 6.2.

Table 6.2 Coefficients of contractions of the Sierpinski Triangle

	a	b	c	d	t_1	t_2
F_1	$\frac{1}{2}$	0	0	$\frac{1}{2}$	0	0
F_2	$\frac{1}{2}$	0	0	$\frac{1}{2}$	$\frac{1}{2}$	0
F_3	$\frac{1}{2}$	0	0	$\frac{1}{2}$	$\frac{1}{4}$	$\frac{\sqrt{3}}{4}$

Again we obtain that a very complex set can be coded with a small number of data.

In the *Dynamica* session in Section 6.6 we will visualize the fact that the attractor set does not depend on the initial set. ∎

Example 6.7 Our next example is Koch's curve which was obtained by repetitive application of four transformations, two of which included rotations. The first transformation F_1 is a contraction by a factor 1/3, the second one F_2 is the combination of a contraction by a factor 1/3, rotation counterclockwise by an angle $\pi/3$, and then translation by the vector $(\frac{1}{3}, 0)^T$.

The third transformation F_3 is similar combination of the transformations as F_2 where the rotation is performed clockwise by $\pi/3$, and translation is done by the vector $(\frac{1}{2}, \frac{\sqrt{3}}{6})^T$.

Finally, the fourth transformation F_4 is a contraction by a factor 1/3 and then translation by the vector $(\frac{2}{3}, 0)^T$.

These four transformations are all affine and can be written in matrix form as follows.

$$F_1\begin{pmatrix}x\\y\end{pmatrix}=\begin{pmatrix}\frac{1}{3}&0\\0&\frac{1}{3}\end{pmatrix}\begin{pmatrix}x\\y\end{pmatrix}$$

$$\begin{aligned}F_2\begin{pmatrix}x\\y\end{pmatrix}&=\begin{pmatrix}\frac{1}{3}\cos\frac{\pi}{3}&-\frac{1}{3}\sin\frac{\pi}{3}\\\frac{1}{3}\sin\frac{\pi}{3}&\frac{1}{3}\cos\frac{\pi}{3}\end{pmatrix}\begin{pmatrix}x\\y\end{pmatrix}+\begin{pmatrix}\frac{1}{3}\\0\end{pmatrix}\\&=\begin{pmatrix}\frac{1}{6}&-\frac{\sqrt{3}}{6}\\\frac{\sqrt{3}}{6}&\frac{1}{6}\end{pmatrix}\begin{pmatrix}x\\y\end{pmatrix}+\begin{pmatrix}\frac{1}{3}\\0\end{pmatrix}\end{aligned}$$

$$\begin{aligned}F_3\begin{pmatrix}x\\y\end{pmatrix}&=\begin{pmatrix}\frac{1}{3}\cos(-\frac{\pi}{3})&-\frac{1}{3}\sin(-\frac{\pi}{3})\\\frac{1}{3}\sin(-\frac{\pi}{3})&\frac{1}{3}\cos(-\frac{\pi}{3})\end{pmatrix}\begin{pmatrix}x\\y\end{pmatrix}+\begin{pmatrix}\frac{1}{2}\\\frac{\sqrt{3}}{6}\end{pmatrix}\\&=\begin{pmatrix}\frac{1}{6}&\frac{\sqrt{3}}{6}\\-\frac{\sqrt{3}}{6}&\frac{1}{6}\end{pmatrix}\begin{pmatrix}x\\y\end{pmatrix}+\begin{pmatrix}\frac{1}{2}\\\frac{\sqrt{3}}{6}\end{pmatrix}\end{aligned}\tag{6.9}$$

$$F_4\begin{pmatrix}x\\y\end{pmatrix}=\begin{pmatrix}\frac{1}{3}&0\\0&\frac{1}{3}\end{pmatrix}\begin{pmatrix}x\\y\end{pmatrix}+\begin{pmatrix}\frac{2}{3}\\0\end{pmatrix}$$

The IFS $\{[0,1], F_1 \cup F_2 \cup F_3 \cup F_4\}$ generates the Koch curve K_C. Obviously, $F(K_C) = K_C$. Similarly, as in the cases of Cantor set and Sierpinski triangle, the Banach contraction principle implies that the limiting attractor set K_C is independent of the initial set, which is in our case $[0, 1]$. Consequently, the Koch curve K_C is generated by four contractions F_1, F_2, F_3 and F_4, all of them affine transformations, and can be coded by the numbers shown in Table 6.3.

Table 6.3 Coefficients of contractions of the Koch curve

	a	b	c	d	t_1	t_2
F_1	$\frac{1}{3}$	0	0	$\frac{1}{3}$	0	0
F_2	$\frac{1}{6}$	$-\frac{\sqrt{3}}{6}$	$\frac{\sqrt{3}}{6}$	$\frac{1}{6}$	$\frac{1}{3}$	0
F_3	$\frac{1}{6}$	$\frac{\sqrt{3}}{6}$	$-\frac{\sqrt{3}}{6}$	$\frac{1}{6}$	$\frac{1}{2}$	$\frac{\sqrt{3}}{6}$
F_4	$\frac{1}{3}$	0	0	$\frac{1}{3}$	$\frac{2}{3}$	0.

Just as before, we have a complex set that may be coded with a small set of data. ∎

6.4 Basic Results on Iterated Function Systems

In this section we present definitions and basic results that make rigorous some of our intuitive reasoning from motivating examples. Results will be given for a general metric space (X, D), although we are especially interested in the case where $X = R^2$ as it contains the fractals we have considered so far.

Let H be the set of all subsets of X that are closed and bounded. We now proceed to define a metric D on H. First define the distance between a point $a \in X$ and a set $B \in H$ as

$$d(a, B) = \inf\{d(a, b) : b \in B\}.$$

For example, $d(7, \{1, 2, 4, 5\}) = 2$ if we define the distance between two points $x, y \in R$ to be $d(x, y) = |x - y|$. If the distance between two sets $A, B \in H$ was defined by

$$d(A, B) = \sup\{d(a, B) : a \in A\}$$

then we would face a problem because

$$d(\{1, 2, 4\}, \{6, 7, 11, 14\}) = 5 \quad \text{and} \quad d(\{6, 7, 11, 14\}, \{1, 2, 4\}) = 10.$$

Thus, even in a simple case, $d(A, B) \neq d(B, A)$. This problem can be easily fixed by taking the maximum of these two distances, which leads to the following definition.

DEFINITION 6.6 *Let $A, B \in H$. The* **Hausdorff distance** *between A and B is*

$$D_H(A, B) = \max\{\, d(A, B)\,,\; d(B, A)\,\}.$$

D_H is also called the **Hausforff metric**.

One can easily check that D_H is the distance in the sense of Definition 1.14. Thus (H, D_H) is a metric space. In order to formulate the Banach Contraction Principle additional notions must be defined.

DEFINITION 6.7 *A sequence $\{x_n\}$ in a metric space X is said to be a* **Cauchy sequence** *if for every $\epsilon > 0$ there exists a positive integer N such that $d(x_n, x_p) < \epsilon$ for every $n, p \geq N$.*

DEFINITION 6.8 *A metric space (X, d) is called a* **complete metric space** *if every Cauchy sequence in X converges to an element in X.*

An important property of (H, D_H) is that it is a complete metric space, see [B] and [F] for a proof. The following is a powerful result for complete metric spaces.

THEOREM 6.1 (Banach Contraction Principle)
Let $F : X \to X$ be a contraction mapping, with a contraction constant c, on a complete metric space (X, d). Then there exists a unique point $x^* \in X$ such that:

(1) x^* is a fixed point of F, i.e., $F(x^*) = x^*$.

(2) $\lim F^n(x_0) = x^*$ for every $x_0 \in X$. That is, x^* is a global attractor.

To apply this theorem successfully we have to check whether the union of contractions is a contraction itself. It turns out that this holds and that in fact there is an explicit formula for the contraction constant in this case.

THEOREM 6.2
Let $F_1, F_2, \ldots, F_m$ be the contractions with contraction constants $c_1, c_2, \ldots, c_m$, on H, then the union mapping $F = \bigcup_{i=1}^{m} F_i$ is a contraction, too, with a contraction constant $c = \max\{c_1, \ldots, c_m\}$.

Now we formulate one of the main results of this chapter.

THEOREM 6.3
Let $F_1, F_2, \ldots, F_m$ be contractions on H. Then, there exists a unique global attractor $A_F \in H$ for the union mapping $F = \bigcup_{i=1}^{m} F_i$, that is,

$$\lim_{n \to \infty} F^n(A) = A_F \tag{6.10}$$

for every initial set $A \in H$. Convergence is considered in the Hausdorff metric, which means that equation (6.10) should be read as

$$D_H(F^n(A), A_F) \to 0, \quad \text{as } n \to \infty.$$

Theorem 6.3 gives a theoretical justification for the statements we made earlier about the independence of the limiting fractal set from the initial set.

We have seen how simple codes can generate complex images of fractals generated by an IFS. It seems that the family of fractals that can be generated in such a way is still a small subset of the set of all closed and bounded sets in the plane, in a way similar to the case where all polynomials are a small subset of the set of continuous functions on a given closed interval. However, every continuous function on a given closed interval can be approximated uniformly by a polynomial, and so we may hope that our fractals possess a similar property. This is the content of the following theorem.

THEOREM 6.4 (The Collage Theorem)
Let $\{B, F_1, F_2, \ldots, F_m\}$ be an IFS, with contraction factors $c_1, \ldots, c_m$, and set

$$c = \max\{c_1, \ldots, c_m\}$$

Let $A_F \in H$ be the unique global attractor for the union mapping $F = \cup_{i=1}^{m} F_i$. If $\epsilon > 0$ is such that

$$D_H(B, F(B)) < \epsilon,$$

then

$$D_H(B, A_F) < \frac{\epsilon}{1-c}.$$

This result is an immediate consequence of one of the formulas in the proof of the Banach contraction principle and so theoretically was well known for a long time. The interesting thing is that Barnsley and Hurd [BHu] have developed a patented algorithm called the IFS compression which effectively implements this result.

6.5 Calculation of Box Dimension for IFS

One of the advantages of using an IFS is that the box dimension of the global attractor is often relatively easy to calculate or estimate in terms of the contraction constants of the generating contractions. For the sake of simplicity of exposition we discuss the special case where the contractions $F_1, F_2, \ldots, F_m$ are **similarities** on R^2, i.e.,

$$d_E(F_i(x), F_i(y)) = c_i\, d_E(x, y), \quad \text{for every } x, y \in R^2, \tag{6.11}$$

where $0 < c_i < 1$ for every $i = 1, \ldots, m$. Here d_E is the usual Euclidean distance

$$d_E(x, y) = d_E((x_1, x_2), (y_1, y_2)) = \sqrt{(y_1 - x_1)^2 + (y_2 - x_2)^2}.$$

A set that is invariant under such a family of similarities is called a **strictly self-similar set** [F]. All sets mentioned in Examples 6.1 through 6.3 are strictly self-similar sets. If the components $F_i(A)$ of A do not overlap "too much," then one can prove that the box dimension $D_B(A) = d$ satisfies a simple algebraic condition. The precise statements of the conditions and the theorem are the following.

DEFINITION 6.9 *A family of similarities $F_1, F_2, \ldots, F_m$ satisfies the* **open set condition** *if there exists a bounded set V such that*

$$F_i \cap F_j = \emptyset \quad \textit{whenever} \quad i \neq j$$

and

$$\cup_{i=1}^{m} F_i(V) \subset V$$

For example, in the case of the Middle Third Cantor Set we can choose $V = (0, 1)$.

THEOREM 6.5 (The Hutchinson Theorem)
Let $\{A, F_1, F_2, \ldots, F_m\}$ be an IFS, with contraction factors $c_1, \ldots, c_m$. Let $A_F \in H$ be a unique global attractor for the union mapping $F = \bigcup_{i=1}^{m} F_i$. Then the box dimension $D_B(A_F)$ satisfies

$$\sum_{i=1}^{m} c_i^d = 1. \tag{6.12}$$

In the case of Examples 6.1 to 6.3 it is clear that all conditions of Theorem 6.5 are satisfied and so we can use the formula (6.12) to calculate the box dimension. See Table 6.4.

Table 6.4

Global Attractor	$\sum_i c_i^d = 1$	d
Middle Third Cantor Set	$\left(\frac{1}{3}\right)^d + \left(\frac{1}{3}\right)^d = 1$	$\frac{\ln 2}{\ln 3}$
Sierpinski Triangle	$\left(\frac{1}{2}\right)^d + \left(\frac{1}{2}\right)^d + \left(\frac{1}{2}\right)^d = 1$	$\frac{\ln 3}{\ln 2}$
von Koch curve	$\left(\frac{1}{3}\right)^d + \left(\frac{1}{3}\right)^d + \left(\frac{1}{3}\right)^d + \left(\frac{1}{3}\right)^d = 1$	$\frac{\ln 4}{\ln 3}$

Example 6.8 Consider the IFS $\{[0, 1], \frac{1}{2}x, \frac{1}{4}x + \frac{3}{4}\}$. This IFS satisfies all conditions of Theorem 6.5 and the box dimension d of its attractor satisfies

$$\left(\frac{1}{2}\right)^x + \left(\frac{1}{4}\right)^x = 1.$$

Thus

$$d = \frac{\ln\left(\frac{1}{2}\sqrt{5} + \frac{1}{2}\right)}{\ln 2} = .69424.$$

∎

In the case where a collection of contractions of an IFS does not consist of similarities, one may use a related result to estimate box dimension of the attracting set. To present the next result we introduce the notions of upper and lower box dimensions.

DEFINITION 6.10 *The* **upper box dimension** *and* **lower box dimension** *of a subset F of R^n are given by*

$$\overline{dim_B} F = \limsup_{\delta \to 0} \frac{\ln N_\delta(F)}{\ln \frac{1}{\delta}}$$

$$\underline{dim_B} F = \liminf_{\delta \to 0} \frac{\ln N_\delta(F)}{\ln \frac{1}{\delta}}$$

where $N_\delta(F)$ is the smallest number of cubes of side δ that cover F.

THEOREM 6.6

Let $\{A, F_1, F_2, \ldots, F_m\}$ be an IFS, with contraction factors $c_1, \ldots, c_m$. Let $A_F \in H$ be a unique global attractor for the union mapping $F = \cup_{i=1}^m F_i$. Then the box dimension $D_B(A_F)$ satisfies

$$\overline{dim_B} A_F \leq d,$$

where d is a solution of the equation

$$\sum_{i=1}^{m} c_i^d = 1.$$

6.6 *Dynamica* Session

Some function systems have been predefined. For example, there is one called `Sierpinski` whose attractor is the Sierpinski triangle. `Sierpinski` is a list of three functions, each of which is a contraction with constant 1/2.

```
In[1]:= << Dynamica'

In[2]:= Sierpinski
Out[2]= {sf1, sf2, sf3}
```

Each function has two components arranged in a list like $\{f(x, y), g(x, y)\}$. The first function is just a scaling by a factor of 1/2 and the others are the same plus a translation.

```
In[3]:= IterateIFS[Sierpinski, {{x, y}}, 1]
Out[3]= {{{0.5 x, 0.5 y}},
         {{0.5 + 0.5 x, 0.5 y}},
         {{0.25 + 0.5 x, 0.433013 + 0.5 y}}}
```

The n-th iterate of the Sierpinski IFS consists of 3^n functions which contract by $1/2^n$.

```
In[4]:= IterateIFS[Sierpinski,
          {{x, y}}, 2]//Simplify
Out[4]= {{{0.25 x, 0.25 y}},
          {{0.25 (1 + x), 0.25 y}},
          {{0.125 + 0.25 x, 0.216506 + 0.25 y}},
          {{0.5 + 0.25 x, 0.25 y}},
          {{0.75 + 0.25 x, 0.25 y}},
          {{0.625 + 0.25 x, 0.216506 + 0.25 y}},
          {{0.25 (1 + x), 0.433013 + 0.25 y}},
          {{0.5 + 0.25 x, 0.433013 + 0.25 y}},
          {{0.375 + 0.25 x, 0.649519 + 0.25 y}}}

In[5]:= Length[%]
Out[5]= 9

In[6]:= IterateIFS[Sierpinski,
          {{x, y}}, 3]//Simplify;

In[7]:= Length[%]
Out[7]= 27
```

Note that the function in the second iterate contracts by a factor of 1/4 and those in the third iterate contract by a factor of 1/8.

This is the Sierpinski IFS with an initial shape of a triangle and only one iteration. There are several predefined initial shapes: **point**, **segment**, **triangle**, and **square**.

```
In[8]:= IFSPlot[Sierpinski, triangle, 1];
```

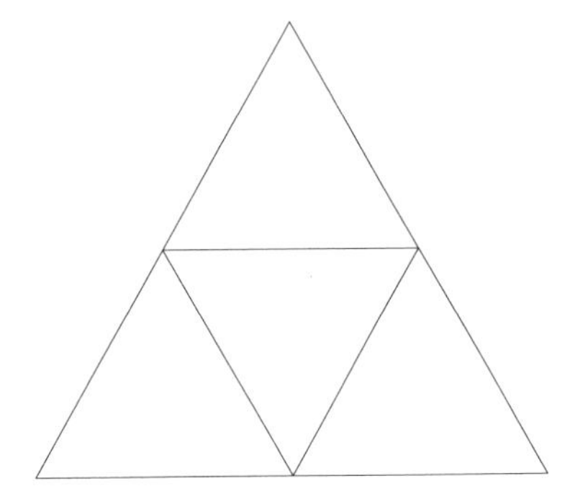

To produce plots of the Sierpinski triangles after n iterations one may type `IFSPlot[Sierpinski,triangle,n]`. The second through the fifth iterates of the Sierpinski triangle are shown in Figure 6.1.

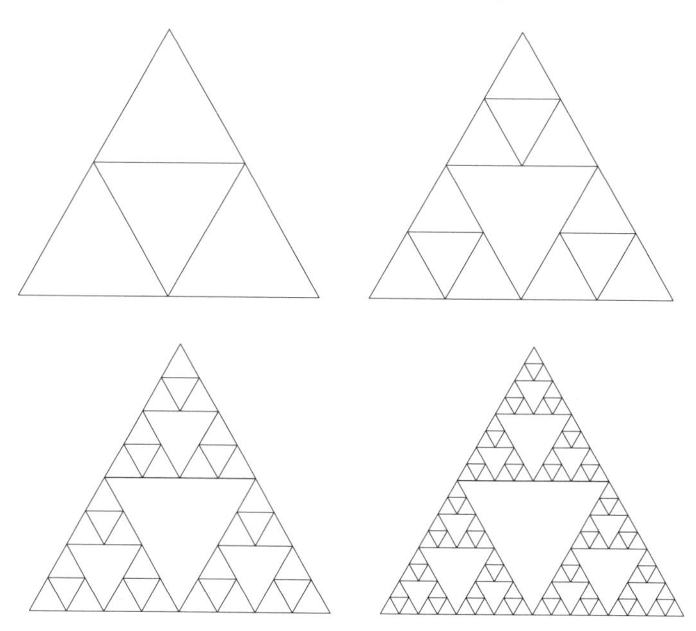

FIGURE 6.1: The next four steps in the construction of the Sierpinski triangle.

The first four steps in the construction of von Koch curve where the starting shape is a segment are shown in Figure 6.2. The *Dynamica* commands used for this construction were of the form `IFSPlot[koch,segment,n]`, for $n = 1, 2, 3, 4$.

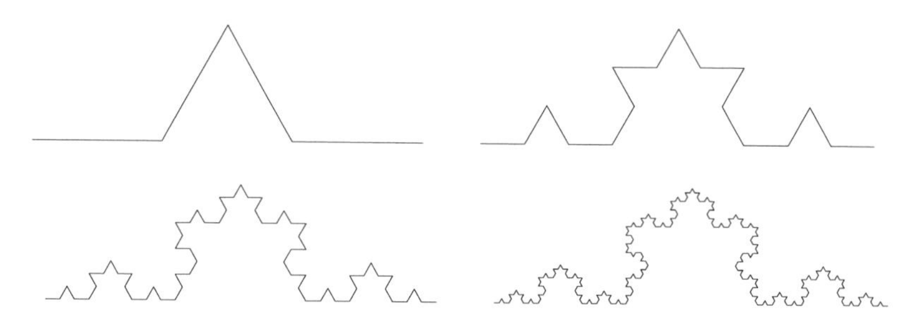

FIGURE 6.2: First four steps in the construction of the von Koch curve.

The Banach contraction principle states that the initial shape does not affect the shape of the attractor. See Figure 6.3 for a visualization of this fact.

FIGURE 6.3: Five iterates of the IFS with Sierpinski triangle attractor with different starting shapes (originators).

Now try out some other IFS. First another Sierpinski gasket. This IFS has 8 functions so its n-th iterate has 8^n functions.

```
In[9]:= IFSPlot[square8, square, 4]
```

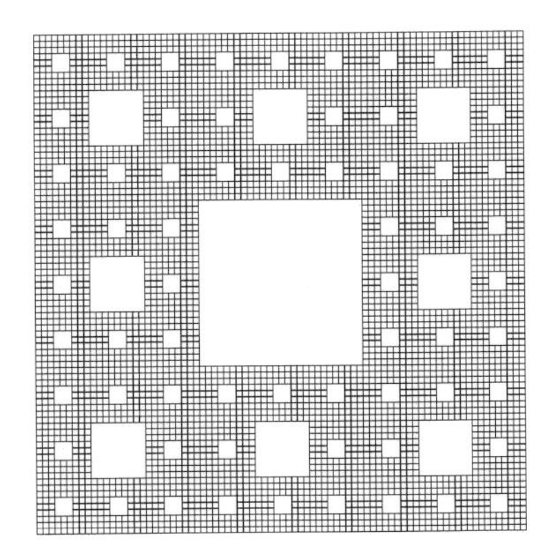

Another gasket with 5 functions.

```
In[10]:= IFSPlot[square5, square, 4];
```

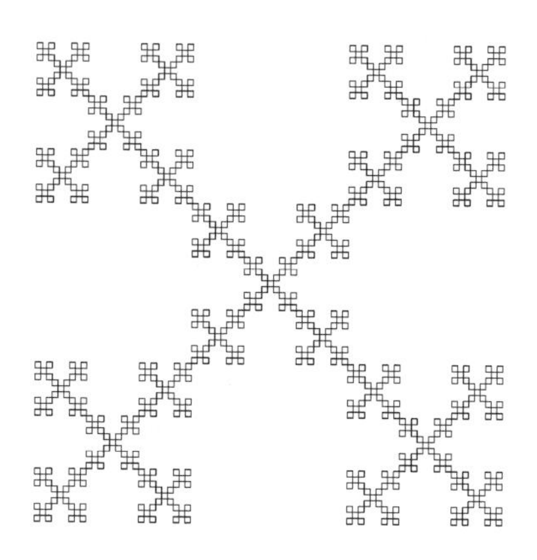

The von Koch curve with starting shape which is the triangle. As before, one may experiment with different initial shapes.

```
In[11]:= IFSPlot[koch, triangle, 4]
```

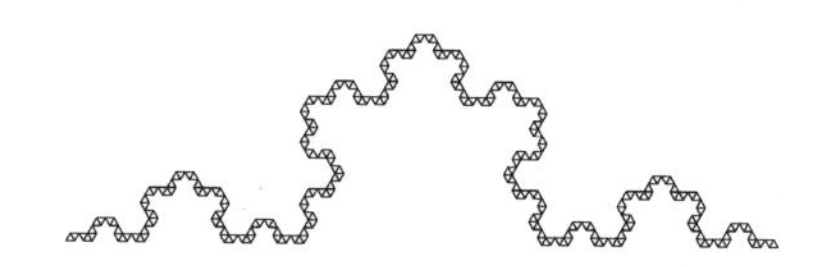

`FractalDimension` computes the fractal dimension of an attractor.

```
In[12]:= FractalDimension[koch]
Out[12]= 1.26186
```

The output of `Fractal Dimension` agrees with formula 6.12. as we may see from the fact that the Koch curve consists of 4 pieces of scale 1/3 so it should have dimension Log[4]/Log[3].

```
In[13]:= N[Log[4]/Log[3]]
Out[13]= 1.26186
```

The function `FractalDimension` works by finding the number of pieces and size of the pieces in the n-th iterate of the IFS. Then these are used to estimate the dimension by a generalization of formula (6.12). For IFS that contain only similarities, it suffices to do one iteration. This is the default value so one can call `FractalDimension` without specifying n.

These are additional examples of fractal dimension calculations.

```
In[14]:= FractalDimension[Sierpinski]
Out[14]= 1.58496

In[15]:= FractalDimension[square8]
Out[15]= 1.89279

In[16]:= FractalDimension[square5]
Out[16]= 1.46497
```

At this point you can design your own fractal by constructing an IFS and plotting it. Certain simple transformations of the plane are provided in *Dynamica* to save typing.

Here we create a similarity with scale factor 1/2 that rotates by 45 degrees and translates by {3/4, 0}.

```
In[17]:= f1 = Similarity[1/2, 3/4, 0, 45];

In[18]:= f1[x, y]
Out[18]= {0.75 + 0.353553 x - 0.353553 y,
          0. + 0.353553 x + 0.353553 y}
```

There is a way to create a transformation that scales the x and y variables by two different factors, followed by rotation and translation.

```
In[19]:= f2 = BiSimilarity[1/2, 1/4, 0, 0, 0];

In[20]:= f2[x, y]
Out[20]= {0. + 0.5 x + 0. y, 0. + 0. x + 0.25 y}
```

Finally, one can create a general affine transformation by specifying a 2×2 matrix and a translation. The matrix should be a contraction, however.

```
In[21]:= f3 = Affine[{{1/2, -1/2}, {0, 1/2}},
          {3/2, 0}];

In[22]:= f3[{x, y}]
Out[22]= {1.5 + 0.5 x - 0.5 y, 0.5 y}
```

Now putting these together into an IFS, the attractor can be plotted.

```
In[23]:= ifs = {f1, f2, f3};
```

```
In[24]:= IFSPlot[ifs, square, 1]
```

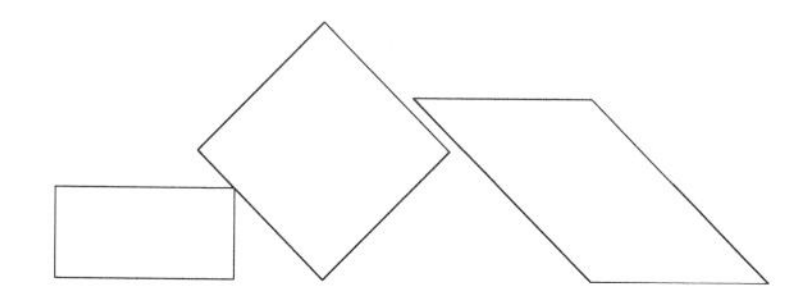

```
In[25]:= IFSPlot[ifs, square, 5]
```

Since this IFS contains maps which are not similarities, the fractal dimension cannot be computed exactly. The `FractalDimension` function can be used with an increasing number of iterations to obtain an approximate dimension.

```
Out[25]= Do[Print[FractalDimension[ifs, i]],
          {i, 1, 7}]

1.1042147453635278
1.1382460606980478
1.1798992901803325
1.2185579293022597
1.2449131041887587
1.2610221613867918
1.2712727763832934
```

A more interesting IFS can be created by adding a rotation to the similarities used for the Sierpinski triangle. These have scale 1/2.

```
In[26]:= f1 = Similarity[1/2, {0, 0}, 60];

In[27]:= f2 = Similarity[1/2, {1/2, 0}, 60];

In[28]:= f3 = Similarity[1/2, {1/4, √3/4}, 60];

In[29]:= ifs = {f1, f2, f3};
```

Here are the first and the sixth iterates of our IFS.

FIGURE 6.4: First and sixth iterates of the IFS. It can give us an idea of the shape of the attractor.

6.7 Exercises

Exercise 6.1 Find a set of four similarity transformations on R for which the Middle Third Cantor Set C is an attractor. Calculate the box dimension $D_B(C)$ of the attractor C.

Exercise 6.2 Use *Dynamica* to generate the first four steps in the construction of an attractor A of the IFS given by $\{[0,1], \frac{1}{2}x, \frac{1}{4}x + \frac{3}{4}\}$. Calculate the box dimension $D_B(A)$ of the attractor A.

Exercise 6.3 Use *Dynamica* to generate the first five steps in the construction of an attractor A of the IFS given by $\{[0,1], \frac{1}{5}x, \frac{1}{4}x + \frac{3}{4}, \frac{1}{3}x + \frac{1}{5}\}$. Calculate the box dimension $D_B(A)$ of the attractor A.

Exercise 6.4 Find an upper bound of the box dimension of an attractor A of IFS given by $\{[0,1], \frac{1}{2}x, \frac{1}{3}x + \frac{1}{3}\}$.

Exercise 6.5 Use *Dynamica* to generate the first five steps in the construction of an attractor A of the IFS given by $\{\square, F_1, F_2, F_3\}$ where F_1, F_2, F_3 are the affine transformations given by Table 6.5. Calculate the box dimension $D_B(A)$ of the attractor A.

Table 6.5

	a	b	c	d	t_1	t_2
F_1	$\frac{1}{3}$	0	0	$\frac{1}{3}$	0	0
F_2	$\frac{1}{2}$	0	0	$\frac{1}{2}$	$\frac{1}{2}$	0
F_3	$\frac{1}{4}$	0	0	$\frac{1}{4}$	$\frac{1}{2}$	$\frac{\sqrt{3}}{4}$

Exercise 6.6 Use *Dynamica* to generate the first five steps in the construction of an attractor A of IFS given by $\{\square, F_1, F_2, F_3\}$, $\{\diamond, F_1, F_2, F_3\}$, and $\{\triangle, F_1, F_2, F_3\}$ where F_1, F_2, F_3 are the affine transformations given in Exercise 6.5.

Exercise 6.7 Use *Dynamica* to generate the first five steps in the construction of an attractor A of IFS given by $\{\triangle, F_1, F_2, F_3\}$ where F_1, F_2, F_3 are the affine transformations given in Table 6.6. Calculate the box dimension $D_B(A)$ of the attractor A.

Table 6.6

	a	b	c	d	t_1	t_2
F_1	$\frac{1}{3}$	0	0	$\frac{1}{3}$	0	0
F_2	$\frac{1}{8}$	$-\frac{\sqrt{3}}{8}$	$\frac{\sqrt{3}}{8}$	$\frac{1}{8}$	$\frac{1}{3}$	0
F_3	$\frac{1}{8}$	$\frac{\sqrt{3}}{8}$	$-\frac{\sqrt{3}}{8}$	$\frac{1}{8}$	$\frac{1}{2}$	$\frac{\sqrt{3}}{8}$

Bibliography

[A] R. Agarwal, *Difference Equations and Inequalities. Theory, Methods and Applications*, Marcel Dekker Inc., New York, 1992.

[Ag] H. N. Agiza, On the analysis of stability, bifurcation, chaos and chaos control of Kopel map, *Chaos Solitons Fractals* 10(1999), 1909–1916.

[AB] L. J. S. Allen and A. B. Burgin, Comparison of deterministic and stochastic SIS and SIR models in discrete time, *Math. Biosciences*, 163(2000), 1–33.

[ACT] A. Arneodo, P. Coullet, and C. Tresser, Possible new strange attractors with spiral structure, *Commun. Math. Phys.*, 79(1981), 573–579.

[AGP] A. Agliari, L. Gardini, and T. Puu, The dynamics of a triopoly Cournot game, *Chaos Solitons Fractals* 11(2000), 2531–2560.

[AHS] L. J. S. Allen, M. K. Hannigan, and M. J. Strauss, Mathematical analysis of a mathematical model for a plant–herbivore system, *Bull. Math. Biol.*, 55(1993), 847–864.

[AKL] A. Amleh, V. Kirk, and G. Ladas, On the dynamics of difference equation, (to appear).

[Al] L. J. S. Allen, Some discrete-time *SI*, *SIS*, and *SIR* epidemic models, *Math. Biosciences*, 124(1994), 83–105.

[ASTL] L. J. S. Allen, M. J. Strauss, H. G. Thorvilson, and W. N. Life, A preliminary mathematical model of the apple twig borer and grapes on the Texas High Plaines, *Ecol. Model.*, 58(1991), 369–383.

[ASY] K.T. Alligood, T. Sauer and J.A. Yorke , *CHAOS An Introduction to Dynamical Systems*, Springer-Verlag, New York, Berlin, Heidelberg, Tokyo, 1997.

[Ar] L. Arriola, Global first integrals for first order difference equations, *J. Differ. Equations Appl.* 4(1998), 523–532.

[As] M. A. Asmussen, Regular and chaotic cycling in models of ecological genetics, *Theor. Pop. Biol.*, 16(1979), 172–190.

[Au] J. P. Aubin, *Optima and Equilibria: An Introduction to Nonlinear Analysis*, Springer, New York, 1993.

[BGT] E. Barbeau, B. Gelbord, and S. Tanny, Periodicity of solutions of the generalized Lyness recursion, *J. Differ. Equations Appl.* 1(1995), 291–306.

[B] M. Barnsley, *Fractals Everywhere*, Academic Press, Boston, MA, 1988.

[BB] J.M. Borwein and P.B. Borwein, *Pi and the AGM,* Wiley, New York, 1987.

[BC] M. Benedicks and L. Carleson, The dynamics of the Henon map, *Annals of Math.* 122(1991), 73–169.

[BF] M. Basson and M. J. Fogarty, Harvesting in discrete-time predator-prey systems, *Math. Biosciences* 141(1997), 41–74.

[BFL1] J. R. Beddington, C. A. Free, and J.H. Lawton, Dynamic complexity in predator-prey models framed in difference equations, *Nature*, 255(1975), 58–60.

[BFL2] J. R. Beddington, C. A. Free, and J.H. Lawton, Characteristics of successful natural enemies in models of biological control of insect pests, *Nature*, 273(1978), 513–519.

[BHu] M. Barnsley and L. Hurd, *Fractal Image Compression,* A. K. Peters, Ltd. Wellesley, MA, 1993.

[BGK] G. I. Bischi, L. Gardini, and M. Kopel, Analysis of global bifurcations in a market share attraction model, *J. Econom. Dynamics. Control*, 24(2000), 855–879.

[BH] S. Bourgault and Y. S. Huang, The existence of continuous invariants for linear difference equations, *J. Differ. Equations Appl.* 6(2000), 739–751.

[BL] D. J. Bernstein and J. C. Lagarias, The $3x + 1$ conjugacy map, *Canadian J. Math.* 48 (1996), 1154–1169.

[BMG] G. I. Bischi, C. Mammana, and L. Gardini, Multistability and cyclic attractors in duopoly games, *Chaos Solitons Fractals*, 11(2000), 543–564.

[BN] J. Bishir and G. Namkoong, Effect of density on the interaction between competitive plant species, (Preprint).

[BlC] L. S. Block and W. L. Coppel, *Dynamics in One Dimension*, Lectures in Mathematics, 1513, Springer–Verlag, New York, 1992.

[BS] J. R. Buchanan and J. F. Selgrade, Constant and periodic rate stocking and harvesting for Kolmogorov–type population interaction models, *Rocky Mountain J. Math.*, 25(1995), 67–85.

[BTR] M. Bernardo, T. T. Truong, and G. Rollet, The disrete Painlevé I equations: transcendental integrability and asymptotic solutions, *J. Phys. A*, 34(2001), 3215–3252.

[BZ] W. Briden and S. Zhang, Stability of solutions of generalized logistic difference equations, *Period. Math. Hung.* 29(1994), 81–87.

[C] A. Cournot, *Recherche sur la Principes Matematiques de la Theorie de la Richesse,* Hachette, Paris, 1838.

[C1] J. M. Cushing, Nonlinear matrix models and population dynamics, *Natural Resource Modeling,* 2(1988), 539–580.

[C2] J. M. Cushing, The Allele effect in age–structured population dynamics, *Mathematical Ecology,* T. G. Hallam, L. J. Gross, and S. A. Levin, eds., World Scientific Publishing Co., Singapore, 1988, 479–505.

[Cl] D. Clark, Second order difference equations related to the Collatz $3n + 1$ conjecture , *J. Differ. Equations Appl.*, 1(1985), 73–86.

[CC] J. H. Conway and H. S. M. Coxeter, Triangulated polygons and frieze patterns, *Math. Gaz.* 57(1973), 87–94 and 175–183.

[CF] D. M. Chan and J. E. Franke, Multiple extinctions in a discrete competitive system, *Nonlin. Anal. RWA,* 2(2001), 75–91.

[CH] H. N. Comins and M. P. Hassell, Predation in multiprey communities, *J. Theor. Biol.* 62(1976), 93–114.

[CK] D. Clark and M. R. S. Kulenović, On a coupled system of rational difference equations, *Comp. Math. Appl.* 2002 (to appear).

[CKi] J. T. Crow and M. Kimura, *An Introduction to Population Genetics Theory,* Harper & Row, New York, 1970.

[CKLV] K. Cunningham, M. R. S. Kulenović, G. Ladas, and S. Valicenti, On the recursive Sequence $x_{n+1} = \frac{\alpha+\beta x_n}{Bx_n+Cx_{n-1}}$, *Nonlin. Anal. TMA,* 47(2001), 4603–4614.

[CKS] D. Clark, M. R. S. Kulenović, and J. F. Selgrade, Global asymptotic behavior of a two-dimensional difference equation (to appear).

[CL1] D. Clark and J. T. Lewis, A Collatz-type difference equation, *Congr. Numer.* 111(1995), 129–135.

[CL2] D. Clark and J. T. Lewis, Symmetric solutions to a Collatz-like system of difference equations, *Congr. Numer.* 131(1998), 101–114.

[D] B. Davies, *Exploring Chaos*, Perseus Books, Reading, MA, 1999.

[D1] R. Devaney, *A First Course in Chaotic Dynamical Systems: Theory and Experiment*, Addison–Wesley, Reading, MA, 1992.

[D2] R. Devaney, A piecewise linear model for the zones of instability of an area-preserving map, *Phys. D* 10(1984), 387–393.

[D3] M. Barnsley, R. Devaney, et al., Fractal patterns arising in chaotic dynamical systems, in *The Science of Fractal Images*, Springer-Verlag, New York, 1988.

[DBG] R. Dieci, G. I. Bischi, and L. Gardini, From bi-stability to chaotic oscillations in a macroeconomic model, *Chaos Solitons Fractals*, 12(2001), 805–822.

[DKL] R. DeVault, V. Kocic, and G. Ladas, Global stability of a recursive sequence, *Dynam. Systems Appl.*, 1(1992), 13–21.

[E1] S. Elaydi, *An Introduction to Difference Equations*, 2nd ed., Springer-Verlag, New York, 1999.

[E2] S. Elaydi, *Discrete Chaos*, Chapman& Hall/CRC Press, Boca Raton, FL, 2000.

[E3] S. Elaydi, A converse to Sharkovsky's theorem, *Amer. Math. Month.* 103(1996), 386–392.

[EGL] H. El-Metwally, E. A. Grove, and G. Ladas, On the recursive sequence $x_{n+1} = \frac{\alpha+\gamma x_{n-1}}{Cx_{n-1}+Dx_{n-2}}$, (to appear).

[EGLR] H. El-Metwally, E. A. Grove, G. Ladas, R. Levins, and M. Radin, On the difference equation $x_{n+1} = \alpha + \beta x_{n-1}e^{-x_n}$, *Nonlin. Anal. TMA*, 47 (2001), 4623–4634.

[EMTW] S. Eubank, W. Miner, T. Tajima, and J. Wiley, Interactive computer simulationand analysis of Newtonian dynamics, *Am. J. Phys.* 57 (1989), 437.

[F] K. Falconer, *Fractal Geometry: Mathematical Foundations and Applications*, Wiley, New York, 1990.

[F1] M. J. Feigenbaum, Quantitative universality for a class of nonlinear transformations, *J. Stat. Phys.* 19(1978), 25–52.

[F2] M. J. Feigenbaum, The universal metric properties of nonlinear transformations, *J. Stat. Phys.* 21(1979), 669–706.

[FB] J. Fraleigh and R. Beauregard, *Linear Algebra*, 3rd ed., Addison-Wesley, New York, 1994.

[FIK] A. S. Fokas, A. R. Its, and A. V. Kitaev, Discrete Painlevé equations and their appearance in quantum gravity, *Commun. Math. Phys.* 142(1991), 313–344.

[FJL] J. Feuer, E. Janowski, and G. Ladas, Invariants for some rational recursive sequences with periodic coefficients, *J. Differ. Equations Appl.* 2(1996), 167–174.

[FY1] J. E. Franke and A-A. Yakubu, Mutual exclusion versus coexistence for discrete competitive systems, *J. Math. Biol.* 30(1991), 161–168.

[FY2] J. E. Franke and A-A. Yakubu, Global attractors in competitive systems, *Nonlin. Anal. TMA* 16(1991), 111–129.

[FY3] J. E. Franke and A-A. Yakubu, Geometry of exclusion principles in discrete systems, *J. Math. Anal. Appl.* 168(1992), 385–400.

[GH] J. Guckenheimer and P. Holmes, *Nonlinear Oscillations, Dynamical Systems, and Bifurcations of Vector Fields*, Springer-Verlag, New York, Berlin, Heidelberg, Tokyo, 1983.

[GJKL] E. A. Grove, E. J. Janowski, C. M. Kent, and G. Ladas, On the rational recursive sequence $x_{n+1} = (\alpha x_n + \beta)/((\gamma x_n + \delta)x_{n-1})$, *Comm. Appl. Nonlinear Anal.* 1(1994), 61–72.

[GKL] E. A. Grove, V. L. Kocic, and G. Ladas, *Advances in Difference Equations* (Veszprm, 1995), 289–294, Gordon and Breach, Amsterdam, 1997.

[GKL1] C. H. Gibbons, M. R. S. Kulenović, and G. Ladas, On the recursive sequence $x_{n+1} = \frac{\alpha+\beta x_{n-1}}{\gamma+x_n}$, *Math. Sci. Res. Hot-Line*, 4(2)(2000), 1–11.

[GKL2] C. H. Gibbons, M. R. S. Kulenović, G. Ladas, and H. D. Voulov, On the trichotomy character of $x_{n+1} = \frac{\alpha+\beta x_n+\gamma x_{n-1}}{A+x_n}$, *J. Differ. Equations Appl.* 8(2002) (to appear).

[GLMT] E. A. Grove, G. Ladas, L.C. McGrath, and C. T. Teixeira, Existence and behavior of solutions of a rational system, *Comm. Appl. Nonlinear Anal.* 8(2001), 1–25.

[GLPL] E. A. Grove, G. Ladas, N. R. Prokup, and R. Levins, On the global behavior of solutions of a biological model. *Comm. Appl. Nonlinear Anal.* 7 (2000), 33–46.

[GM] I. Gumowski and C. Mira, *Recurrences and Discrete Dynamic Systems*, Lecture Notes in Mathematics, Springer, Berlin, 1980.

[H] M. Henon, A two dimensional mapping with a strange attractor,*Comm. Math. Phys.*, 50(1976), 69–77.

[HaW] G. H. Hardy and E. M. Wright, *An Introduction to the Theory of Numbers*, Oxford University Press, Oxford, 1980.

[HBQC] F. A. Haggar, G. B. Byrnes, G. R. W. Quispel, and H. W. Capel, k-integrals and k-Lie symmetries in discrete dynamical systems. *Physica A* 233(1996), 379–394.

[HC] M. P. Hassell and H. N. Comins, Discrete time models for two-species competition, *Theor. Pop. Biol.* 9(1976), 202–221.

[HCM] M. P. Hassell, H. N. Comins, and R. M. May, Spatial structure and chaos in insect population dynamics, *Nature* 353(1991), 255–258.

[Hg] C. S. Holling, Some characteristics of simple types of predation and parasitism, *Can. Ent.* 91(1959), 385–398.

[HK] J. Hale and H. Kocak, *Dynamics and Bifurcations*, Springer-Verlag, New York, Berlin, Heidelberg, Tokyo, 1991.

[HM] M. P. Hassell and R. M. May, Stability in insect host-parasite models, *J. Anim. Ecol.* 42(1973), 693–726.

[HT] Phase portraits for a class of difference equations, *J. Differ. Equations Appl.* 5(1999), 177-202.

[HW] R. Hilborn and C. J. Walters, *Quantitative Fish Stock Assessment*, Chapman & Hall, London, 1992.

[Ho] R. A. Holmgren, *A First Course in Discrete Dynamical Systems*, second edition. Universitext, Springer-Verlag, New York, 1996.

[Hu] J. Hutchinson, Fractals and self-similarity, *Indiana Univ. J. Math.*, 30(1981), 713–747.

[HuW] J. H. Hubbard and B. H. West, *Differential Equations: A Dynamical Systems Approach, Part I*, Springer-Verlag, New York, Berlin, Heidelberg, Tokyo, (1991).

[J] N. Joshi, Singularity analysis and integrability for disrete dynamical systems, *J. Math. Anal. Appl.*, 184(1994), 573–584.

[Ja] O. L. R. Jacobs, *Introduction to Control Theory*, Clarendon Press, Oxford, 1974.

[JLV] E. J. Janowski, G. Ladas, and S. Valicenti, Lyness-type equations with period two coefficients, *Advances in Difference Equations*, Veszprm, Hungary, August 7-11, 1995. 327–334, Gordon and Breach Science Publishers, Amsterdam 1997.

[JP] F. Jones and J. Perry, Modelling populations of cyst-nematodes (nematoda: heteroderidae), *J. Appl. Ecology,* 15(1978), 349–371.

[JR] H. Jiang and T. D. Rogers, The discrete dynamics of symmetric competition in the plane, *J. Math. Biol.* 25(1987), 573–596.

[K] H. Kocak, *Differential and Difference Equations through Computer Experiments, with diskette containing PHASER,* second edition, Springer-Verlag, New York, 1989.

[Kc] H. von Koch, Sur une corbe continué sans tangente, obtenue par une construction géometrique élémentaire, *Arkiv fur Mathematik* 1(1904), 681–704.

[KH] V. Kaitala and M. Heino, Complex non-unique dynamics in ecological interactions, *Proc. R. Soc. London B* 263(1996), 1011–1013.

[KHG] V. Kaitala, M. Heino, and W Getz, Host-parasite dynamics and the evolution of host immunity and parasite fecundity strategies, *Bull. Math. Biol.* 59(1997), 427–450.

[KJ] H. J. Korsch and H.-J. Jodl, *Chaos, A Program Collection for the PC,* Springer-Verlag, New York, Berlin, Heidelberg, Tokyo, 1994.

[KK] S. Kalabušić and M. R. S. Kulenović, On the recursive sequence $x_{n+1} = \frac{\gamma x_{n-1}+\delta x_{n-2}}{Cx_{n-1}+Dx_{n-2}}$, (to appear).

[KKLT] W. A. Kosmala, M. R. S. Kulenović, G. Ladas, and C. T. Teixeira, On the recursive sequence $y_{n+1} = \frac{p+y_{n-1}}{qy_n+y_{n-1}}$, *J. Math. Anal. Appl.* 251(2000), 571–586.

[KL] M. R. S. Kulenović and G. Ladas, *Dynamics of Second Order Rational Difference Equations,* Chapman & Hall/CRC, Boca Raton, London, 2001.

[KL1] M.R.S. Kulenović and G. Ladas, Linearized oscillations in population dynamics, *Bull. Math. Biol.* 49 (5) (1987), 615–627.

[KLMR] M. R. S. Kulenović, G. Ladas, L. F. Martins, and I. W. Rodrigues, On the dynamics of $x_{n+1} = \frac{\alpha+\beta x_n}{A+Bx_n+Cx_{n-1}}$ facts and conjectures, *Comput. Math. Appl.,* 2002(to appear).

[KLP1] M. R. S. Kulenović, G. Ladas, and N. R. Prokup, On the recursive sequence $x_{n+1} = \frac{\alpha x_n+\beta x_{n-1}}{1+x_n}$, *J. Differ. Equations Appl.* 6(5) (2000), 563–576.

[KLP2] M. R. S. Kulenović, G. Ladas, and N. R. Prokup, On a rational difference equation, *Comput. Math. Appl.* 41(2001), 671–678.

[KLR] V. L. Kocic, G. Ladas, and I. W. Rodrigues, On the rational recursive sequences, *J. Math. Anal. Appl.* 173(1993), 127–157.

[KLS] M. R. S. Kulenović, G. Ladas, and W. S. Sizer, On the recursive sequence $x_{n+1} = \frac{\alpha x_n + \beta x_{n-1}}{\gamma x_n + C x_{n-1}}$, *Math. Sci. Res. Hot-Line* 2(1998), 1–16.

[KLTT] V. L. Kocic, G. Ladas, G. Tzanetopoulos, and E. Thomas, On the stability of Lyness' equation, *Dynam. Contin. Discrete Impuls. Systems*, 1(1995), 245–254.

[KN1] M. R. S. Kulenović and M. Nurkanović, An example of a construction of the invariant interval for the system of difference equations, (to appear).

[KN2] M. R. S. Kulenović and M. Nurkanović, Global asymptotic behavior of a two dimensional system of difference equations modelling cooperation, *J. Differ. Equations Appl.* 8(2002), to appear.

[Ko] M. Kopel, Simple and complex adjusment dynamics in Cournot Duopoly model, *Chaos Solitons Fractals* 7(1996), 2031–2048.

[KoL] V. L. Kocic and G. Ladas, *Global Behavior of Nonlinear Difference Equations of Higher Order with Applications*, Kluwer Academic Publishers, Dordreht/Boston/London, 1993.

[KP] W. G. Kelley and A. C. Peterson, *Difference Equations,* Academic Press, Boston, MA, 1991.

[KR] W. Krawcewicz and T. D. Rogers, Perfect harmony: the discrete dynamics of cooperation, *J. Math. Biol.* 28(1990), 383–410.

[Ku] M. R. S. Kulenović, Invariants and related Liapunov functions for difference equations, *Appl. Math. Lett.* 13(2000), 1–8.

[Kv] A. N. Kolmogorov, Local structure of turbulence in an incompressible liquid for very large Reynolds numbers, *C. R. Acad. Sci. USSR* 30(1941), 299–303.

[KYH] V. Kaitala, J. Ylikarjula, and M. Heino, Non-unique population dynamics: basic patterns, *Ecol. Modelling* 135(2000), 127–134.

[L] D. G. Luenberger, *Introduction to Dynamic Systems, Theory, Models, and Applications*, Wiley, New York, 1979.

[L1] R. C. Lyness, Note 1581, *Math. Gaz.* 26(1942), 62.

[L2] R. C. Lyness, Note 1847, *Math. Gaz.* 29(1945), 231.

[La] H. Lauwerier, *Fractals,* Princeton University Press, Princeton, 1991.

[La1] H. A. Lauwerier, One-dimensional iterative maps, in *Chaos*, A. V. Holden, Ed., Manchester Univ. Press, Manchester, 1986, 39–57.

[La2] H. A. Lauwerier, Two-dimensional iterative maps, in *Chaos*, A. V. Holden, Ed., Manchester Univ. Press, Manchester, 1986, 58–97.

[Lg] J. C. Lagarias, The $3x+1$ problem and its generalizations, *Amer. Math. Monthly* 92(1985), 3–23.

[Lop1] G. Ladas, Invariants for generalized Lyness equations, *J. Differ. Equations Appl.* 1(1995), 209–214.

[Ly] S. Lynch, *Dynamical Systems with Applications using Maple*, Birkhauser, Boston, 2001.

[LL] H. Levy and F. Lessman, *Finite Difference Equations*, Dover, New York, 1992.

[LT] V. Lakshmikantham and D. Triggiante, *Theory of Difference Equations*, Academic Press, Boston, MA, 1988.

[LTT] G. Ladas, G. Tzanetopoulos, and A. Tovbis, On May's host parasitoid model, *J. Differ. Equations Appl.* 2(1996), 195–204.

[LY] T. Y. Li and J. A. Yorke, Period three implies chaos, *Amer. Math. Monthly* 82(1975), 985–992.

[M] J. D. Murray, *Mathematical Biology*, second edition, Springer-Verlag, Berlin, 1993.

[M1] R. M. May, Biological populations with nonoverlapping generations: stable points, stable cycles and chaos, *Science* 186(1974), 645–647.

[M2] R. M. May, Host-parasitoid systems in patchy environments: a phenomenological model, *J. Animal Ecol.* 47(1978), 833–843.

[M3] R. M. May and M. P. Hassell, The dynamics of multiparasitoid-host interactions, *The American Naturalist* 117(1981), 234–261.

[Ma] S. Maeda, The similarity method for difference equations, *IMA J. Appl. Math.* 38(1987), 129–134.

[Mc] R. Moeckel, Dynamical Systems Lab, *Mathematica* package for simulation of Dynamical Systems, http:// www. math. umn. edu/rick /mathematica /Lab.5535.s01.3.nb.

[Mi] C. Mira, *Chaotic Dynamics*, World Scientific, Singapore, 1987.

[Ml] M. Martelli, *Introduction to Discrete Dynamical Systems and Chaos*, Wiley-Interscience, New York, 1999.

[Mo] P. A. B. Moran, Some remarks on animal population dynamics, *Biometrics* 6(1950), 250–258.

[Mr] J. Moser, *Stable and Random Motions in Dynamical Systems*, Princeton University Press, Princeton, 1973.

[MK] R. S. MacKay, *Renormalisation in Area-Preserving Maps*, World Scientific Publishing Co., River Edge, NJ, 1993.

[MSHZ] C. Meier, W. Senn, R. Hauser, and M. Zimmermann, Strange limits of stability in host-parasitoid systems, *J. Math. Biol*, 32(1994), 563–572.

[Mt] B. Mandelbrot, *The Fractal Geometry of Nature*, W. H. Freeman, New York, 1982.

[MT] A. Milne-Thompson, *The Calculus of Finite Differences*, McMillan and Co., London, 1933.

[My1] P. J. Myrberg, Iteration der reellen polynome zweiten grades, I *Ann. Acad. Sci. Fennicae* 256(1958), 1–10.

[My2] P. J. Myrberg, Iteration der reellen polynome zweiten grades, II *Ann. Acad. Sci. Fennicae* 268(1959), 1–18.

[My3] P. J. Myrberg, Iteration der reellen polynome zweiten grades, III *Ann. Acad. Sci. Fennicae* 336(1963), 1–10.

[My4] P. J. Myrberg, Sur l'itration des polynomes rels quadratiques, *J. Math. Pures Appl.* (9) 41 (1962), 339–351.

[MW] N. MacDonald and R. R. Whitehead, Introducing students to nonlinearity: Computer experiments with with Burgers mappings, *Eur. J. Phys.* 6(1985), 143.

[NB] A. J. Nicholson and V. A. Bailey, The balance of animal populations, *Proc. Zool. Soc. Lond.*, 1935, 551–598.

[NY] H. Nusse and J. A. Yorke, *Dynamics: Numerical Explorations*, second edition, Springer-Verlag, New York, Berlin, Heidelberg, Tokyo, 1998.

[O] P. Olver, *Applications of Lie Groups to Differential Equations*, Graduate Texts in Mathematics, second edition, Springer-Verlag, New York, Berlin, Heidelberg, Tokyo, 2000.

[OC] J. D. Opsomer and J. M. Conrad, An open-access analysis of the northern anchovy fishery, *J. Environ. Econ. Manag.* 27(1994), 21–37.

[P] T. Puu, *Nonlinear Economic Dynamics*, Springer, New York, 1997.

[P1] T. Puu, Chaos in duopoly pricing, *Chaos Solitons Fractals* 1(1991), 573–581.

[P2] T. Puu, The chaotic duopolists revisited, *J. Econom. Behavior & Organization* 33(1998), 385–394.

[PJS] H-O, Peitgen, D. Saupe, H. Jurgens, *Chaos and Fractals : New Frontiers of Science*, Springer, New York, 1992.

[PS1] G. Papaschinopoulos and C. J. Schinas, Stability of a class of nonlinear difference equations, *J. Math. Anal. Appl.*, 230(1999), 211–222.

[PS2] G. Papaschinopoulos and C. J. Schinas, Generalized invariants for systems of difference equations of rational form. *Neural Parallel Sci. Comput.* 7(1999), 379–404.

[PT] J. Palis and F. Takens, *Hyperbolicity and Sensitive Chaotic Dynamics at Homoclinic Bifurcations*, Cambridge University Press, Cambridge, 1993.

[QRT1] G. R. W. Quispel, J. A. G. Roberts and C. J. Thompson, Integrable mappings and soliton equations I, *Phys. Lett.* A 126 (1988), 419–421.

[QRT2] G. R. W. Quispel, J. A. G. Roberts, and C. J. Thompson, Integrable mappings and soliton equations II, *Physica D* 34 (1989), 183–192.

[QS] G. R. W. Quispel and R. Sahadevan, Lie symmetries and integration of difference equations, *Phys. Lett.* A 184 (1993), 64–70.

[R] C. Robinson, *Stability, Symbolic Dynamics, and Chaos*, CRC Press, Boca Raton, 1995.

[Ra] D. Rand, Exotic phenomena in games and duopoly models, *J. Math. Econ.* 5(1978), 173–184.

[Ri] W. E. Ricker, Stock and recruitment, *J. Fish. Res. Bd. Can.* 11(1954), 559–623.

[RQ] J. A. G. Roberts and G. R. W. Quispel, Chaos and time-reversal symmetry. Order and chaos in reversible dynamical systems, *Phys. Rep.* 216(1992), 53–117.

[Ro] D. J. Rogers, Random search and insect population models, *J. Anim. Ecol.* 41(1972), 369–383.

[Roy] T. Royama, A comparative study of models for predation and parasitism, *Res. Popul. Biol. Ecol.* (1971), 1–91.

[S] W. Sierpinski, Sur une corbe Cantorienue qui contient une image biounivoquet et continué detoute corbe donée, *C. R. Acad. Paris*, 162(1916), 629–632.

[S1] H. L. Smith, Planar competitive and cooperative difference equations, *J. Differ. Equations Appl.*, 3(1998), 335–357.

[Sa] T. L. Saaty, *Modern Nonlinear Equations*, Dover, New York, 1981.

[Sc1] C. J. Schinas, Invariants for some difference equations, *J. Math. Anal. Appl.*, 212(1997), 281–291.

[Sc2] C. J. Schinas, Invariants for difference equations and systems of difference equations of rational form, *J. Math. Anal. Appl.*, 216(1997), 164–179.

[Se] J. F. Selgrade, Using stocking or harvesting to reverse period-doubling bifurcations in discrete population models, *J. Differ. Equations Appl.*, 4(1998), 163–183.

[Sf] J. T. Sandefur, *Discrete Dynamical Systems. Theory and Applications*, Clarendon Press, Oxford University Press, New York, 1990.

[Sh] A. N. Sharkovsky, Co-existence of cycles of a continuous mapping of a line into itself, *Ukranian Math. Z.* 16(1964), 61–71.

[Si] D. Singer, Stable orbits and bifurcation of maps of the interval, *SIAM J. Appl. Math.*, 35(1978), 260–267.

[Sk] A. G. Sivak, Periodicity of recursive sequances and the dynamics of homogeneous piecewise linear maps of the plane, *J. Differ. Equations Appl.*, 8(2002), to appear.

[SH] A. M. Stuart and A. R. Humphries, *Dynamical Systems and Numerical Analysis*, Cambridge University Press, Cambridge, 1996.

[SM] C. L. Siegel and J. Moser, *Lectures on Celestial Mechanics*, Springer, New York, Berlin, Heidelberg, 1971.

[SMR] A. N. Sharkovsky, Yu. L. Maistrenko, and Yu. Romanenko, *Difference Equations and Their Applications*, Kluwer Academic, Dodrecht, 1993.

[SN1] J. F. Selgrade and G. Namkoong, Stable periodic behavior in a pioneer-climax model, *Nat. Resource Model.*, 4(1990), 215–227.

[SN2] J. F. Selgrade and G. Namkoong, Population interactions with growth rates dependent on weighted densities, *Differential Equations* in *Models in Biology, Epidemiology and Ecology*, S. Busenberg and M. Martelli, eds., Lect. Notes in Biomath. 92, Springer-Verlag, Berlin, 1991, 247–256.

[SQ] R. Sahadevan and G. R. W. Quispel, Lie symmetries and linearisation of the QRT mapping, *Physica A*(1997) 234, 775–784.

[SR1] J. F. Selgrade and J. H. Roberds, Lumped-density population models of pioneer-climax type and stability analysis of Hopf bifurcations, *Math. Biosciences* 135(1996), 1–21.

[SR2] J. F. Selgrade and J. H. Roberds, Period-doubling bifurcations for systems of difference equations and applications to models in population biology, *Nonlin. Anal. TMA* 29(1997), 185–199.

[SZ] J. F. Selgrade and M. Ziehe, Convergence to equilibrium in a genetic model with differential viability between the sexes, *J. Math. Biol.*, 25(1987), 477–490.

[T] A. D. Taylor, Aggregation, competition, and host-parasitoid dynamics: stability conditions don't tell it all, *The American Naturalist* 141(1993), 501–506.

[TB] E. S. Thomas and S. Bourgault, A note of the theoretical existence of invariants: part I, *J. Differ. Equations Appl.* 6(2000), 681–693.

[To] H. Todd, Private communication with R.C. Lyness, 1941–1945.

[VB] M. Vellekop and R. Berglund, On intervals: Transitivity → chaos, *Am. Math. Month.* 101(1994), 353–355.

[W] H. M. Wilbur, Multistage life cycles, in *Population Dynamics in Ecological Space and Time*, University of Chicago Press, Chicago, 1996, 75–108.

[Wi] J. Wimp, *Computation with Recurrence Relations*, Pitman Advanced Pub. Program, Boston, 1984.

[WZ] W. Wang and L. Zhengyi, Global stability of discrete models of Lotka-Volterra type, *Nonlin. Anal. TMA* 35(1999), 1019–1030.

[Y1] A-A. Yakubu, The effect of planting and harvesting on endangered species in discrete competitive systems, *Math. Biosci.* 126(1995), 1–20.

[Y2] A-A. Yakubu, A discrete competitive system with planting, *J. Differ. Equations Appl.*, 4(1998), 213–214.

[Z] E. C. Zeeman, Geometric unfolding of a difference equation, http://www. math. utsa. edu/ecz/gu.html.

Index